Universal Design

Principles and Models

Universal Design

Principles and Models

R O B E R T A   N U L L

CRC Press
Taylor & Francis Group
Boca Raton London New York

CRC Press is an imprint of the
Taylor & Francis Group, an **informa** business

CRC Press
Taylor & Francis Group
6000 Broken Sound Parkway NW, Suite 300
Boca Raton, FL 33487-2742

First issued in paperback 2017

© 2014 by Taylor & Francis Group, LLC
CRC Press is an imprint of Taylor & Francis Group, an Informa business

No claim to original U.S. Government works

ISBN-13: 978-1-4665-0529-2 (hbk)
ISBN-13: 978-1-138-07600-6 (pbk)

Visit the Taylor & Francis Web site at
http://www.taylorandfrancis.com

and the CRC Press Web site at
http://www.crcpress.com

Contents

Preface

This book revision represents my networking experiences in the journey of telling the story of universal design (UD). I have learned to appreciate the contributions of a wonderful group of design professionals and professional groups who share the goal of making the world a better place for everyone.

In the late 60s, I started teaching housing and design classes at Purdue University. As part of my professional development as a new faculty member, I joined and became an active member of the American Association of Housing Educators (now called Housing Education and Research Association [HERA]). Other important influences included Bill Sims, one of my PhD program advisors at Ohio State and an expert in environmental design research, and the Environmental Design Research Association (EDRA). In 1979, at the first national meeting of EDRA that I attended, I won the student design competition for my PhD research on residence hall environments. I met Ron Mace—the Father of Universal Design—at a "Handi-tap" seminar in Columbus, Ohio, in 1980. His common sense approach provided a unique strategy for building supportive environments for handicapped individuals.

When I went to teach Housing and Interior Design at San Diego State University (SDSU), one of my students' assignments—a Kitchen Design Project—started my work with the San Diego Center for the Blind. I designed three working kitchens with the involvement of my SDSU students and the American Society of Interior Designers (ASID). Our project won an ASID national award and led to the very successful community project described in Chapter 3. Telling the story of this project provided opportunities for giving presentations at professional meetings (American Society on Aging [ASA], "I'm Old and I Cannot See") and writing articles for professional publications.

In 1993, while teaching at Miami of Ohio, with colleagues Barbara Flannery and Ken Special, I entered a Universal Design Education Project competition.

Our proposal, "Strategies for Teaching Lifespan Issues to Future Designers," was one of the 14 chosen as a model curriculum in UD. We did a two-day charrette that was described in the book *Strategies for Teaching Universal Design*—1995, Polly Welch, editor, Adaptive Environments, Boston.

Several years after the passage of the Americans with Disabilities Act, the textbook that I wrote with Ken Cherry, *Universal Design: Creative Solutions for ADA Compliance* (Belmont, California, Professional Publications, Inc., 1996), was published. The book opened doors for my recognition as a UD professional. In 1998, when a Japanese businessman asked Ron Mace to recommend a UD text that would be a good choice for translation into Japanese, Ron gave him a copy of my book.

That spring marked the first International Universal Design meeting at Hofstra University in New York State. At that landmark meeting, there were more than 400 people attending—and of that group, one-third were from Japan. Ron Mace gave the keynote address at the meeting and was recognized for his leadership in the field of UD. Sadly, Ron died unexpectedly soon after the meeting, but through his research and writing, he had set the foundation for the UD movement.

Later that year (1998), because of my writing on UD, I was invited to present a lecture on UD for the 2nd Universal Design Consortium in Tokyo. I then flew to Korea for the meeting, "East Meets West—Housing for People of Diverse Cultures," sponsored by the Korean Association and the HERA, August 6–8. It was a unique networking opportunity created by Dr. Yeun Sook Lee from Yonsei University in Seoul. I also presented a UD lecture—"Universal Design as a Major Concept for the 21st Century." My presentation, along with that of Amos Rappoport, author of *House Form and Culture*, was sponsored by the Korean Institute of Architecture.

In the following year, 1999, I was invited to attend the "Unlimited by Design" exhibit at the Cooper–Hewitt National Design Museum in New York City. The exhibit provided a powerful tool for promoting UD.

In November 1999, I presented an invited lecture, "Universal Design for the 21st Century," at the 3rd Interior Design Scientific conference in Kuwait (sponsored by the Public Authority for Applied Education and Training).

In 2000, Dr. Yeun Sook Lee planned the World Congress on Environmental Design for the New Millennium in Seoul, Korea, November 8–22. I did a design charrette for Korean design students and served as a reviewer for educational submissions for the Congress. From November 13–17, I went from Korea to Japan for a presentation to Misawa (second largest home builder in Japan) and then lectured on UD in office environments for the Kokuyo Company in Yokohama.

In 2001, I was one of four women invited to give a keynote talk at the first UD symposium in Kumamoto, Japan (the other speakers were Valerie Fletcher, Pattie Moore, and Molly Story). The following year, the 3rd Universal Design Conference was held in Yokohama, Japan. I participated in six poster presentations and later that week went to Korea where I gave a keynote speech, "UD for an Aging Population," at an Aging-in-place Symposium at Yonsei University in Seoul, along with John Christophersen, an architect from Norway. I have continued to attend and give presentations at

UD meetings and meetings that featured UD components. Because UD is invisible, marketing efforts need to identify good examples and tell why they are good examples.

Exhibits (like the one at the Cooper–Hewitt museum) have helped to promote UD. UD pioneer, Dr. Yeun Sook Lee initiated a UD exhibit and symposium at the Hangaram Design Museum Art Gallery in Seoul in 2004. Dr. Lee worked with Satoshi Nakagama from Japan in planning the program and exhibit. I was fortunate to be one of 10 international speakers at the symposium (emphasis was on product design). Dr. Lee created a book (in Korean and English) that showed pictures of the exhibit and described the new paradigm of UD that she had developed (Chapter 1).

In 2006, I was given the opportunity to develop a UD exhibit in the Art Gallery on the Medical Campus of the University of Southern California (USC). The exhibit, titled "Lifespan Collaborative Strategies," was in place from March 12 to June 28. We held a competition for Interior Design students and the winning posters became part of the exhibit. The faculty from the USC Andrus School of Gerontology worked with us in planning several events at the gallery. Early in June, Dr. Lee brought a group of her graduate students from Yonsei University to the gallery for a seminar and tour on their way to the Environmental Design Research Association (EDRA) meeting in Atlanta, Georgia. When we were at the EDRA meeting, I took the students to meet with Rebecca Stahr from LifeSpring Environments in Atlanta and we got to see the ASID–UD demonstration house (pictures in Chapter 9). One of the problems with the demonstration homes is that, because of cost considerations, they are only open and staffed for short periods.

The demonstration facility developed on The Ohio State University farm campus in 2006 was unique because it is part of a permanent exhibit on UD. Susan Zavotka worked with Lowe's to develop kitchen, bath, and UD exhibits. These have been featured in Lowe's' monthly design magazine and the facility is available for tours.

In 2006, faculty members from The Ohio State University, Jack L. Nasar, Jennifer Evans-Crowley, and Scott Lissner, received funding from the National Endowment for the Arts, as part of its Annual Universal Design Leadership Initiative, for an International Conference on Universal Design and Visitability and an edited book. The book *Universal Design and Visitability: from Accessibility to Zoning* was published in 2007 in Columbus, Ohio. The conference was attended by 200 people (125 on site and 75 online) from as far away as Finland and Japan. Steven Jacobs, president of IDEAL Group, arranged to put the whole conference online for active distance participation. We were able to ship about 30 posters from the UD exhibit "Lifespan Collaborative Strategies" to Ohio State for a gallery display in the registration area. At the meeting, I received the USC Morton Kesten Summit Award of Excellence in Universal Design Education.

In 2010, I received the Irma Dobkin Universal Design Grant of $2000 from the IFDA Educational Foundation for a Universal Design Teacher's Manual to accommodate this textbook *Universal Design: Principles and Models* (2013, CRC Press). The manual *Teaching and Evaluation Strategies for Universal Design* will also serve as an independent resource for design professionals.

This second edition of the UD text includes definitions of UD, applications, examples of best practices, case studies of successful UD projects, and trends and resources that will enable UD to progress into the future to make life more meaningful to countless new generations.

Author

Dr. Roberta Null holds degrees from South Dakota State University, the University of Minnesota, and The Ohio State University. She has taught housing and interior design courses at Purdue University, San Diego State University, and most recently at Miami University of Ohio. She received the 1986 ASID Environmental Design Award for design of training kitchens at the San Diego Center for the Blind. She has been invited to lecture on UD in the United States, Korea, Japan, and Kuwait and is the author of a textbook titled *Universal Design: Creative Solutions for ADA Compliance*, published in 1996. In 2006, Dr. Null received the USC Morton Kesten Summit Award for Excellence in Universal Design Education. Dr. Null is retired and lives in Whittier, California.

Contributors

Allsteel designers wrote sections "Modern Office Ergonomics: Encouraging Healthy Movement from the Seated Position" and "Ergonomics and Your Bottom Line" in Chapter 6.

ASID and USGBC leaders wrote "REGREEN Residential Remodeling Guidelines 2008" in Chapter 10.

Connie Barker of the Environmental Health Network wrote sections "How Universal Can It Be If It's Not Green? How Green Can It Be If It's Not Universal?" and "Ecology House" in Chapter 10.

Julia O. Beamish cowrote section "Center for Real Life Kitchen Design" in Chapter 3.

Franklin Becker cowrote section "Non-Territorial Office: Cornell University's International Facility Management and Workplace Studies Program" and "Offices That Work: Balancing Cost, Flexibility, and Communication" in Chapter 6.

Michael Braungard, North Point Press, a Division of Farrar, Straus and Giroux, New York: Cowrote "Cradle to Cradle: Remaking the Way We Make Things" in Chapter 10.

Margaret P. Calkins, President, I.D.E.A.S., Inc., Kirtland, Ohio: Wrote section "Evidence-Based Long-Term Care Design" in Chapter 7.

James C. Canestaro, AIA, cowrote "Case Study: Facility Management: A Cultural Universal Design Perspective" in Chapter 4.

Nancy C. Canestaro, PhD, cowrote "Case Study: Facility Management: A Cultural Universal Design Perspective" in Chapter 4.

Ken Cherry cowrote "Case Study: San Francisco Redevelopment Housing; Mendelsohn House: Milestone in Low-Income Housing for the Elderly" in Chapter 3.

Owen J. Cooks wrote sections "Case Study: Purdue University's Americans with Disabilities Act Compliance Plan" and "Remodel of the Football Stadium at Purdue University" in Chapter 4.

Clare Cooper-Marcus, Professor Emeritus, Department of Architecture and Landscape Architecture, University of California at Berkeley; Principal, Healing Landscapes in Berkeley: Wrote section "No Ordinary Garden: Alzheimer's and Other Patients Find Refuge in a Michigan Dementia-Care Facility" in Chapter 7.

Mary Lou D'Auray, University of Southern California, cowrote "Case Study: Andrus Gerontology Center Bathroom Renovation Project" in Chapter 3. She also wrote the section "Universal Design and Green Design, an Imperative" in Chapter 9.

Carolyn J. Deardorff wrote section "Defining Universal Design" in Chapter 1.

Brian Donnelly, Industrial designer, wrote "Case Study: Ghost Ranch Evaluation" in Chapter 3. He cowrote section "Industrial Design Curriculum at San Francisco State University" in Chapter 5.

Joan M. Eisenberg wrote section "Technique: Universal Bath Design" in Chapter 8.

JoAnn M. Emmel cowrote section "Center for Real Life Kitchen Design" in Chapter 3.

Mark Epstein, ASLA, wrote section "Building an Alzheimer's Garden in a Public Park" in Chapter 7.

Gilbert Geis cowrote section "Americans with Disabilities Act in Action: Legal Rulings in Sports Arenas" in Chapter 4.

GenShift 2011 researchers wrote section "Lifestages: Redefining the Kitchen—A Comprehensive Study about Generational and Societal Influences on Kitchen Design" in Chapter 8.

Kim Gibbons cowrote "Case Study: San Diego Center for the Blind: Design as a Team Effort" in Chapter 3.

Ricardo Gomes cowrote section "Industrial Design Curriculum at San Francisco State University" in Chapter 5.

Alan Harp wrote section "AWPL's Autumn Chair Features Principles of Universal Design" in Chapter 3.

Sandra C. Hartje wrote research section "Developing an Incentive Program for Universal Design in New Single-Family Housing" in Chapter 3.

Leona Hawks wrote "Case Study: The Utah House—Teaching/Research Demonstration Facility" in Chapter 3.

healthyindoorair.org wrote Chapter 10's "Healthy Indoor Air for America's Homes".

Arlena Hines, Lansing Community College, Lansing, Michigan, cowrote section "Kitchen Design for a Family of Cooks: Kitchen Design Strategies to Increase Home Dining for All Family Members" in Chapter 8.

Karen Hirsch wrote section "The Oral History Interview" in Chapter 3.

InPro wrote section "Bariatric Design 101—An Introduction to Design Considerations" in Chapter 7.

Mark Johnson wrote sections "Principles of Universal Design/Creating Accessible, Equitable Kitchens" in Chapter 2 and "3-D Computer Design Application" in Chapter 9.

Louise Jones, PhD, wrote section "A Universally Designed Academia" in Chapter 10.

Mac Kennedy wrote section "An Award-Winning Universally Designed Home (Senior Housing)" in Chapter 8.

Bradley A. Knopp, from the HTM Group wrote "Case Study: Airport Interior Design: An Integrated Approach" in Chapter 7.

Virginia W. Kupritz, PhD, University of Tennessee, wrote "Case Study: The Effects of Workplace Design Features on Performance for Different Age Groups" in Chapter 6.

Carol Lamkins wrote section "Universal Design Features for Aging in Place—the Bath" in Chapter 8.

Cherie Lebbon, Helen Hamlyn Research Centre, Royal College of Art, cowrote section "Obstacles and Solutions to Inclusive Design" in Chapter 5.

Yeun Sook Lee, PhD, Professor, Yonsei University, Seoul, Korea, wrote section "Universal Design Paradigm and 21st Century Culture—A New Wave of Design for Humanity toward World Community and Future" in Chapter 2.

Tom Lent, Healthy Building Network, wrote section "California 01350 and Indoor Air Quality" in Chapter 10.

LogiSon Sound Masking System wrote section "ABC's of Effective Acoustics" in Chapter 6.

Susan Mack, OTR/L, Homes for Easy Living wrote section "Award-Winning Universal Design Home (Senior Housing)" in Chapter 8.

Joseph A. Maxwell, from the HTM Group wrote "Case Study: Airport Interior Design: An Integrated Approach" in Chapter 7.

Sanjoy Mazumdar, PhD, cowrote section "Americans with Disabilities Act in Action: Legal Rulings in Sports Arenas" in Chapter 4.

Janetta Mitchell McCoy, PhD, Mazumdar, PhD wrote section "A Summer Camp Where Disabilities Are the Norm: Implications and Guidelines for Design" in Chapter 7.

William McDonough, North Point Press, a Division of Farrar, Straus and Giroux, New York: Cowrote section "Cradle to Cradle: Remaking the Way We Make Things" in Chapter 10.

Lee Meyer wrote section "Residential Redesign for Accessibility" in Chapter 8.

Daniel D. Mittleman wrote section "Facilitating Virtual Meetings: Lessons Learned" in Chapter 4.

Ruth Morrow, School of Architecture, Sheffield University: Cowrote section "Obstacles and Solutions to Inclusive Design" in Chapter 5.

Satoshi Nakagama wrote sections "Tripod Exhibit of the Design Process" and "Toyota Universal Design Showcase (Examples of over 400 Products Developed in Japan and Abroad)" in Chapter 9.

Kerry A. Nelson, ASID, IDEC, wrote section "Defining and Designing Public Places" in Chapter 7.

Ann Warble Nienow cowrote section "Environmental Programming" in Chapter 3.

Eunice Noell-Waggoner, LC, wrote section "Light—A Universal Need" in Chapter 6.

Tracy F. Ostroff wrote section "Chicago Universal Design Architecture Competition" in Chapter 9.

Julie Overton, University of Southern California: wrote section "National Home Modification Coalition" in Chapter 9.

Kathleen R. Parrott, Virginia Tech, cowrote section "Center for Real Life Kitchen Design" in Chapter 3.

Mary Jo Peterson wrote section "Universal Design and an Aging Population", "How to Design for Aging in Place" and "Design Trends to Follow for Aging in Place" in Chapter 2 and the section "How to Design for Aging in Place—Kitchens and Baths" in Chapter 8.

Jillianne Pfeifer cowrote "Case Study: One Restaurant's Efforts at ADA Compliance" in Chapter 11.

Joy Potthoff, Bowling Green State University, Bowling Green, Ohio, cowrote section "Kitchen Design for a Family of Cooks: Kitchen Design Strategies to Increase Home Dining for All Family Members" in Chapter 8.

Karim Rachid, Furniture Designer, wrote section "A New Role for the Designer" in Chapter 5.

Victor Regnier, FAIA, USC, wrote section "Design for Assisted Living: Guidelines for Housing the Physically and Mentally Frail" in Chapter 7.

Susan Russell cowrote "Case Study: Re-Creating the School Lunch Kit: Research in Universal Product Design" in Chapter 5.

John P. Salmen wrote section "Home Design for a Lifetime" in Chapter 8.

Elizabeth B.N. Sanders led the research team and also wrote "Case Study: Re-Creating the School Lunch Kit: Research in Universal Product Design" in Chapter 5.

Henry Sanoff, PhD, AIA: wrote "Case Study: Redesigning a Child Development Center" in Chapter 7.

Arricca Elin SanSone wrote section "OXO: Universal Design Innovator" in Chapter 5 and reviewed section "Universal Design Principles Can Help Make Your Home More Functional and Fashionable—Now and in Years to Come" in Chapter 8.

Charles M. Schwab, Architect, AIA, CAPS, CGP, wrote sections "Strolling through the Universal Designed Smart Home" in Chapter 1, "The Universal Designed Smart Home Office" in Chapter 6, which was published in *Special Living Magazine*, Spring 2006 Article, "Technique: Strolling through the Universal Designed "Smart" Home" in Chapter 8, and "Achieving Clean Indoor Air" in Chapter 10.

Anne Seltz, MA, Audiologist, wrote sections "ANSI Classroom Acoustics Standard: Let the Word Be Heard" in Chapter 4, "An Update from the BOSTI Group" in Chapter 6, "Design for Acoustic Environments: Let the Word Be Heard" in Chapter 7, and "Acoustics, the ADA, and ANSI Standards" in Chapter 11.

Aaron Shamberg, MLA, cowrote section "The Occupational Therapy Accessibility Specialist: Consultation for Implementing Universal Design and ADA Compliance" in Chapter 4.

Shoshana Shamberg, OTR/L, MS, cowrote section "The Occupational Therapy Accessibility Specialist: Consultation for Implementing Universal Design and ADA Compliance" in Chapter 4.

William Sims cowrote section "Non-Territorial Office: Cornell University's International Facility Management Program" in Chapter 6.

Dai Sogawa, Editor of the Japanese Universal Design magazine contributed images on UD products shown in Chapter 5.

Timothy J. Springer, PhD, wrote section "Universal Design Implications for Facility Management" in Chapter 4.

Steelcase wrote "Case Study: Wellness in the Workplace" in Chapter 6.

Sylvia Sullivan, Universal Design/Development, Inc., Thousand Oaks, California, wrote section "Hillside Home with Elevator and Other Universal Design Features Residential Elevators: Value Added for a Hillside Home" and cowrote section "Kitchen Design for a Family of Cooks: Kitchen Design Strategies to Increase Home Dining for All Family Members" all in Chapter 8.

Meg Teaford, The Ohio State University, cowrote section "Teaching Universal Design through Community Building Services Learning Techniques" in Chapter 9 and with Lowe's developed at The Ohio State University the section "Case Study: Permanent Teaching/Research Facility" in Chapter 3.

Sandra S. Thurlow, PhD, wrote section "Using Customer Feedback for New Product Design: A Study of Appliance Controls" in Chapter 5.

Jim Tobias, MA, Inclusive Technologies, cowrote section "Barriers, Incentives, and Facilitators for Adoption of Universal Design" in Chapter 5.

Sharon Toji, Access Communications, wrote section "Universal Sign and Wayfinding Design" in Chapter 7.

Gregg Vanderheiden, PhD, Trace R&D Center, University of Wisconsin–Madison, cowrote section "Barriers, Incentives, and Facilitators for Adoption of Universal Design" in Chapter 5.

Linda Welch, CKD, Lansing, Michigan, cowrote section "Kitchen Design for a Family of Cooks: Kitchen Design Strategies to Increase Home Dining for All Family Members" in Chapter 8.

Betty Jo White, **PhD**, Kansas State University, College of Human Ecology: wrote "Case Study: Universal Design Teaching Facility" in Chapter 3.

Nancy Wolford wrote section "Surveying Professionals regarding Universal Design" in Chapter 3.

Noriko Yamamoto wrote section "CHAMP House, a Unique Child Care Facility in Japan" in Chapter 7.

Susan Zavotka, **PhD**, The Ohio State University, developed the section "Case Study: Permanent Teaching/Research Facility" in Chapter 3 with Lowe's at The Ohio State University; cowrote sections "Teaching Universal Design through Community Building Services Learning Techniques" in Chapter 9 and "Case Study: One Restaurant's Efforts at ADA Compliance" in Chapter 11.

John Zeisel wrote sections "Alzheimer's and Environment-Behavior Research" and "Alzheimer's Facility and Attached Garden designed by Martha Tyson" in Chapter 7.

Introduction

1

PIONEERS OF UNIVERSAL DESIGN

The term *universal design* was first used in the 1970s by the staff at the Center for Accessible Housing at North Carolina State University. Its earliest and most important promoter was Ron Mace, Director of the Center and a wheelchair user since a childhood bout with polio. Ron, an architect, product designer, and educator, said that one of the most important changes brought about by the use of universal design was the elimination of the special needs label.

In the introduction to a Department of Housing and Urban Development (HUD) publication on universal design (see Figure 1.1), Ron stated: "Too often older or disabled people live limited lives or give up their homes and neighborhoods prematurely because the standard housing of the past cannot meet their current needs. While a "truly universally usable" house is a goal for the future, many features in houses today already can be made "universally usable." Application of the universal design concept in home construction increases the supply of usable housing available to the public. This is accomplished by including universal features in as many houses as possible. These added features also allow people to remain in their homes as long as they like because Universal Design provides for easy adjustment to changing needs which accompany aging."

Ron was a pioneer, both in his own work and through his encouragement of others. He wrote the introduction to the first edition of this book and later, when asked by a Japanese healthcare provider to recommend the best Universal Design book to be translated and used in Japan, he presented him with a copy of the first edition. The Japanese, with their expanding numbers of aging adults, realized the importance of Universal Design to meet the specialized needs of this huge segment of its population.

Before his untimely death in 1998, Ron was able to see many of his ideas come to fruition on both a national and international scale. "Designing for the 21st Century: An International Conference on Universal Design" was held at Hofstra University in June 1998 (Mace 1998). (Over a third of participants at this meeting came from Japan.) In a presentation at that conference, Ron commented that he was often asked about the terminology, the definitions, and the differences between barrier-free design, universal design, and assistive technology. He said:

> "First, I think it's important that we know the differences between these three things so we can go out and help industry and other people understand some of the subtle but important distinctions between them. When they get muddled, the message becomes vague...."

FIGURE 1.1   House diagram, HUD brochure, 1988.

Barrier-free Design is what we used to call the issue of access. It is predominantly a disability focused movement. Removing architectural barriers through the building codes and regulations is barrier-free design. The ADA Standards are barrier-free design because they focus on disability and accommodating people with disabilities in the environment. In fact, the ADA is now the issue of access in this country. So, what is the difference between barrier-free and universal? ADA is the law, but the accessibility part, the barrier-free design part, is only a portion of that law. This part, however, is the most significant one for design because it mandates what we can do and facilitates the promotion of universal design. But it is important to realize and remember that ADA is not Universal design. I hear people mixing it up, referring to ADA and universal design as one in the same; this is not true....

Universal Design broadly defines the user. It's a consumer market driven issue. Its focus is not specifically on people with disabilities, but all people. It actually assumes the idea that everybody has a disability and I feel strongly that's the case. We all become disabled as we age and lose ability, whether we want to admit it or not. It is negative in our society to say, "I am disabled" or "I am old." We tend to discount people who are less than we popularly consider to be "normal." To be normal is to be perfect, capable, competent, and independent. Unfortunately, designers in our society also mistakenly assume that everyone fits this definition of normal. This just is not the case....

Now, ***assistive technology*** to me is really personal use devices, those things focused on the individual, things that compensate or help one function with a disability.... Another example of assistive technology is my wheelchair (see Figure 1.2). I need it as an individual. It is not a consumer product. It's for me. It's an assistive technology device. Assistive technology really started in the medical industry with durable medical equipment. Here again, people needing equipment are discounted as not being whole people. We are considered to be "patients." We should be grateful to have an oxygen system that keeps us breathing or a wheelchair that provides mobility. Whether or not the product looks nice, is easy to live with, or is available at a marketable price is unimportant to those developing and providing it, *or to those of us who have to use it.*

So, if you could separate barrier-free, universal, and assistive technology distinctly, they would look like this: assistive technology is devices and equipment we need to be functional in the environment; barrier-free, ADA, and building codes are disability mandates; and universal design is design

FIGURE 1.2   Ron Mace cartoon (manga) by Dai Sogawa.

for the built environment and consumer products for a very broad definition of the user that encourages attractive, marketable products that are more usable by everyone. The reality, however, is that the three blend and move into each other.

It is critical for all designers, educators, researchers, and advocates to really understand this relationship between barrier-free, universal, and assistive technology in order to develop and implement truly universally usable designs.

(These are excerpts from a presentation made by Ronald L. Mace, FAIA, at "Designing for the 21st Century: An International Conference on Universal Design" on June 19, 1998. Edited by Jan Reagan for publication in UD Newsline, Quarterly Newsletter of THE CENTER FOR UNIVERSAL DESIGN, Volume 1, Number 4/Volume 2, Number 1, August 1998.)

DEFINITION OF UNIVERSAL DESIGN

The general acceptance of the term *universal design* was itself a big step forward and is exemplified in the following definition from a resolution adopted by the Committee of Ministers of the Council of Europe at their February 2001 meeting:

> Universal design is a strategy that aims to make the design and composition of different environments and products useable for everyone. It attempts to do this in the most independent and natural manner possible, without the need for adaptation or specialized design solutions. The intent of the universal design concept is to simplify life for everyone by making the built environment, products, and communications equally accessible, useable, and understandable at little or no extra cost. The universal design concept emphasizes user-centered design by following a holistic approach to accommodate the needs of people of all ages, sizes, and abilities. It provides for the changes that all people experience throughout their lives. Consequently, universal design is becoming an integral part of the architecture, design, and planning of the built environment.

INTRODUCTION TO RESIDENTIAL DESIGN

STROLLING THROUGH THE UNIVERSAL DESIGNED "*SMART*" HOME

As we move into the 21st century, we are experiencing the phenomenon of multi-generations moving into homes under the same roof. At the same time, people are living beyond 100 years of age. Twenty years ago, I myself was a boomerang grandchild. I moved in with my grandfather after he had a series of strokes. I became a family caregiver overnight.

Architect Charles Schwab (2009) has identified findings from his research that have influenced his home design. I clearly remember him saying over and over, "I want to stay in my own house as long as possible" and jokingly he would quote from the *Wizard of Oz*, "There's no place like home Charlie." It was then that I realized that most homes in the United States did not meet the needs of the elderly and certainly not anyone who used any kind of mobility device. The split-level two-bedroom home did not work for my grandfather and was unmanageable for me as the caregiver. He used a walker regularly and sometimes had need for a wheelchair. **I wish I knew then what I know now**.

Research brought me to what was back then a new concept called Universal Design (UD). Experience was teaching me that this would be the housing choice of the not-so-distant future. I learned that even back then, Medicare and Medicaid funds were declining as the boomers were aging and people are living longer. Now, 15 years later, 10,000 people are turning 65 every day! At current funding rates in the United States, Medicare, which helps pay for nursing home care, is estimated to be without funds by the year 2017. That's only 4 years away! Medicaid is also bankrupting many states. Since hospital stays and out-of-home long-term care are among the largest costs, it is clear that there are changes coming our way.

Long-term care will begin to occur in homes, and homes need to be properly designed in order to accommodate this. "Smart" UD in housing on a massive scale is the solution.

While researching the problem, I also discovered that there were really no home plans that were available to the general public on an easily available basis. What did elderly or disabled people do? Where were they to live with peace of mind and dignity? Thus began the motivation for designing a home plan book based on the principles of UD. As people inquired, they also asked for energy-efficient and sustainable details and specifications. I thought the two were a natural fit. While researching for the book, I discovered that breathing diseases represented the largest disability among children and the fourth largest disability among adults. Clean indoor air is often the second most popular feature that attracts people to "Green" building, ranking immediately behind energy efficiency and conservation.

Thus, clean indoor air became the common denominator in the UD Smart Home as many people with physical disabilities also have allergies and need clean indoor air. Thus became the rationale for the *Universal Designed Smart Home. Smart* in this case refers to "smart" energy-efficient construction. The two form a symbiotic, yin-and-yang natural relationship. After all, what is the point of having an energy-efficient home if you can't get into the front door or use the kitchen or bathroom? And why have an accessible home if you can't pay the utility bills or the poor air quality makes you ill?

I then discovered that the true definition of sustainability is the use and implementation of resources in a manner that does not defer and deplete from future generations. I realized that the standard American home, since it is unusable for many elderly and disabled, is unsustainable in and of itself. Homes NOT designed for everyone require more time from younger generations as caregivers and the cost of nursing home and hospital care is draining financial coffers for future generations. A 2001 Harvard Joint Center for Studies report states that the ratio of taxpayers to non-taxpayers was 5:1; by 2050, it will be 2:1, mostly due to the elderly population.

Multi-generational living is already a popular reality and will continue to be so. And as a result, homes that feature UD will be a great asset and will only increase in value. There are 76 million baby boomers who will live longer than in the past, not to mention, owing to medical technological advances, more children are being born with disabilities than ever. There are currently about 3 million people who use wheelchairs. Because of the cost of war, since 2001, the number of disabled veterans has jumped 25% to 2.9 million. There is also the obesity epidemic and obesity can often lead to physical disabilities. Let's face it, there is no way government programs will be able to support these populations as their needs for long-term housing grow. The time to build UD housing is now.

UD is not simply accessible design but is "design for all." It is truly inclusive, and when properly designed, it functions just as well for people of tall and short stature, people with visual or audible impairments, overweight people, or anybody who uses mobility devices. It is appreciated by people of all ages. A person pushing a baby stroller will agree that no step entries make it easier for strolling and an elderly person will enjoy being able to live in his/her own home longer and "age in place." A child or seated person can help prepare meals in the kitchen owing to varying height countertops. Wider doors and hallways will be appreciated on moving day. Paramedics and firemen will be able to professionally do their job (without turning a gurney sideways to get someone through the door). Ease of use, safety, and convenience are the by-products of UD. It is not a series of codes and it does not use signs or other designations to identify it. The purpose of UD, like assistive technology, is essentially the same: to reduce the physical and attitudinal barriers between people with and without disabilities. A well-conceived UD should be stealthy and can be invisible; it is quite simply good and economical design. Another approach to defining UD was a Master's thesis research conducted by Carolyn Deardorff at Colorado State University (Deardoff and Birdsong 2003).

DEFINING UNIVERSAL DESIGN

The topic of universal design has become more prevalent in educational and consumer literature over the past decade. However, some sources use multiple terms synonymously with universal design, creating confusion about the meanings of all of the terms. This study examined the confusion in the literature between the definition of universal design and accessible design, adaptable design, barrier-free design, lifespan design, and transgenerational design.

During the spring of 2000, questionnaires were mailed to 55 experts in the field of universal design. These individuals were current and former members of the National Advisory Council at the Center for Universal Design, other leaders in universal design organizations, authors found in the literature, and authors referenced in the literature for their knowledge on the issues. Twenty-seven questionnaires were returned, four of which were incomplete and not included in the analysis. The sample consisted of 23 usable questionnaires, representing a 42% response rate. Frequencies, correlation, and T testing were used to analyze the data.

The findings revealed a general sense of agreement with the most prevalent definition of universal design. However, several terms had a stronger degree of agreement than others. Written comments amended as well as augmented definitions. Three of the term definitions, adaptable, barrier-free, and universal design, met the baseline of 80% agreement and no modification was suggested. Based on the comments of the experts and the descriptives used in the dictionaries, three of the definitions could be further clarified.

The definition of lifespan design had 61% agreement among the respondents. Eighteen percent disagreed with the definition. Respondents who disagreed felt it was not different from transgenerational or it sounded like barrier-free. Several concluded that it met the needs of people through changing abilities from birth to death. The dictionary description "over the longest period over which life may extend" speaks to this concept. The literature and experts' comments suggest the definition of lifespan design be changed to "products and environments that consider the needs and abilities of people from childhood to late life." This accounts for needs from birth to death inclusive of changing abilities in this process.

The definition of transgenerational design found agreement from 61% of the experts. Nine percent felt strong disagreement with the definition. Several respondents stated the confusing nature of the term and a sense of similarity or overlap with lifespan design. There were no entries in the dictionary for this term. The findings confirmed the literature in terms of identifying lifespan and transgenerational as being very similar. Revision on the basis of the findings state that the terms transgenerational and lifespan are synonymous. Lifespan considers a birth-to-death timeline. The term transgenerational is repetitious in that all age segments are included at some point in the lifespan. Composition of a more concise group of terms associated with universal design suggests that the term transgenerational should not be used.

Accessible design had the greatest diversity in response to the definition. Sixty-one percent agreed with the definition. Twenty-two percent of the respondents disagreed with the definition. The experts did not want accessibility tied into code requirements; however, codes mandate minimums for access (emphasis added), which makes it an unavoidable use of the term. It describes design that is accessible. The descriptives used in the dictionary are "easy to approach, reach, enter, speak with, or use." The literature and responses from the experts suggest revising the accessible design definition to "products and environments meeting requirements for use by people with disabilities." However, the dictionary descriptives give a more general encompassing definition. It was suggested that accessible be defined more succinctly with dictionary descriptives as "products and environments that are easy to approach, reach, enter, or use."

A major difficulty in the promotion of Universal Design has been the identification and location of excellent examples to illustrate how well the concept can work. This difficulty is exacerbated by a general misunderstanding of how environmental settings affect the way we live and work. The concept of Universal Design promotes the creation of environments that are usable by everyone to the greatest extent possible ... every faucet, light fixture, telephone, bathroom, and entrance. Universal Design is both convenient and profitable. It is a philosophy of design that removed distinctions among varying abilities by adhering to four major principles, identified by Ron Mace as follows:

1. *Universal design is supportive*: it makes environments work for the individual, stressing ease of use and maintenance.
2. *Universal design is adaptable*: it serves a wide range of users whose needs change over time (Figure 1.3).

FIGURE 1.3   Adjustable office chair.

3. *Universal design is accessible*: the everyday comforts and conveniences that "normal" individuals enjoy are provided to all users of the environment. Codes and ADA (Americans with Disabilities Act) guidelines for accessibility are minimal and the interpretation of accessibility is frequently limited to providing access to buildings for people with impaired mobility (Figures 1.4a and 1.4b).
4. *Universal design is safe*: it not only provides environments and tools for the presently disabled but also actually anticipates and prevents disabilities such as repetitive strain injuries (Figure 1.5).

These principles have provided a standard against which products and environments can be measured. As the movement has grown, Universal Design principles have continually been evaluated and refined. In Japan, where the aging of the population has reached crisis proportions, the idea of Universal Design is being recognized as an essential approach. In a presentation on the necessity of Universal Design for Japan, Dr. Satoshi Kose when he was with the Building Research Institute of the Ministry of Land, Infrastructure & Transport, and an expert on Universal Design, stated that: "By the year 2020, 28% of Japan's population will be aged 65 or older, making it one of the world's oldest populations. The idea of Universal Design will become a necessity in this situation." Dr. Kose went on to say that "The … Principles of Universal Design are primarily guidelines for designers." On the basis of these principles, Dr. Kose raised the following six general areas as necessary conditions: "Safety, Accessibility, Usability, Appropriate pricing, Durability and Aesthetics. The first three of these are close to the idea of barrier-free, while the three additional considerations go beyond this, and also the process is an important element of Universal Design" (UDF News).

The Center for Universal Design (formerly the Center for Accessible Design), represented by UD advocates Bettye Rose Connell, Mike Jones, Ron Mace, Jim Mueller, Abir Mullick, Elaine Ostroff, Jon Sanford, Ed Steinfeld, Molly Story, and Greg Vanderheiden, developed an expanded list of Universal Design principles:

1. *Equitable use*. The design does not disadvantage or stigmatize any group of users.
2. *Flexibility in use*. The design accommodates a wide range of individual preferences and abilities.

FIGURE 1.4   (a) No-step accessible entrance at the home of Ruby Trow, Whittier California. (b) Easy access ramp to no-step entrance built in 1990.

3. *Simple, intuitive use.* Use of the design is easy to understand, regardless of the user's experience, knowledge, language skills, or current concentration level.
4. *Perceptible information.* The design communicates necessary information effectively to the user, regardless of ambient conditions or the user's sensory abilities.
5. *Tolerance for error.* The design minimizes hazards and the adverse consequences of accidental or unintended action.
6. *Low physical effort.* The design can be used efficiently and comfortably, with a minimum of fatigue.
7. *Size and space for approach and use.* Appropriate size and space is provided for approach, reach, manipulation, and use, regardless of the user's body size, posture, or mobility.

Not all plans for universally designed facilities have been built, but they can still serve as realistic learning experiences for design students. Several years ago, students of Dr. Phyllis Markussen

FIGURE 1.5   Sharp microwave in a drawer.

(University of Nebraska–Kearny) designed a campus conference facility that featured model senior housing apartments in addition to a universally designed conference center (Chapter 7). An example of a project that was actually completed involved a kitchen design by students at San Diego State University that evolved into the creation of a prototype rehabilitation facility (San Diego Center for the Blind Case Study, Chapter 3).

The facility plans described (University of Nebraska–Kearny and San Diego Center for the Blind) make extensive use of the Universal Design principles, both in their planning (Nebraska) and in their implementation (San Diego).

The expanded list of Universal Design principles will be used throughout the text to illustrate the completeness and correctness of designs described in the case studies that follow.

REFERENCES

Deardorff, C. D., & Birdsong, C. (2003). Universal design: Clarifying a common vocabulary. *Housing and Society, 30*(2), 119–138

Mace, R. L. (Director). (1998, June 19). A perspective on universal design. *Designing for the 21st Century: An International Conference on Universal Design*. Lecture conducted from Hofstra University, Hempstead, New York.

Schwab, C. M. (2009, July 1). America's first universal design home. *Exceptional Parent, 39*(7), 24–28.

What Is Universal Design?

CONTENTS

WHAT IS UNIVERSAL DESIGN?

Over the past few years, the term *universal design* has been showing up in advertisements for a variety of products, has been cited in the text of the Americans with Disabilities Act (ADA) and its accessibility guidelines, and has appeared in the course offerings at many design and building programs in colleges and universities. There has been a wellspring of interest in the concept, but at the same time, there has been a clear lack of any fully realized definition of what exactly is meant by universal design.

Some people define universal design as simply "good" design. This only replaces one word with an equally imprecise term. In the broadest terms, universal design is "design for all people." Universal design, also known as *life span design*, seeks to create environments and products that are usable by children, young adults, and the elderly. They can be used by people with "normal" abilities and those with disabilities, including temporary ones. Still, like many generalities, "designing for all people" provides a nice slogan but doesn't do enough to further understanding and use of the quite remarkable, indeed revolutionary, concept behind the mere definition. (See Chapter 1, "Introduction," for other definitions of universal design.)

Ronald Mace, the architect who coined the term *universal design*, said that one of the most important changes brought about by the use of this term was the elimination of the label "special needs" from segments of the population who are working to maintain or gain their independence. Universal design and the ADA both ask that people be viewed as equal in nature, as having similar rights and obligations, and as deserving of equal opportunity in every facet of society. The approach used by both is "people first," which is the guiding principle of this book and which is reflected in the now-accepted method of referring to people with disabilities: person as a noun followed by disability as an adjective (Story et al. 1998).

People First: Using Language that Dignifies	
Instead of Saying	**Use**
* Handicapped person	* Person with a disability
* Mute, dumb, deaf, blind person	* Person who cannot speak, has a hearing impairment, visual impairment, and so on
* Palsied, CP, or spastic	* Person with cerebral palsy
* Mongoloid	* Person with Down syndrome
* Cripple	* Person who has a physical disability
* Retarded, crazy, mental, defective	* Person who has a mental disability
* Epileptic	* Person who has epilepsy

One of the problems with the phrase "special needs" is that the disability is given more attention than the person (special needs implies that "they" are lacking something "we" have). This results in the individual being further discriminated against by being made to feel separate, different, in need.

Universal design features are good for almost everyone, and as they become incorporated into the everyday world, the similarities between people, as well as their needs for similar products and environments, will become more readily apparent.

Historically, design has met the needs of people with varying abilities by creating specialized (and thus expensive), rather unattractive products and environments to make up for a "missing" ability (prosthetic design) or by removing a barrier to access (accessible, or barrier-free design). Universal design incorporates the features of both of these design styles but goes a step further by looking at people with a more encompassing eye. It defines ways of thinking about and designing environments and products that work for the greatest number of people possible, regardless of their range of ability, body size, or age. One easily recognizable example is the use of levers instead of round knobs on doors. Levers can be used by people with arthritis, small children, and anyone who has confronted a closed door while holding two armfuls of groceries in a sudden downpour of rain (see Figure 2.1).

Universal design asks that designers create spaces and products that adapt to people as individuals and that strengthen their sense of themselves as capable and independent, or even as their needs change, or even if they have a disability that historically would have severely limited their ability to work, play, or do much more than simply exist in the world.

A prevalent example of such adaptable design is the ergonomic office chair that adjusts for height and for forward and backward leaning support (see Figure 1.3). (Ergonomic design is

FIGURE 2.1   Lever handle by Hafele.

discussed in detail in Chapter 6.) This type of design can also be more fully utilized in the home, where people also come in different sizes and deserve decent support. Throughout this book, illustrations of products for both office and home are offered as examples of universal design. As home offices become increasingly accepted, more of the same careful design currently being used in office furnishings will be applied to the home.

FOUR CORNERSTONES OF UNIVERSAL DESIGN

Florida Interior Designer, Susan Behar, American Society of Interior Designers (ASID), views universal design as "an enhancing business strategy needed for survival and renewed opportunities for design professionals. The four A's—accessibility, adaptability, aesthetics, affordability—address the education and design values necessary for incorporating universal design into our environment." Similarly, but with a slightly different focus, this book posits that the following four underlying principles that Ron Mace developed be considered essential for creating a universal design. Universal design must be

1. Supportive
2. Adaptable
3. Accessible
4. Safety oriented

These four interrelated aspects of a design provide useful standards for the measurement and evaluation of new and existing products and environments, and were described in more detail in Chapter 1.

SUPPORTIVE DESIGN

The first test of universal design is that it must be *supportive*: It should provide a necessary aid to function, and it must not, in providing such aid, create any undue burden on any user (see Figure 2.2).

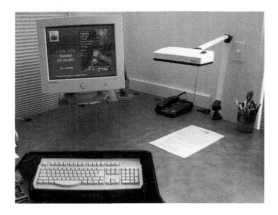

FIGURE 2.2   Supportive design; desk lamp for task lighting.

Consider the lighting used to illuminate a work surface or space. Lack of appropriate lighting can actually lead to decreased visual acuity. And as people grow older, they need more light to see as well. Depending upon the environment (home, work, windowed, enclosed), people need to be able to adjust for different levels and directions of light to support everyday activities. Or consider a kitchen countertop, which should be glare-free and easy to clean. If it lacks these supportive features, it will actually add stress to day-to-day living.

ADAPTABLE DESIGN

Adaptable means that a product or environment should serve a majority of individuals who have a variety of changing needs. One example was mentioned earlier: the ergonomic chair. Adjustable workstations are another example of design that adapts to meet a variety of needs (see Figure 2.3). Desks that adjust in height, with wraparound or detachable surfaces, meet the test of adaptability. Adjustable stands for keyboards and monitors also meet this requirement, as do software programs that allow a computer to display text in varying fonts and sizes. Products such as these are useful for people with visual impairments and for anyone whose eyesight "isn't what it used to be," one of the most common complaints of aging.

FIGURE 2.3   Workstations with desks with rounded edges and personalized task lighting, from Herman Miller.

ACCESSIBLE DESIGN

Accessibility means removing barriers. For universal design and the ADA, such barriers are both attitudinal and physical. By encompassing a broader range of human abilities, universal design subtly empowers individuals, changing a physical environment that currently hinders or harms many people unnecessarily. Universal design promotes accessibility because barriers (to mobility, communication, or well-being) inhibit most people. For example, curb cuts work for bicyclists and parents pushing carriages as well as for people using wheelchairs. However, universal design suggests looking closer at the design of curb cuts by also considering how they affect people with visual impairments. Once this is done, designers may decide to use placement and texture, or a contrasting color or pattern, to alleviate possible accidents that may arise from visual limitations.

Examples of accessible design include placing wall sockets at an 18-inch height from the floor for ease of reach from a wheelchair, using wider, standardized doors, and creating a travel path free of obstacles. These are all features that would benefit everyone—higher wall sockets mean less bending from a standing position, wider doors provide more room for maneuvering packages and furnishings, and a clear travel path helps prevent accidents. Accessible design means rethinking space and equipment to better enable use by all people (Figures 2.4 and 2.5).

FIGURE 2.4   ARJO's Freedom Bath is easy to access; door open with view of seat, grab bar, handheld shower head, and control panel.

FIGURE 2.5 ARJO's Freedom Bath is easy to access, featuring watertight door.

SAFETY-ORIENTED DESIGN

Safety-oriented design promotes health and well-being. It is corrective and preventative. Using contrasting colors or patterns to mark changes in floor level helps protect against tripping injuries (see Figure 2.6). Desks and cabinets with rounded edges are safer than those with sharp edges. Redundant alarms that have both audible and visual signals are safer than those that use only a single cue. A smoke alarm that also provides a light source can save time in exiting a burning building and can also mark the path of exit.

There is more to safety, however, than overcoming physical threats. Safety also entails a sense of psychological well-being, of belonging, of self-esteem and self-worth. Any environment affects both the physical and the psychological, and design must be directed toward both. Safe design must recognize and deal with both physical and psychological challenges.

Products and spaces that allow their users to gain a high level of competence support a state of psychological health. They protect individuals from the loss of independence as they cope with changes that occur naturally as they age. When people no longer work as well within a given environment because of changes in their physical capabilities, they should not have to curtail their

FIGURE 2.6 Contrasting colors on stairs from Eunice Noel-Waggoner.

activities or lower their expectations about what they can accomplish. Rather, the environment should be flexible enough to accommodate changing human needs and abilities.

OTHER BENEFITS OF UNIVERSAL DESIGN

Universal design has several beneficial features besides the four already noted. First, universal design is economical. It does not focus on creating products and environments for an individual disability (since each person manages a disability in a unique way), but instead goes beyond specialization, not only by utilizing existing products in different ways but also by standardizing those things that can be beneficial to everyone. Here, the call for a standard door width (3 feet) stands out because it not only provides access for people using wheelchairs and walkers but also saves time and money for builders, designers, and manufacturers. A wider door is also more convenient for people who need to move furniture through doorways.

The narrowest doors in a house go into bathrooms because early builders thought that no one would be moving furniture in and out of one. Unfortunately, this standard has remained, even though more and more people use wheelchairs and walkers and consequently must have a wider access into the bathroom. Wider halls also need to become a standard within housing. Anyone who is ever in need of emergency service that requires a stretcher will find his or her care severely hampered by narrow hallways that don't allow turns into and out of rooms. Universal design is also aesthetically pleasing; products and environments do not stand out as different or necessary. Designers for "special needs" have frequently given little consideration to appearance, and so people with certain disabilities are surrounded by institutional-looking products. Often the world at large thinks all products that meet varying abilities need to be cumbersome and ugly, a bit like living in a stereotypical sterile hospital ward. The products themselves add to the problem by calling unwelcome attention to the disability at the expense of the individual's need for an aesthetically pleasing environment. Universal design, on the other hand, adapts products that are already accepted by the population at large or creates ones that will be pleasing to everyone.

Finally, universal design is marketable. Millions of Americans want to buy what universal design can provide. As the baby boom generation has changed, so have the main areas where their money is spent. When "boomers" had their own babies, lots of money was spent on products for children. And as boomers grow into later adulthood, more money will be spent on products and environments that allow them to maintain their independence.

Another important marketing consideration is that the ADA will provide access to a fuller lifestyle for millions of people with disabilities who have been kept out of the mainstream. The ADA will bring millions of people who must be accommodated through assistive devices and accessible environments into the workplace. And it will bring these same people into the social world of restaurants, grocery stores, theaters, and so on. Virtually all owners of businesses or services need to consider how universal design can help make environments accessible to these individuals.

Products will have to be developed and provided, and environments will have to be created or adapted, that provide access for people who have been shut out because of a physical or mental impairment. As people in general begin to interact more fully with a broader population that has been socially isolated, more people will want to live in home environments that not only meet changing personal needs but also make it possible to entertain friends who have disabilities. There will be a huge demand for designs and products that provide such opportunities. Universal design products that fulfill all of these standards will be the best alternative to mere prosthetic design and will be among the most highly demanded. As the demand increases, so will the production levels, leading to lower costs and greater availability.

SEVEN PRINCIPLES OF UNIVERSAL DESIGN

The four original principles of universal design have been expanded into seven principles of universal design. These have provided a standard against which products and environments can be measured. As the movement has grown, universal design principles have continually been evaluated and refined. A leadership team (identified in Chapter 1, page 9) was formed at the Center for

Universal Design in 1997 at North Carolina State University. This group developed a list of seven principles under a grant from the National Institute on Disability and Rehabilitation Research.
Specific illustrations of the seven principles are as follows:

1. *Equitable use.* The design does not disadvantage or stigmatize any group of users (Figure 2.7).
2. *Flexibility in use.* The design accommodates a wide range of individual preferences and abilities (Figures 2.8 and 2.9).
3. *Simple, intuitive use.* Use of the design is easy to understand, regardless of the user's experience, knowledge, language skills, or current concentration level (Figure 2.10).
4. *Perceptible information.* The design communicates necessary information effectively to the user, regardless of ambient conditions or the user's sensory abilities (Figure 2.11).
5. *Tolerance for error.* The design minimizes hazards and the adverse consequences of accidental or unintended action (Figure 2.12).
6. *Low physical effort.* The design can be used efficiently and comfortably, with a minimum of fatigue (Figure 2.13).
7. *Size and space for approach and use.* Appropriate size and space is provided for approach, reach, manipulation, and use, regardless of the user's body size, posture, or mobility (Figure 2.14).

Equitable use—This principle was listed as the first priority for Walton D. Dutcher, Jr. because he is a wheelchair user. The premise comes from logic that unless you can get into the home, traverse through it, and have sufficient maneuvering space in each room or area, then everything else is meaningless (Johnson 2008).

Dutcher's approach has been to establish a basic set of features under the term "Life Span Design (Is It Marketable?)." The designer should also be aware of what features are priorities for women.

FIGURE 2.7   Equitable use: design application side-by-side refrigerator, from ASID Atlanta Show Home.

FIGURE 2.8   Flexibility in use: 48″ work aisles, multiple counter heights, 48″ work aisles, D-shaped pulls, a Susan Mack kitchen.

FIGURE 2.9   Flexibility in use, design application, drawer storage adjusts for a variety of pan and dish sizes, from ASID Atlanta Show Home.

FIGURE 2.10   Simple, intuitive: single-lever faucet from ASID Atlanta Designer Show Home.

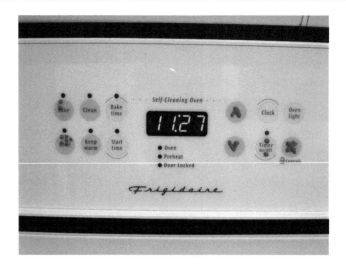

FIGURE 2.11 Perceptible information: microwave oven controls using indicator lights with added fluorescent Braille markings, San Diego Center for the Blind.

FIGURE 2.12 Error tolerance: magnetic induction cooktop is not hot to the touch, from ASID Designer Show Home.

FIGURE 2.13 Low physical effort: design application, remote window blind control, from ASID Designer Show Home, Atlanta, GA.

FIGURE 2.14   Size and space: knee space at sink and cooktop with retractable cabinet doors, by Susan Mack.

Principles of Universal Design/Creating Accessible, Equitable Kitchens

Kitchen & Bath Business (KBB) and Residential Design/Build

Universal design concepts need to be applied to kitchen planning so the kitchen will function for, and benefit, all residents and visitors.

The term *universal design* is sometimes inaccurately used as the politically correct description of compliance with the Americans with Disabilities Act (ADA) and other access codes or guidelines. Universal design is a broader approach that incorporates the needs of all users, not one specific group. Universal design is an ideal whereas code compliance is simply following a dictate.

Understanding the principles of universal design is fundamental to creating environments that ensure the end user's well-being. Universal design is inclusive and equitable, meeting the needs of a variety of people. It is much more than the misconception that it is design limited to medical solutions or access challenges.

Because it is typically used by all occupants of the house, the kitchen is the primary focus of universal design applications.

Following is the Center for Universal Design's Seven Principles of Universal Design with applications that apply to kitchens and the Life Span Design features for the entire house. These principles could be used as a checklist of additional criteria during the design process. The checklist follows:

1. EQUITABLE USE

Design is useful and marketable to people with diverse abilities.

- Provide the same means of use for all users: identical whenever possible; equivalent when not.
- Avoid segregating or stigmatizing any users.
- Provisions for privacy, security, and safety should be equally available to all users.
- Make the design appealing to all users.

Design applications

- Rocker light switch
- Motion sensor lighting, ventilation, or faucets
- Side-by-side refrigerator (Figure 2.7)

Life span design features for equitable use

- A 4-foot-wide walkway from sidewalk or driveway
- No step entries
- Americans with Disabilities Act Accessibility Guidelines (ADAAG)–compliant thresholds
- Thirty-six-inch-wide doors throughout
- Minimum 44-inch-wide hallways
- Electrical outlets and telephone jack 18 inches from the floor
- Switches 42 inches from the floor
- Environmental controls 48 inches from the floor
- Access to the circuit breaker panel; topmost breaker at maximum reach of 48 inches

2. FLEXIBILITY IN USE

Design accommodates a wide range of individual preferences and abilities.

- Provide choice in methods of use.
- Accommodate right- or left-handed access and use.
- Facilitate the user's accuracy and precision.
- Provide adaptability to the user's pace.

Design applications

- Knee spaces with door and storage options, allowing for seated or standing use
- Forty-eight-inch work aisles, ensuring either a perpendicular or parallel approach to appliances (Figure 2.8)
- Multiple counter heights
- Movable (portable) storage
- Deep drawers with or without divider pegs (Figure 2.9)
- Storage for an optional stool

Life span design features for flexibility use

- Blocking for grab bars and shower-seat installations
- Roll-in showers offering adequate maneuvering room for wheelchairs
- Shower system including temperature set/pressure balance single-handle control, diverter valve, and handheld shower
- Side or front transfer access space to commodes
- Single-lever kitchen and bathroom faucets

3. SIMPLE, INTUITIVE

Design is easy to understand, regardless of the user's experience, knowledge, language skills, or current concentration level.

- Eliminate unnecessary complexity.
- Be consistent with user expectations and intuition.

- Accommodate a wide range of literacy and language skills.
- Arrange information consistent with its importance.
- Provide effective prompting and feedback during and after task completion.

Design applications

- Operation of single-lever faucet that moves left for hot and right for cold (Figure 2.10)
- Use of red to indicate hot and blue to indicate cold
- One-step controls on a microwave for preprogrammed recipes

Life span design features for simple and intuitive use

- Thermostats with intuitive features and directive notations or symbols large enough to read and with sufficient color contrast

4. PERCEPTIBLE INFORMATION

Design communicates necessary information effectively to the user, regardless of ambient conditions or the user's sensory abilities.

- Use different modes (pictorial, verbal, tactile) for redundant presentation of essential information.
- Provide adequate contrast between essential information and its surroundings.
- Maximize "legibility" of essential information.
- Differentiate elements in ways that can be described (i.e., make it easy to give instructions or directions).
- Provide compatibility with a variety of techniques or devices used by people with sensory limitations.

Design applications

- Digital temperature control on faucets or ovens that sound and blink when limits are reached
- Lighting controls that light up in the off position and go dark when on
- Smoke detectors with sound and light alarms
- Cooking controls that use numbers and pictures to indicate cooking mode/process (Figure 2.11)
- Use of color contrast

Life span design features for perceptible information

- Contrasting colors of floor materials delineating traffic passages
- Energy-saving illumination
- Various floor materials, all of which comply with the Federal Housing Accessibility Design Guidelines, and colors to different areas

5. ERROR TOLERANCE

Design minimizes hazards and the adverse consequences of accidental or unintended actions.

- Arrange elements to minimize hazards and errors. Most-used elements, most-accessible; hazardous elements eliminated, isolated, or shielded.
- Provide warnings of hazards and errors.
- Provide failsafe features.
- Discourage unconscious action in tasks that require vigilance.

Design applications

- GFCI outlets that reduce risk of shock
- Temperature-limiting faucets that prevent accidental scalding
- Timed automatic shutoff on faucets or ventilation
- Induction cooktops (Figure 2.12)

Life span design features for error tolerance
- Low volatile organic compound materials and finishes
- Fire extinguisher mounted on base cabinet next to the range/cooktop

6. PHYSICAL EFFORT
Design can be used efficiently, comfortably, and with minimum fatigue.
- Allow user to maintain a neutral body position.
- Reasonable operating forces used.
- Minimize repetitive actions.
- Minimize sustained physical effort.

Design applications
- Lever handles
- Remote window controls (Figure 2.13)
- Remote controls for cooktop ventilation
- Motion-activated appliances and controls
- D-pulls on cabinetry
- Conveniently located storage and appliances (raised dishwashers, counter height microwaves and ovens)

Life span design features for low physical effort
- Lever handles on all swinging doors
- Handles that accommodate grasp on all sliding or folding doors
- Kitchen, bathrooms, and other cabinet doors fitted with D-shaped or other styles of handle that facilitate grasp and are ergonomic
- Garage door opener

7. SIZE AND SPACE
Appropriate size and space are provided for approach, reach, manipulation, and use regardless of user's body size, posture, or mobility.
- Provide a clear line of sight to important elements for any seated or standing user.
- Make the reach to all components comfortable for any seated or standing user.
- Accommodate variations in hand and grip size.
- Provide adequate space for the use of assistive devices or personal assistance.

Design applications
- Split double ovens at comfort height
- Storage accessories installed within the universal reach range (15 to 48 inches above finished floor)
- Movable (portable) storage
- The 30-inch × 48-inch clear floor space in front of all appliances
- Knee space at a sink, cooktop, work counters or adjacent to tall appliances (Figure 2.14)

Life span design features for size and space
- Sixty-inch turning radius in bathrooms and kitchen
- Lazy Susan cabinets in kitchen where indicated
- Pull-out shelves in kitchen base cabinets
- Front controls on the range or cooktop
- Switches for garbage disposal installed in the front apron of the sink's base and range/cooktop exhaust fan/light switch installed in the base cabinet next to the range

Life Span Design Features are needed because of the following:

- The aging of the population
- The advancement of medical technology increasing the capability of recovery at home, after treatments for traumatic injury or disease, which also increases the potential for disability
- The need to address the Federal Budget deficit with proposed solutions forcing either a tax increase, which is unlikely, or the diminishment in social services and health care owing to a reduction in legislation advancing community-based services and support
- Potential for increases in the number of disabilities on the basis of obesity, the nation's number one health care issue, and the potential for a greater number of disabilities originating in an increasing number of low birth-weight babies
- Parents living with their children because of the rising cost of alternative housing or long-term care facilities

UNIVERSAL DESIGN PARADIGM AND 21ST CENTURY CULTURE, A KOREAN VIEW

A NEW WAVE OF DESIGN FOR HUMANITY TOWARD WORLD COMMUNITY AND FUTURE

According to Interior Design Professor Yeun Sook Lee, Universal Design is the 21st century's creative paradigm that enables us to realize human dignity and equality. Universal design is the reconstruction of the meaning "for human" in that design work is for making human life more rich and convenient. Universal design means design of environment or products that satisfy the customers' needs as much as possible. It is also defined as a process to make life pleasant for all people by having more of them conveniently use the products or environment.

The origin of universal design started from barrier-free design, as the position of the weak in the society, who were overlooked during the 20th century, became recognized. Barrier-free design was upgraded to universal design as a more comprehensive society became necessary. Universal design will evolve into a concept with a dynamic and aesthetic value that respects each individual's characteristics in the new age of actual diversity where everyone in the general population is considered. In other words, universal design has paid attention to the weak in the society (e.g., the handicapped, the aged, females, low-income groups, and children) from the beginning to the present. From now on, however, it will develop to an even broader meaning for those temporarily under a physical handicap, common people who have potential handicaps, and eventually all people with different individualities.

The 21st century's design culture will grow as universal design for the time being and eventually develop into a culture described by the term "design for human being." If the 20th century is characterized with material and machine-based civilization and thus a standardized gray culture, the 21st century will be a movement to a culture that attempts to recover humanity, rediscover the value [of a human being], and increase it.

Some 21st century design enables society members to make selections for their lifestyles and preferences. Examples include children's furniture systems that can be easily transformed to match the stages of a child's growth, office system furniture that can be freely arranged to conform to the nature of tasks and preference, and DIY (do it yourself) furniture that can be created to fit the resident's lifestyle.

The fundamental concept of universal design can be described as "user-oriented design in the post-industrial society" that intends to escape from the standardization of the early industrial society based on the mass production system. One can also infer the basic attributes from various terms many people use to express the concept of universal design: flexibility, variable design, life span (design that easily accommodates changeable stages of life), transgeneration design, adaptable design where environment itself can vary with needs, and additive/expandable design that gradually accommodates change of needs. Analysis of these terms reveals that they more broadly accommodate ranges in time, situation, customer needs, and user types.

UNIVERSAL DESIGN AND PRINCIPLES

The concept and principles of universal design have been gradually developing and differentiating for application to designing actual products and environments and as standards for environmental evaluation in the future. When the term *universal design* showed up the first time, it just meant "excellent" design and, in a broader meaning, "design for all." Such expressions, however, are not enough to understand and use the highly excellent concepts hidden in universal design.

Early advocates in the field have interpreted universal design to make it more easily understood through the use of identifying "principles." The following were the four principles in the initial phase: supportive design, adaptable design, accessible design, and safety-oriented design.

These principles were later deemed to be too abstract and limited to explain the characteristics of universal design. As a result of efforts to present a more concrete description, therefore, seven principles were proposed: equitable use, flexibility in use, simple and intuitive, perceptible information, tolerance for error, low physical effort, and size and space for approach and use (Figures 2.7 through 2.14). In Dr. Lee's opinion, these principles, however, are also insufficient to express the functionality and beauty of universal design. That is, the principles are mainly terms against negation to relatively emphasize universal design in comparison with the 20th century's design. The seven phrases, in particular, are too concrete to properly express the potential characteristics actually contained in universal design. It is expected that a checklist will be developed beyond the seven principles from the viewpoint of the whole.

As one of such possibilities, a 2006 exhibit at a Korean art gallery presents, as shown below, the nature and future direction of the 21st century's universal design using the 24 characters of "Beautiful Universal Design" as the initials of the words. It is necessary, prior to discussing the topic, to estimate the direction of change by summarizing how the meaning of the 21st century's design is changing, how it will change, and if it must change eventually. The 21st century's design will pursue the characteristics of Deep, Ethical, Sensible, Integrative, Gentle, and Nourishing in its description.

FIGURE 2.15   Dr. Lee's 21st Century Paradigm for Beautiful Universal Design © 2006.

The design of the 20th century's industrial society was shallow in focus, as it became immersed in mass production and commercialism without any time for serious and careful consideration on the impact the result would exercise on the human, societal, and ecological systems. To recover this, the work of the 21st century should be reborn as a deeper design that fills the hollow center (Figure 2.15).

The 21st century, in addition, should contain the characteristics listed below.

- It should include ethics that cherish the human itself and participate in helping satisfy the social morality.
- It should revive the sense and emotion of the human, who has been once considered as senseless, and make them interact.
- It should be based on an integrative approach less satisfying or emphasizing one aspect that might result in loss of others.
- It should have a gentle characteristic that treats and leads the users more comfortably.
- It should add not only visible convenience but also a positive image of life. That is, it should nourish the life.

Deep: Full of deep thought and sound philosophy, not just with a thin wrapper
Ethical: Based on ethics that respect public interest and the ecosystem
Sensible: Having a combination of human senses function and being positively appealing
Integrative: Having overall versatility that a design should maintain and not exclude or isolate users
Gentle: Interacting with users in a more comfortable relationship
Nourishing: Nourishing life like food with plenty of ingredients

Over the meaning of the 21st century's design is added that of "UNIVERSAL." The nine characters of "Universal" represent Useable, Normalizing, Inclusive, Versatile, Enabling, Respectable, Supportive, Accessible, and Legible. The description of each of the words is as follows:

Useable: Able to be easily used without causing any inconvenience or interruption of use
Normalizing: Having users naturally exist in the group without being excluded or discriminated against
Inclusive: Inclusively satisfying a broader range of user group, rather than a limited one
Versatile: Having versatile characteristics, rather than one function or characteristic, and is thus able to be complementarily or selectively used
Enabling: Enabling, rather than having users become frustrated, give up, or become depressed
Respectable: Having users maintain dignity and self-esteem without damaging their self-esteem or making them feel inferior
Supportive: Having users easily adapt to the daily life by supplementing their physical or mental limitation
Accessible: Able to be easily accessed and have users easily access information
Legible: Clearly informing users so that they can understand

The said universal design should be recognized as "Beautiful Universal Design," in order to develop more successfully. Characteristics necessary to be satisfied in addition for such development can be presented with the nine characters of "BEAUTIFUL": Benign, Enhancing, Adorable, User-friendly, Touching, Inspiring, Flexible, Useful, and Loving. These attributes are not the condition of universal design but characteristics required to realize the potential to ensure the possibility of success in life.

Benign: Occupying an advantageous position in the market economy, attracting consumers
Enhancing: Making the user's life more pleasant and leading it to the pursuit of a better quality of life
Adorable: Attracting the user's eyes and mind
User-friendly: Easily accommodating the user's sensitive needs

Touching: Causing the spirit of impression and gratitude beyond simple consumption and use

Inspiring: Inspiring enthusiasm and enhancing the spirit

Flexible: Able to be flexibly changed or selectively used for a given situation and atmosphere

Useful: Efficiently satisfying the purpose that the user intends

Loving: Having users feel a serious attitude, enthusiasm, and love during the process to create the designed output

Figure 2.15 shows the combination of the words from "Beautiful Universal Design," the 21st century's design described above. In summary, the 21st century's "Design" characteristics should be satisfied first for a design to be "Universal," and the design should additionally satisfy the "Beautiful" characteristics to develop to a more successful and more popular universal design.

How should today's people foresee this overwhelming change and get prepared for it wisely? A paradigm is a thought for looking at things and phenomena, which is formed by itself and explained as a cultural phenomenon, but which can also be predefined and developed in a short while as a directional guide to the future society.

The design of the new era that creates all the environment and objects must play the role of deconstructing the industrial society itself, as well as creating a paradigm of consciousness that designs the area of thoughts that are not visible.

Universal design is a paradigm that explains the various phenomena we are experiencing now and it is the characteristic of the civilization that will come in the future and the principle that creates the future society. It is a fast-growing paradigm that is enough to declare "Freedom by Design" since the "Unlimited by Design" exhibit at the Cooper Hewitt Museum in New York in 1998.

In conclusion, "universal design" foresees the flow of the society to suggest the direction of design that contributes to the improvement in quality of life, and, also, this can be called a meta-design exhibition that makes and consolidates the flow of changing times and life.

Universal design is a design paradigm to comprehend the diverse requirements and characters of modern consumers. While the Renaissance highlighted "human" and rediscovered the value of human culture that had been hidden under the veil of religion during the Middle Ages, universal design is a second Renaissance movement; thought and design converge to rediscover human reality and living value that have been overlooked and hidden by industrialization and reestablish the relationship between persons who pursue quality of life and the artificial creation that is the setting for human life.

Universal design is the orientation and destiny of designs of the 21st century, a century that is requested to comprehend diverse, complicated, and multilateral human needs, users of diverse categories, and dynamic changes.

The products of the 21st century based on the efficiency of mass production need to transform as the market pursues quality in the age of globalization. In addition to the very small number of products that have succeeded in the international or domestic markets, there are many more products that are struggling for qualitative growth, and the success of a national economy depends on them. As long as products are targeted at human users, the "pro-human" and "user-oriented" designs will remain the way to the orbit of success. Because the products that attract and delight consumers through good design can contribute to the industrial development of a nation, governments should take up universal design with a strategic approach and keen interest.

How to educate the next generation entails the problem of how to shape the human resources of a nation to meet the needs of the changing population. The most important thing is to make designers read and adapt to the trends of social changes and give them an opportunity to challenge their own thinking. During the past, the institutional education has gone through the strains of educational reforms for a short period, but still has many limits in helping teachers and students communicate and digest the latest information in the fast-changing digital information society. In a society where the gaps between generations, social classes, and the powerful and the weak have broadened, universal design will help students understand the relationships between members of the society in the "Age of Diversity" from a more healthy point of view, prepare them to cope with the future, and complement the institutional education of future designers.

The 21st century is the Age of Diversity, and for the various members of society to give and receive the respect for each other's personality and individuality, it is essential to breed the culture of coexistence. Even though individualism may get deeper and the awareness of the community may become weaker than those of the 20th century, they are different since they will develop on the foundation of the culture of coexistence and fusion. Universal design will provide us with the basis of a united society where people can coexist in mutual respect. Migration, accelerating with globalization, together with the aging society, will bring about more tourists, and our understanding of them and the construction of urban foundations will be additional factors influencing the national economy. Designers are needed to facilitate the development of support systems to maximize the benefits to society of these changes.

As the index of aging scores go higher, the atmosphere of society will become more depressed, and the productivity of people will suffer. Besides, the individuals devoid of activities will sense that they are aging faster. In the future society, while it is important to nurture the culture of leisure to comfort the long period of old age, it is more important to create the environment for the older citizens to participate in productive activities. We should reorient the characteristics of the environment favorable to young workers to those for the old population. Japan already began many years ago building more accessible, easy-to-use, safe, and pleasant environments to accept the aged workers in factories avoided by young workers. In this regard, universal design can be viewed as an issue related to the rearrangement of the future industrial environment that must be taken seriously and actively supported.

Universal design is not only a transitory trend in one field but also a massive trend that can have impacts on all the areas of our society.

UNIVERSAL DESIGN AND AN AGING POPULATION

Achieving universal design is a complex process that has been further complicated by lack of recognition of need by an increasingly aging population. Older people must first risk the stigma of being considered "disabled" after a lifetime of relative independence and competence. Few are willing to admit that their bodies are changing and that their demands on their environments are also changing, despite the availability of design information and assistive devices. The autonomy one gains in youth creates a sense of internal control over one's life; however, in old age, decreased income, role losses, and diminished vigor and social status combined leave many with a sense of being controlled by external circumstances. To correct the imbalance, residential environments for the elderly must be designed so that they ensure safety, health, comfort, convenience, and, most importantly, independence—features that provide a psychological feeling of internal control over the environment.

Why should designers be concerned with the relationship of universal design to aging? One reason is that a whole new population of older adults has resulted from people living longer because of improved health care and nutrition. Because Americans over 60 will number almost 88 million by 2030, the aging of the population has colossal implications for our society. Organizations wishing to survive in the 21st century will need to rethink their marketing strategies, products, and services to meet this burgeoning segment of consumers. Not only are individuals living longer, the 76 million boomers have now started joining this older population.

Marketing to a specialized "elderly" group is like trying to reach a nonexistent segment of the population. After all, we generally consider "elderly" to be 15 years older than we are, whether we are 15, 55, or 85 years old!

No one wants to be labeled as "old" or "disabled," so any products or services that are targeted for this group are sure to meet with a lack of interest. Universal design is a general approach that provides maximum appeal and benefits for all age groups, rather than to a niche market such as the frail elderly or disabled/wheelchair users. We need to take a universal design approach to ensure safe, comfortable, convenient, and accessible dwellings for people of all ages, sizes, and abilities, not just for the elderly.

Because universal design is invisible and inclusive, it will meet the needs of an aging population (and all others). This is not only an advantage but also a challenge. It means that we need to show good examples of universally designed products and environments and explain why and how they incorporate universal design. We also need to be able to evaluate whether products and environments can, to the greatest extent possible, be used by everybody. Some basic questions related to the four universal design ideas originally developed by Mace (Chapter 1, page 6 and Chapter 2, page 3) would include the following:

- *Is it easy to use and take care of?* A bad example would be the inclusion of highly reflective, shiny, dark-colored granite countertops.
- *Is it adaptable?* Can it be used in a variety of ways to accommodate different users? Wide doorways are great for wheelchair access, as well as for moving furniture and equipment into rooms.
- *Is it accessible?* For example, no-step entrances are easy for wheelchair access and also provide easy access for walkers, strollers, and rolling luggage.
- *Is it safe?* A microwave oven placed over a gas range—too high for established kitchen design guidelines and not safe because of the presence of an open flame—is a fire danger, especially for older persons.

The American Association of Retired Persons' (AARP's) annual member surveys consistently reveal a strong preference by seniors to remain in their homes—to "age in place." In 10 years of surveys, over 80% of respondents expressed this preference. Household members of all ages have roots in their communities and strong emotional ties to their homes. They prefer to remain where they are. Few people want to move solely because their homes no longer fit their needs (AARP 2013).

Despite the growing need for supportive home environments, few consumers seem to be requesting universally designed houses or modifications of their existing homes. Research has shown that the people who would most benefit from modifications were not demanding access features. The National Home Modifications Action Coalition Steering Committee suggested three major reasons for the lack of demand for these services.

First, the demographics are skewed. The population of people with disabilities is, in reality, a very diverse group. The group encompasses those with physical, sensory, and cognitive disabilities. It also includes those with mobility impairments, grasp and reach limitations, and vision and hearing problems. Members of the group represent all ages, income levels, family types, and residential locations. When one adds the complications of disabilities related to aging, which can change from day to day, it becomes clear that this is not a large homogenous group that can be targeted with a single marketing or design concept. The home modifications and products they are seeking span an extremely wide range.

Second, it is a mistake to confuse the population of people who could benefit from more universally designed housing with those who recognize the function of universally designed living environments and who will seek them out in the marketplace. Relatively few people, at any point in time, have serious enough impairments to seek specially designed products or a housing change. (A chart developed by Dr. Gill from the Helen Hamlyn Research Centre in London shows percentages of persons with various disabilities [Figure 2.16].) It helps clarify the small percentage of persons with serious impairments (i.e., wheelchair users). Although wheelchair users represent a fraction of the population of people with disabilities, these individuals and others with severe impairments are the most likely to recognize their needs and act to make changes. The independent person who might benefit from a grab bar or stair railing is not acutely aware of specific personal needs. For an aging population concerned with maintaining independence, there is frequent denial of need for products and home modifications designed to create a supportive environment for people with disabilities. One client told her health care provider that she would rather fall than have someone come to her house and see a grab bar in her bathroom.

Recent developments in information dissemination have helped to educate users and caregivers to the potential offered by universally designed products and environments. Instant messaging through the Internet, recognition of the importance of showing beautiful examples, and a more enlightened design community have all combined to emphasize the need for universal design. There is a growing awareness of the importance of universal design, partly because many

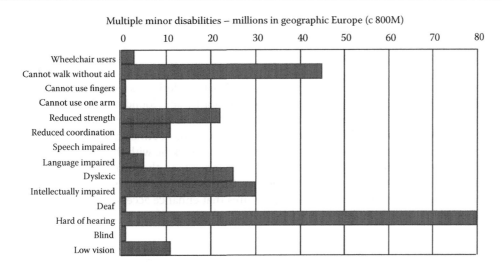

Multiple minor disabilities – millions in geographic Europe (c 800M)

FIGURE 2.16   Chart of incidence of disabilities by type in Europe.

baby boomers, who are responsible for the care of aging parents, recognize the need for supportive environments and are demanding universally designed products. The national meeting of the AARP held October 14 to 16, 2004, was a good example of the trend to show beautiful examples of universal design. In the exhibit hall, the Generations Home, part of the Universal Design Series for Advantage Homes (www.advantagehomes.com), was one of the most popular exhibits. Designed by AARP architect Laurence A. Weinstein of Shared Solutions America, the home featured many good examples of universal design. Larry gave scheduled presentations titled "Good Ideas for Better Living in Universal Design Featured Homes." Each presentation was followed by a tour of the house in which he would explain the universal design concepts incorporated in the design. A colorful folder included a floor plan, a listing of universal design features, and a description of the goals for the house, "to save energy, simplify housekeeping and maximize living to make a house a home through all life brings for all stages of life" (Figure 2.17). The work of Larry Weinstein in bringing the universal design concept to the attention of AARP members is an example of a trend to the gained acceptance of universal design as essential to the creation of supportive housing for everyone.

FIGURE 2.17   Computer rendering of the exterior view of a house with a step-free entrance at Manorwood Homes by Larry Weinstein. Also see the step-free entrance in Figures 1.4 and 1.5.

PIONEERING GROUPS

THE AMERICAN SOCIETY OF INTERIOR DESIGNERS (ASID)

ASID has been a leader in promoting the use of research data by designers (evidence-based design). The study "Aging in Place: Aging and the Impact of Interior Design" found that thinking about the home and how it factors into their futures was as important to the people surveyed as planning for financial security or health care. According to the study, most Americans (82%) want to live in their homes even as they require assistance and care. In working with the baby boomer population, designers have an opportunity to help clients understand the long-term impact of their decisions and to advise them on how to combine aesthetics and function. Interior designers from around the country mentioned the following features that enhance access, mobility, and ease of use without detracting from appearance (ASID 2012).

- People want a house that is easier to maintain (57%) and easier to get around in (40%). Nearly a quarter of those surveyed (23%) felt that downsizing would make the home easier to maintain and access. Those choices influence selection of everything from flooring, wall, and counter surfaces to organizer/storage units.
- A significant number of those surveyed want upgrades and changes that enhance lifestyle. Appliances (from compact washer/dryer units to top-of-the-line refrigerators/freezers) and fixtures were high on the list.
- Nearly two-thirds (62%) planned for social activities that can influence design decisions if they intend to entertain or host guests.

SPECIAL DESIGN FEATURES

In addition to the features mentioned in the previous list, design researchers identified the following special design criteria.

- Locate the master bedroom and bathroom on the ground floor, but remember that increased security is a desire of an aging population. Ground floor bedrooms/master suites need to have security features that provide a safe feeling for users.
- Create good traffic flow with few or no step-ups or step-downs between rooms.
- Select furniture that is easy to move and to get into and out of.
- Reduce the number of pieces of furniture to make it easier to get around.
- Use smaller kitchen appliances that are more lightweight and easier to handle.
- Eliminate soft padding under carpet. A hard commercial padding gives a more sure-footed surface.
- Replace basement laundry rooms with smaller washers and dryers that fit into the bathroom or a utility closet.
- Install an elevator or stair lift.
- Employ color contrasts as an aid to visual acuity.

(A Vision for Aging in Place, produced by the ASID Aging in Place Council)

THE NATIONAL KITCHEN AND BATH ASSOCIATION (NKBA)

Because kitchens and baths create the most problems for an aging population that are best addressed by the use of universal design, NKBA has taken a leadership role in providing training and resources for designers (NKBA 2013). Resources include yearly national meetings (KBIS) and trade publications such as *Kitchen and Bath Business* and *Kitchen & Bath Design News*.

The best known advocate for universal design in NKBA is Mary Jo Peterson. Mary Jo has been recognized nationally and internationally for her expertise in kitchen and bath and universal design. Her regular column in *Kitchen & Bath Design News* has become a "must read" for kitchen and bath designers. She has consistently emphasized universal design in her writing and design work. Her case study, "How to Design for Aging in Place," builds on her background experience as a designer and educator. It is followed by excerpts from recent columns in *Kitchen & Bath Design News*.

HOW TO DESIGN FOR AGING IN PLACE

Often when I have a conversation with other designers regarding design for aging, they talk about parents or grandparents who found that neither they nor their homes were prepared for changes that occurred with age. Yet, many of these changes are predictable, and we should, as designers, be able to incorporate concepts into our designs that accommodate these changes.

Ten years ago, my incredible Irish grandmother died at age 99, and her living situation in the last years of her life propelled me to pay closer attention to design as it pertains to aging. Although basically healthy and alert at age 95, her strength and hearing had diminished to a point that she could no longer safely live in the home she'd known since before I was born. This began my commitment to creating environments that would enable older adults to live comfortably in their homes for the duration of their lives.

Likewise, personal experiences prompted New York–based interior designer Rosemary Bakker to focus her efforts on these same considerations. In a *New York Times* article, Bakker related that when her mother returned home from hip surgery, she was faced with trying to maneuver a walker through narrow doorways, over area carpets and raised thresholds, and into a kitchen where she couldn't bend to get food out of the refrigerator or reach pots and pans. Additionally, there was no bathroom on the first floor.

As the article said, "Suddenly, the house that had suited her for 42 years was a time bomb waiting to go off." Out of this experience came Rosemary's book, *Elderdesign*, a resource for designing and furnishing homes for later years. She also has a recent book, titled *Revitalizing Your Home, An AARP Guide to Beautiful Living for the Second Half of Life* (Bakker 2010). In that book, she has listed the universal design features that she sees as essential. She emphasized that design for a lifetime was not about spending a lot of money, but thinking about the most important changes you can make to enjoy a healthy, long, and fulfilling life. Rosemary's list addressed the three main barriers to aging in place: difficulty getting in and out of the home, difficulty getting around the home, and an accessible bathroom. Her list of top universal design features included the following:

1. A no-step entry to your home
2. Wider doorways and hallways
3. A bedroom and bathroom on the same floor
4. Walk-in no-threshold showers
5. Reachable, rocker-style light switches
6. Lever-style door handles and faucets
7. Kitchen appliances with automatic shutoff features
8. Nonslip flooring, especially in the kitchen and bathroom
9. Abundant and even lighting
10. Grab bars in bathing areas
11. Comfortable furniture and furnishings
12. Telephones and doorbells with a low-frequency tone

As you can see, the two lists on universal design features share many of the same goals.

As kitchen and bath designers, we have an opportunity and a responsibility to design flexibility, access, and support into each project we approach. For the first time in history, there are more people over age 65 than under age 25, and many of the homes we live and work in were not designed for this new longevity.

AARP surveys show that more than 80% of the people over age 60 want to remain in their homes. Accessible kitchens and baths are critical to this desire. To this end, designers must address issues of safe movement throughout the home, as well as efficient yet accessible use of the spaces we design.

Visitability Ordinances had been passed in several cities in the US. This means that new housing built in that area must meet a "Visitability" standard. To meet the Visitability standard, the home must have at least one entrance that is accessible, wide enough passage through the main floor, and at least one main floor bathroom that is designed for use by people of varying abilities. The standard frequently also recommends at least one bedroom on the main floor (or a room that can be converted to a bedroom).

As we advance in the aging process, our senses decline, and our flexibility, balance, stamina, and reflexes diminish. These are often compounded by side effects of medications and chronic or injury-related conditions such as arthritis or limited recovery from broken bones. Rather than reacting with denial or depression, we can design to accommodate and support these changes.

Both the kitchen and bath begin with the entry, where the clearance at the opening, maneuvering space around the door swing and threshold must be examined. Sometimes, just reversing a door swing and installing a swing-clear hinge and lever handle to the door will improve the situation (see swing-clear hinges and lever handles in Valinda Martin's home, Figure 2.18). An important break with tradition is to replace the raised threshold at the door with a flush conversion at the entry.

Once in the kitchen or bath, lighting is a critical element to reduce risk. We all realize that generous amounts of task and ambient light are important. In addition, we must avoid glare and use contrast appropriately to guide the way. If we increase the bath lighting, we must also carefully light the path to the bathroom, perhaps with a motion-activated system, as aging eyes will be blinded by a quick change from darkness to bright light, or the reverse.

Criteria for selecting flooring should include slip resistance and some forgiveness for dropped items, or to prevent serious harm in the case of a fall. Pattern or contrast should be gentle and can

FIGURE 2.18   Expanding hinges and lever handles in Valinda Martin's home.

be used to help guide the eye. Area rugs should be taped to the floor or, better, eliminated. The point at which flooring materials change should be flush.

Clear floor space for maneuvering is relatively easy to accomplish in the kitchen, but often difficult in the bathroom. Pocket doors or reversed door swing help in the bath, as do vanity designs that increase open space below. Particularly in traditional 5′ × 8′ bathrooms, converting from a tub to a roll-in shower will also help.

In the kitchen, planning retractable doors to conceal an open knee space will open up the clear space and provide a storage spot that easily converts to a place to sit while working (see Figure 2.14). A big consideration for storage is that our height decreases as we age, and for many of us, it becomes less comfortable to bend or climb. Design that provides generous storage between 24″ and 45″ off the floor eliminates the need to do either.

This means that the backsplash area in the kitchen becomes valuable for storage, and at least some wall cabinets might be lowered. Rolling storage in either the kitchen or bath can provide flexible clear floor space and storage that moves to the point of use as desired. Open or glass door storage help accommodate changes in memory.

Support in the form of railings or grab bars is essential as we age, yet this is often distasteful to both clients and designers. With the broad offerings of grab bars today, many coordinated to match accessories, the challenge is minimized.

While this is only the tip of a very big iceberg, it offers food for thought. If my grandmother's home had been designed to support her, she might have stayed comfortably at home. If Rosemary's mother's home had been originally designed to be supportive, her trauma might have been reduced. A 45-year-old couple and their teenage children might not seem to need "aging in place" design, but their parents or friends might. If we can design beautifully and incorporate solutions respectful of our elders, why wouldn't we?

Universal design and access don't have to be the only focus of our efforts. Rather, we can make them an integral part of every project we design.

DESIGN TRENDS TO FOLLOW FOR AGING IN PLACE

1. People are looking at ways to create a level entry and generally more open floor space for easier maneuvering.
2. People are asking for plans that include a master suite on the main floor, so "upstairs" becomes guest or other flex space.
3. Easy maintenance is cited as a top priority.
4. In the kitchen, fewer wall cabinets are key.
5. Appliances are being placed at comfortable heights
6. Drawers should be called the great equalizer. All of us benefit from bringing things closer to us without straining. Even most moderately priced cabinetry offers drawers. Today's drawer appliances—dishwashers, microwave ovens, refrigerators, to name a few—are in demand from people, and particularly those in the boomer segment.
7. In both the kitchen and the bath, not only drawers but also doors that go away are a strong trend. Whether they fold to the side, swing up, recess in, or otherwise open, getting them out of the way while one is accessing what's behind them is good. Hardware has been created and improved so that there have never been more options.
8. In the bath, let's start with the vanity area and talk knee spaces. People are requesting designs that include the option of sitting for at least some of the tasks at hand. Today's lavatory designs invite an open knee space and they are, at last, a strong trend.
9. Have you ever seen more choices in toilets? The trend is definitely to comfort—or right height seats and, given the choice, plan more than one height when doing a whole house, to accommodate changing needs and varied user heights. Although this trend is still designer instigated, clients are responding strongly to the many additional options becoming available, including heated and self-closing seats, personal hygiene, dual-flush, and so on.

FIGURE 2.19 Drive-in shower in Valinda Martin's house.

10. No-threshold showers have begun to take hold with not just designers, but with builders and consumers as well. When containment of water is planned carefully based on the size, position, direction, and amount of water flowing, the extent of waterproof layer, the slope of floor, type and location of drains, and the plan for doors, curtains, or open entries, this is a wonderful choice (Figure 2.19).

11. Through applications of technology new to many of us, we have found a home modification system that linked the homeowner to both community and health care. This system allowed a homeowner or resident to take his/her blood pressure or check blood sugar levels or otherwise communicate with the health care provider. As we move from our current population of 35 million over aged 65 to an expected 70 million in 2030, the benefits of staying at home, where we wish to be, are immeasurable.

(Kitchen & Bath Design News, January 2007, p. 29, "Planning and Design: Ten Design Trends to Follow for Aging-in-Place")

HOUSING EDUCATION AND RESEARCH ASSOCIATION (HERA)

A major contribution to universal design has come from housing and design educators in family and consumer sciences programs. Prototype facilities and programs have been developed by housing and design educators, both in Cooperative Extension and in academic programs. Their emphasis has been in the identification of needs related to housing environments. One example is a 1992 bulletin published by the North Carolina Cooperative Extension Service and written by Wilma Hammett, Extension Interior Design Specialist. The bulletin titled Life-cycle Housing:

Furnishing a User-friendly Home, included a list of nearly 40 design guidelines and suggestions such as "If carpet is used, select a low-level loop pile. Pile height should be no greater than 1/4 inch. Use thin padding underneath or no padding and glue the carpet down directly to the subfloor." Such design guidelines provided specific information for home builders and others involved in the design and renovation of housing.

Members of the Housing Education and Research Association (HERA) (formerly the American Association of Housing Educators) have been instrumental in establishing prototype universal design facilities. An early success story involved the kitchen design project by students at San Diego State University that evolved into the creation of a prototype rehabilitation facility at the San Diego Center for the Blind (Null 1985).

As the universal design movement has progressed, there has been an increased recognition of the importance of universal design in the creation of supportive kitchen and bath environments, especially in residential settings. Kitchen and bath design were emphasized in the first edition of the universal design book, *Universal Design, Creative Solutions for ADA Compliance,* and the choice of case studies and appendix material for this edition continue this emphasis.

At Virginia Tech, Julia Beamish and her colleagues have established a teaching/research kitchen center (described in detail in Chapter 3, "The Design Process"). It includes the "real-life" kitchen designed by Mary Jo Peterson (Dobkin and Peterson 1999). The Kitchen Center is used for teaching kitchen design courses (Virginia Tech has an endorsed college program from the National Kitchen and Bath Association, NKBA) and also for conducting research (the design team has received research funding from NKBA and a recent project involves revising the NKBA Kitchen and Bath Design Guidelines). In an article titled "Universal Design in Residential Spaces," DeMerchant and Beamish (1995) identified many specific features and recommendations. They listed over 30 features related specifically to kitchen design and described how these features should be incorporated into successful universally designed kitchens. Because a universally designed kitchen has invisible solutions that enable family members to be self-sufficient within this living environment, DeMerchant and Beamish highlighted universal design features and detailed how they work.

Leona Hawks, extension housing specialist at Utah State University, was instrumental in the design and construction of the Utah House, a prototype facility that incorporates universal design and green design concepts. B. J. White, housing professor at Kansas State University, developed a complete universal design research and teaching facility in a former household equipment laboratory at Kansas State University. Mary Yearns, extension housing specialist at Iowa State University, developed universal design prototype facilities that have been shown at the state fair and other locations. These exhibits have recently been installed in a classroom at Iowa State University so they can be used for resident design instruction as well as for community outreach. Marilyn Bode, extension housing specialist at The University of Georgia, created a design checklist for residential design when she was at Kansas State University. This checklist has provided valuable guidelines for designers of senior housing.

Susan Zavotka at The Ohio State University has established a program for training Cooperative Extension professionals, students, and sales associates from Lowe's Home Improvement stores in universal design concepts. There exist many other examples of HERA's efforts in promoting universal design (HERA 2003).

OTHER ORGANIZATIONS

The Center for Universal Design in Raleigh, NC, was started by Ron Mace and continues to be a leader in the dissemination of information on universal design. One of the main contributions of the Center and USC is the operation of the listserv:

HOMEMODIFICATIONS-LIST@LISTSERVE.BUFFALO.EDU

The listserv has provided information on universal design and home modifications to a national and international group of designers, occupational therapists, and gerontologists. It has also provided a forum for discussion and a means of communicating with members (AOTA 1999–2013; ASA 2013; IDEC 2013).

John Salmen, of Universal Designers and Consultants, Inc., developed and publishes *The Universal Design Newsletter*. Ed Steinfelt founded the IDEA (Center for Inclusive Design & Environmental Access) at the State University of New York at Buffalo.

Other professional organizations provide computer access to divisions within their main structure; for example, after 9/11, Danny Mittleman, a member of the Work Environments division of the Environmental Design Research Association (EDRA), presented information about a Computer Command Center his design group had created for the navy. Several weeks after 9/11 (when there was a great deal of concern about flying), Mittleman sent a message to the Work Environments Listserv, in which he shared what he and his design colleagues had learned regarding the design of videoconferencing spaces. One of the design directives suggested a light gray wall color as the best background for a multicultural, multicolored group. He also mentioned that a backlit glass board with fluorescent markers (like restaurants use to show their daily menus) was more effective than the usual whiteboards for videoconferencing. This trend to the free sharing of design ideas and information has been a major contribution of the Internet to the acceptance of universal design (EDRA 2012).

Another pioneering group is the National Resource and Policy Center for Housing and Supportive Services at USC–Andrus Gerontology Center (Jon Pynoos). They have been very involved in the Home Modifications network and universal design education. A recent renovation of the Center bathrooms will serve as a prototype of universal design applications in a commercial building. You may view images of the renovated bathrooms at www.usc.edu/dept/gero/hmap/homemods/pages/bathroom/ and in Chapter 7 of this textbook.

ENVIRONMENTAL PROGRAMMING

Environmental programming is a flexible research and design tool that lends itself successfully to universal design techniques. In the first stage, there is development of a detailed user profile. Research at this stage requires learning as much as possible about the group or groups for which the design will be done. The concept of universal design implies that design recommendations should consider four basic categories of people: adults, children, older adults experiencing age-related changes, and people of all ages with disabilities. These groups provide a starting place for exploration into recommendations for universal design for residential spaces (DeMerchant and Beamish 1995; Null 2003).

Creating a detailed user profile will help housing and design professionals and students see beyond the aesthetics or perceived functionality of their designs to empathize with people who will actually use the spaces. It is essential to develop sensitivity in order to understand what it is like to have less-abled bodies, use prosthetic devices, and have difficulty interpreting and accessing spaces. Designer/design educator Janetta McCoy from Arizona State University needed to include all the basic categories of users in her design of a summer camp for children with disabilities. She had to consider the needs of each disability group, that is, autistic, AIDS, cerebral palsy, and low vision. The cabins and communal spaces all needed to be supportive environments for the children, their caregivers, parents, counselors, and other adults involved with the camp operations. She spent a great deal of time in the first stage of environmental programming. Her research and development of user profiles for each group helped establish design directives/criteria that could be used by design and construction professionals. Students and professional designers need to incorporate into their housing and design philosophies a sensitivity to people, rather than maintain a "codes and checklists" mentality.

The relation of the elderly to their physical environment has been widely explored and documented. Knowing as much as possible about each user group leads to the most successful user profile. For example, industrial designer Joseph Koncelik has described many of the physical changes that take place during the aging process. With all normal age-related losses, the rate of onset is different for different individuals. Changes in human strength, flexibility, hearing, vision, and mobility all come as part of human aging. In discussing changes in vision, his listing includes yellowing of the lens and a reduction in the ability to discriminate closely related colors (Koncelik 1996). When developing design criteria for living environments to be used by older persons, these changes in vision would determine color schemes that feature contrasting bright colors instead of muted tones and monochromatic color harmonies. Because the rapidly increasing aging population presents major problems in terms of residential environments, it is important to translate research knowledge into design criteria for homes.

One of the trends is the publicity for universally designed environments in the popular press. For example, in summer 2003, there was an entire page in *USA Today* that showed an accessible playground.

Another trend has been for the adoption of universal design standards as part of a community-wide effort. A brochure that contained cartoon representations of universal design applications throughout the community in the Shizuoka Prefecture (Japan) included the statement "Shizuoka Prefecture promotes Universal Design to help create a 'Comfortable Shizuoka' where all residents and visitors can be comfortable. We are adopting the concept of Universal Design in various areas so that all People, regardless of age, gender, or being disabled or not, can act freely and live active lives." In the United States, a group of builders adopted a voluntary certification to create "Easy Living Homes" with the goal of a more livable and more visitable home for everyone. Following the Visitability standards originated by Eleanor Smith, these homes featured at least one No-step entrance, Easy passage—a 32″ width for every interior passage door and Easy use—no less than one bedroom, a kitchen, some entertainment area, and at least one full bathroom with designated maneuvering space—all on the main floor. The coalition for advancement of the Easy Living Home was located in Decatur, Georgia. Despite some administrative problems that led to the discontinuation of the Easy Living Home organization, the residences that were built continue to serve the needs of their owners.

Universal design is truly the promise for the future. Its relevance was recently highlighted in a report in a popular news weekly that listed "universal design architect" as one of the top 20 hot job tracks for the 21st century. Actually, that listing just touches the surface. Universal design will create career opportunities not only for architects but also for housing educators and researchers, builders, urban planners, product developers, interior designers, facility managers, and gerontologists (AIA 2013).

REFERENCES

American Association of Retired Persons (AARP). (2013). *AARP*. Retrieved June 11, 2013, from www.aarp.org.

American Institute of Architects (AIA). (2013). *The American Institute of Architects, AIA*. Retrieved June 11, 2013, from www.aia.org.

American Occupational Therapy Association, Inc. (AOTA) (1999–2013). *The American Occupational Therapy Association, Inc.* Retrieved June 11, 2013, from www.aota.org.

American Society on Aging (ASA). (2013). *American Society on Aging | Developing leadership, knowledge, and skills to address the challenges and opportunities of a diverse aging society*. Retrieved June 11, 2013, from www.asaging.org.

American Society of Interior Designers (ASID). (2012). *American Society of Interior Designers*. Retrieved June 11, 2013, from www.asid.org.

Bakker, R. (2010). *AARP guide to revitalizing your home: Beautiful living for the second half of life: Reimagine, redesign, remodel*. New York: Lark Books.

Dobkin, I. L., & Peterson, M. J. (1999). *Universal interiors by design: Gracious spaces*. New York: McGraw-Hill.

EDRA | The Environmental Design Research Association. (2012). *The Environmental Design Research Association*. Retrieved June 11, 2013, from www.edra.org.

HERA | Housing Education and Research Association. (2003). *Housing Education and Research Association*. Retrieved June 11, 2013, from www.housingeducators.org.

IDEC | Interior Design Educators Council. (2013). *IDEC | Interior Design Educators Council*. Retrieved June 11, 2013, from www.idec.org.

Johnson, M. (2008, November 18). Creating accessible, equitable kitchens. *Residential design + build, November*. Retrieved November 7, 2008, from www.forresidentialpros.com/article/10355175/creating-accessible-equitable-kitchens.

NKBA | National Kitchen and Bath Association. (2013). National *Kitchen and Bath Association*. Retrieved June 14, 2013, from www.nkba.org.

Null, R. L. (1985). Environmental programming: Case study of using this versatile design teaching tool. *Journal of Interior Design, 11*(2), 139–168.

Null, R. L. (2003). Commentary on universal design. *Housing and Society, 30*(2), 109–118.

Story, M. F., Mueller, J. L., & Mace, R. L. (1998). *The universal design file designing for people of all ages and abilities* (Rev. ed.). Raleigh, N.C.: School of Design, the Center for Universal Design, NC State University.

The Universal
Design Process

3

CONTENTS

INTRODUCTION

When people engage in any creative activity, they are, in effect, designing. And this range of creative human activity is enormous. Making a decision is creating—it is choosing from a vast array of possibilities to meet the present needs. Decisions surrounding the outfitting of a home office with a workstation, chair, and storage equipment require a creative act. One has to consider the space the items will occupy, the present color scheme, the general layout of existing furnishings, the lighting, and the types of uses the office will serve (reading, writing, drafting, computer use). Preferences of style and materials (and perhaps the desires of others who might use it) must be considered and then balanced with budgetary constraints. Finally, the furnishings can be purchased.

Once everything is in place, one has to evaluate how successful the design was: Does the color scheme work? Do the desk and cabinets fit within the room size and furnishings? Is it comfortable for long-term work? This evaluation may even lead one to return the desk or chair and begin the process over again.

All these processes will be going on either covertly or overtly. Design is often done as part of a team, enlisting the aid of potential users or a professional designer or at least asking the sales representatives advice. A checklist can be used to evaluate the success of the design. Or one may have the good fortune of doing everything right the first time without any serious, conscious planning at all. However, as the design process becomes more complex, as projects demand more expertise from the designer, accountability increases. The designer must be able to draw on a wealth of knowledge from a variety of sources.

When designers and other professionals decide to be involved in working toward a universal design, they must welcome (and even ask for) new supportive technologies such as wheelchairs that climb stairs and other barriers (thus truly empowering their users), computers that make it possible for people with virtually any disability to interact more fully with their world, and air quality control devices that make the unseen world healthier. Designers also welcome innovations in established policies including companies that empower their workers by giving them the choice of working out of home offices, or that focus on responsibility and accomplishment (not just on adherence to established routine), and that take seriously human rights laws, such as the Americans with Disabilities Act (ADA), that provide protections against discrimination.

Each of these design innovations may make the world work better. To the extent that they accomplish that goal, they will gain recognition for their success, and their principles will be adopted throughout the culture. The creativity they display will then serve as a model, leading to further innovations. Nearly 10 years ago, an automobile designer from Volvo (Otto Sterner) worked with wheelchair user Dennis Sharp to develop a prototype car design (see interview results under "Interviews" later in this chapter). These are all examples of universal design (Price et al. 2004).

EMPATHY AND UNIVERSAL DESIGN

Consider again the "people first" approach introduced in Chapter 2. A firm grasp of this concept will be driven by one's capacity to *empathize* with people in a variety of circumstances. More than anything else, universal design is defined by empathy. Empathy is the foundation from which all good design is built. It is the quality that makes each individual's self-worth something that can be nurtured. The ability to empathize and act on the awareness it awakens is the critical factor in creating a universal design. But empathy is not a quality that suffuses American life. The work world is not generally considered an empathetic environment; workers are often considered to be present to fulfill the employer's needs, with little regard given to their own needs. The same is true of many social agencies.

One of the subtlest requirements of the ADA is its demand that people reevaluate how they interact with each other, in addition to a shift in how people are managed at work, how services and entertainment are provided, and how people treat one another. If the ADA is merely seen as a requirement to make environments accessible, the main point of the legislation is missed since compliance can often be met in the physical environment by following the ADAAG (ADA Accessibility Guidelines) to the letter and making any minor changes that may be requested by a qualified person with a disability.

But what is central to the ADA and universal design movement is a belief that the built and imagined world does not work for many. Universal design can be further defined as "informed, empathic, creative activity focused on altering the known environment." The known environment is not only physical space and objects but also human beings interacting. As such, universal design encompasses every discipline. It's a unifying circle within which designers, architects, lawyers, sociologists, psychologists, educators, and managers interact. The universal design process begins with empathy, while the techniques needed to accomplish whatever goal one has in mind will follow (as the Mendelsohn House case study in this chapter will show).

Everyone is doing some kind of design most of the time. But often, that design means simply going through the motions, repeating what is already there because it's what is known. The universal design process is not just the methodological design of building a house or tinkering with a few specifications to make a slightly different version of an existing environment: Universal design asks for the design of an entirely new creature. Designers are being asked to embrace the chaos of discovery, to put imagination before skill—and in the process re-create the world.

UNIVERSAL DESIGN TECHNIQUES

In addition to asking whether a design meets the criteria described in Chapter 2 (supportive, adaptable, accessible, and safe), the designer can also utilize several techniques throughout the process to help create a universal design: participatory design, modeling and role-playing, and post-occupancy evaluations.

PARTICIPATORY DESIGN

The universal design process is essentially participatory design. The basic steps are outlined briefly here and will appear again in many of the case studies throughout the book. These studies reveal a variety of approaches to participatory design—all of which are useful, none of which should be taken as the only approach.

No matter what type of project one is involved in, once the tasks to be completed have been identified, there are certain key techniques to help define goals and realize solutions. Henry Sanoff illustrates his expertise in participatory design in the case study on school design in Chapter 7, "Redesigning a Child Development Center," and his research on school design.

In the design community, this stage is often termed *programming* and includes the following activities:

- *Establishing goals*: working with all involved parties to determine the general parameters of the project
- *Conducting research*: learning about the people and spaces involved, and also educating everyone in relevant areas of study
- *Uncovering concepts*: identifying the pieces of the conceptual framework that are guiding the project
- *Determining needs*: stating the constraints on the project (financial, spatial, time, and so on) to help define the proper approach
- *Stating problems*: drawing on all of the above to divide the project into logical components that can be solved through design directives

These activities are further delineated in the "Environmental Programming" project described in the Teachers' Manual Chapter Environmental Programming reveals how techniques that develop design criteria (such as those put forth in the strategy sections of Chapter 2) can greatly simplify the design process. Environmental Programming has been an integral part of the work done by designers in the Advanced Wood Products Laboratory (AWPL) at Georgia Tech University. Examples of such products are the Autumn Chair (Figure 3.1) and a complete kitchen and reception desk (Figures 3.2 and 3.3), using universal design strategies. Perhaps the most important, and often most neglected, step in programming is *research*, which includes several methods proven to work well for gathering information.

- Background reading
- Interviews
- Surveys and questionnaires
- Observation
- Focus groups

AWPL'S AUTUMN CHAIR FEATURES PRINCIPLES OF UNIVERSAL DESIGN

The Autumn Chair Project was developed through research performed at Georgia Tech's Center for Assistive Technology and Environmental Access (CATEA) through the initiative of CATEA Director Joseph Koncelik. "The motivation for this project came from a need to demonstrate the capability of advanced wood processing machinery and also to create a product that exemplified the mission of the center," says Koncelik. "The Autumn Chair is a clear demonstration of product development drawn from the relationship between aging and disability. Not only is the chair drawn from the functionality related to meeting requirements of moderate disability, the chair has a marvelous and unique aesthetic that would not have been achieved without careful attention to human need."

Originally designed in 1997 for the ambulatory elderly, the chair incorporates many features to accommodate this population and is also a success in universal design. "Universal Design is defined as the design of products and environments to be usable by all people, to the greatest extent possible, without the need for adaptation or specialized design," says Alan Harp, industrial designer for AWPL. "By designing for comfort and ease of ingress and egress for the elderly, the chair has been found to be extremely comfortable to the average population."

The specific features of the chair (shown in Figure 3.1) that speak to the ideals of universal design and the benefit it provides are as follows:

- Extended armrests with elbow relief: The extended armrest provides a solid grasp of the chair as one approaches or leaves the chair, an important feature to prevent falls and to assist in egress. The elbow relief prevents pinching of the ulnar nerve while seated.
- Lower seat height: seat height is 16″ from floor, 1 inch lower than typical. This provides a height that can accommodate a much greater percentage of the population.
- Sculpted seat pan: The formed seat disperses the pressure points to create a very comfortable sitting area.

FIGURE 3.1   Autumn Chair.

- ■ Wide footprint: The 23″ wide × 24″ deep footprint makes the chair extremely difficult to tip.
- ■ Integral lumbar support and curved back: The design of the back with its deep curve and radiused part design cradles the back and provides great support for the lower back.

To highlight these principles of universal design, the Autumn Chair was chosen to be featured at the 1998 Unlimited By Design Exhibition at the Smithsonian's National Museum of Design in New York.

An aspect of the chair that garners attention at first glance is the leaf inlay in the seat of the chair. The image is of a maple, oak, and cherry leaf fashioned in their respective species of wood. "As with many successful designs, the leaf inlay was created as a method to disguise a flaw," says Harp.

During construction of one of the original hand-built prototypes, a large knot appeared as the seat was being sculpted. Lacking the resources to make another seat blank, the design team created the inlay to cover the knot so as to preserve the work already completed. The inlay has now become one of the most eye-catching elements of the Autumn Chair.

Harp completed the initial CNC (computer numerically controlled) prototype in August 2001 as the subject of his Master of Science degree through Georgia Tech's Industrial Design program. A short test run utilizing six different species of wood also was completed to investigate the relationship of wood properties as it relates to CNC machining. The species used were red oak, black cherry, yellow poplar, southern yellow pine, soft maple, and plantation-grown mahogany. It was found that cherry and oak performed the best overall, with mahogany and maple following closely in terms of machinability without tear-out and splintering. The poplar and pine both proved to be problematic.

During the early stages of the small test run, the idea of the rocking chair was brought up by popular demand of people touring AWPL. According to Harp, "The creation of the rocking chair is a good example of the advantage of designing in CAD (Computer Aided Design) and producing parts on CNC machinery," says Harp. Harp was able to design and fully prototype the rocking chair in less than one day because of the ability to quickly adapt drawings already on file.

"The direct translation from a 3D Solid Model to an actual part creates highly accurate parts unmatched by conventional methods," says Harp. "The CNC technology can also manufacture parts with extremely high accuracy and repeatability. Creating the chair parts on the CNC Router was about 12 times faster than traditional methods."

The mission of Georgia Tech's Advanced Wood Products laboratory is to move US production of finished products using wood and wood composite materials into an internationally competitive position. The three components of the mission are research and development, education and training, and demonstration. Other examples of universal design demonstration projects are a complete kitchen in Figure 3.2 and office furniture (reception desk in Figure 3.3). For more information on these products, visit the AWPL web site at www.arch.gatech.edu/AWPL/alan.html.

FIGURE 3.2   Kitchen.

FIGURE 3.3   Reception desk.

BACKGROUND READING

There will always be information to be gathered that applies specifically to each project. However, some broad areas of study should be considered that affect the design of all environments including, but certainly not limited to, the trade journals for design, aging, and disabilities. Research available on the elderly is especially useful when one is interested in the design needs of people with disabilities. As noted in Chapter 2, the population is aging rapidly, and as people age, they encounter many of the disabilities that must be accommodated under new laws. A great deal of data exists on the specific needs of this population group as their bodies change, all of which is pertinent to the creation of a universal environment. The following are computer resources that provide research summaries from a variety of design publications: Design Research Connections—EDRA (Environmental Design Research Association) (subscription) and Eye on Design—ASID (American Society of Interior Designers) (subscription available at no cost to members of ASID). These summaries can provide valuable information to designers because they represent the most important, credible research available. They are tools that will provide designers with the research information needed to help them establish reliable design directives for their projects.

INTERVIEWS

The more people one can interview about the project—their specific needs and especially current dissatisfactions—the better. Interviews should include as many of the people involved in a project as possible: current and prospective users, architects and designers, psychologists and sociologists (if pertinent), financiers, and so on. Karen Hirsch, a designer, discussed the importance of personal history interviews in the Strategy section later in this chapter. Her article offers very useful advice on ways to get needed information from prospective users.

A designer might interview a person with a disability as in Otto Sterner's questioning of wheelchair user Dennis Sharp for the Volvo Prototype Car (Figure 3.4).

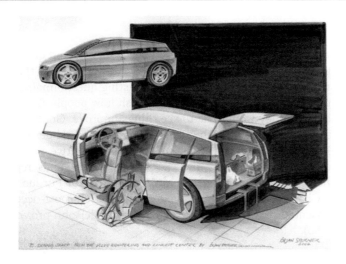

FIGURE 3.4   Volvo prototype car.

The following is the summary of an interview.

RE: Universal Design concept car

Dennis Sharp was invited by Volvo to participate in a project for the 2002 Auto Show at the San Diego Convention Center. Mr. Sharp was asked to describe his "dream car." Volvo designers then created a concept drawing based on his "wish list" that was displayed at the show.

Mr. Sharp is a professional graphic designer who formerly worked as an art director in the space program. He is a wheelchair user who now devotes much of his energy to advocating for persons with disabilities and consulting on issues of access for the disabled.

The car as described includes many features that make it a good example of universal design:

Low body to ground height provides easy access
Auto body has height leveling system
All doors are self-closing and remotely controlled
Wheelchair access ramp and storage
Driver's seat swivels out
Large interior storage
Driver hand controls; dual mode for other operators
Leather ergonomic seating
Full perimeter roll protection
All-around air bags
Voice command navigation and audio system
X-design seat belts

In today's market, wheelchair users have few choices for personal transport. Vans that are converted for wheelchair access are extremely expensive to purchase and maintain. This type of car could provide a more comfortable and economical alternative choice from the traditional van for many consumers.

SURVEYS AND QUESTIONNAIRES

It is important to create a general profile of the audience. Surveys and questionnaires are one way of doing this. Designers need to do research on creating questions that do not point an audience to a specific response as well as modifying existing questionnaires that have proven effective in other areas. Questions should try to establish how people are currently using the environment and the ways they would like to see things improve. Good examples of the use of surveys and questionnaires appear in the case studies of universal design research done by Sandra Hartje (2004) and Nancy Wolford ("Developing an Incentive Program for Universal Design in New Single-Family Housing" and "Surveying Professionals regarding Universal Design") included here. In addition, Owen Cooks' description of Purdue University's ADA Compliance Plan in Chapter 4 and the post-occupancy research (post-occupancy evaluation [POE]) that was an essential part of the planning for the Mendelsohn project are described in this chapter (Figure 3.5).

OBSERVATION

Oftentimes one can gain a better understanding of a project and its audience by observing people in action, either within an existing environment for a renovation or within a similar environment if it's something new. For example, time-lapse photography could be used to build an observation profile of patterns of use for an airport waiting room, or for the use of a social space at a retirement community. Such observations can serve as useful supplements to the feedback given through the survey/questionnaire.

OPINIONS, AWARENESS AND USE OF UNIVERSAL DESIGN FEATURES IN SINGLE-FAMILY HOUSING

Universal design -- attractive spaces, features and products that are functional for most people throughout their lifespan, regardless of ability or disability.

Q1. This question is intended to measure your awareness, use and cost of features that people often want in their homes. *Please circle one response from 1 (very aware) to 4 (unaware) for the "awareness" section. Please circle one response from 1 (very often) to 4 (never) for the "use" section. Please circle one response from 1 (yes, there is added cost), 2 (no added cost) in the "cost" section.*

UNIVERSAL DESIGN FEATURE	AWARENESS OF FEATURE				USE OF FEATURE				ADDED COST OF FEATURE	
	VERY AWARE		UN AWARE		VERY OFTEN		NEVER		YES	NO
A. Single story, no steps between areas	1	2	3	4	1	2	3	4	1	2
B. 5'×5' clear turn space in major activity areas										
1 living area	1	2	3	4	1	2	3	4	1	2
2 one bedroom	1	2	3	4	1	2	3	4	1	2
3 kitchen	1	2	3	4	1	2	3	4	1	2
4 one bathroom	1	2	3	4	1	2	3	4	1	2
C. New/existing multi-story: space for eating, sleeping, laundry, and bathing on ground level	1	2	3	4	1	2	3	4	1	2
D. 36″ wide doorways	1	2	3	4	1	2	3	4	1	2
E. Lever handles on doors	1	2	3	4	1	2	3	4	1	2
F. Thresholds flush or no higher than 1/2″	1	2	3	4	1	2	3	4	1	2
G. Halls minimum 42″ wide	1	2	3	4	1	2	3	4	1	2
H. One entrance at ground level, no steps	1	2	3	4	1	2	3	4	1	2

FIGURE 3.5   Wolford Opinion Questionnaire: universal design features.

Case Study: Ghost Ranch Evaluation

Industrial designer, Brian Donnelly, conducted a participatory design evaluation of the Ghost Ranch Resort in Arizona. He chose a focus group of disabled users to evaluate several parts of the resort setting. Photos of the participants in the research group are included here (Figures 3.6 and 3.7).

Figures 3.8 and 3.9 illustrate the proposed design changes that are based on the research process.

FIGURE 3.6   Photograph of participants in the cafeteria line.

FIGURE 3.7   Participants on Ghost Ranch path.

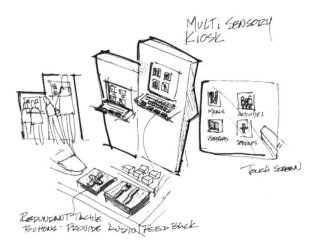

FIGURE 3.8   Proposed design changes for vending machine.

FIGURE 3.9   Proposed changes to outside path.

FOCUS GROUPS

By holding meetings with a cross section of an audience, focused on specific aspects of the project, designers can generate creative solutions to known problems and further identify the issues under consideration.

The approach outlined here has been called *participatory design*, which requires designers to allow the people who will be affected by their decisions to play a significant role in shaping the project. This approach may also lead to a partnership of sorts with a wide range of professionals and laypeople. For some projects, one may want to actively involve psychologists, architects, designers, engineers, facility managers, project managers, people with specific disabilities, and so on. The more knowledge gathered up front, the less chance there will be for unnecessary problems to develop after the project has been completed. The case study on the San Diego Center for the Blind in this chapter is another example of participatory design. Once enough background data have been collected, one will be able to develop a number of design concepts to reach the goals set for the project. Design concepts set down specific strategies for solving individual aspects of a project; for example, if privacy is an issue, one of the design concepts will be to identify ways to ensure that concern is met while keeping within the overall design plan.

MODELING AND ROLE-PLAYING

Once design concepts have been developed, goals have been set, and a plan of action has been created, the designer may feel ready to begin the actual implementation of the design. However, there are a few further steps needed to ensure against unexpected failures. The first of these is *modeling*.

A designer does this to some extent when putting the design onto paper or creating a mock-up of the environment as it is projected to be when finished. Consider taking this process one step further by doing a full-scale working model of one representative space (e.g., a typical patient's room in a health care facility). By doing so, aspects of the environment that may not have been planned for can be identified: for example, furniture management and lighting problems. Graphic designer Jill Mitchell, who had worked in the architectural office of Charles and Ray Eames, told me that they always created a model of the buildup they were designing—these were used to check lighting patterns, and so on, but were not ever shown to clients because the client would be so impressed with the dollhouse-like model that they liked everything!

A second technique to consider is *role-playing*. This can be as simple as taking part in the regular routine of the intended audience. For example, hospital personnel can carry out typical procedures in the model patient room. New technology is available that puts one within an environment before it's been constructed. Computer-aided design programs with full-motion animation and virtual reality have both proven very useful for this purpose to designers with computer expertise.

As work is done to implement a design, there is frequently a need to reevaluate goals and design concepts. Through such activities as modeling and role-playing, one may discover that preliminary designs are not as effective as they could be. Perhaps the project as a whole may be redefined, and in the process, this particular environment will be brought closer to a fully supportive, universal design.

POST-OCCUPANCY EVALUATION

Designers evaluate each project in some manner from the first day they begin to consider what it will demand. Evaluation is an ongoing process. However, a specific form of evaluation that too often gets ignored needs to be focused on here: post-occupancy evaluation (POE). A POE uses many of the same techniques that constitute participatory design. Observation, interviews, questionnaires, surveys, and focus groups can also be used to get feedback on the design after it has been executed. From such feedback, shortcomings can be corrected in any of the techniques used; each of the design directives can be tested against the results, and the designer can thus learn to better define and resolve issues in the future. Basically, the designer must ask, "Did my design work? How well? What things could I improve on? What have I learned from the process?" A well-known research designer speaks and writes eloquently on the importance of evaluation, defining a POE as a study of how well a newly designed interior environment supports the behavior, performance, and satisfaction of its users. Design is considered a hypothesis about how a future environment will affect people's behavior and feelings. Therefore, a POE is the *verification* of that hypothesis.

Case Study: San Francisco Redevelopment Housing

About 30 years ago, urban renewal bulldozers cleared a downtown neighborhood in San Francisco of nearly 4000 low-income housing units to make way for a new convention center. Many of the displaced people were seniors who subsequently went to court and won a landmark settlement requiring the city to provide 1500 units of replacement housing, to be built by nonprofit developers (see Figure 3.10). The court settlement, made in 1973, arranged for construction funding by raising the city's hotel tax by only 0.5%, principally taxing those who would benefit from the new convention center. Since that time, three replacement housing developments for seniors have been designed by Herman Stoller Coliver Architects for nonprofit developer Tenants & Owners Development Corporation of San Francisco: Woolf

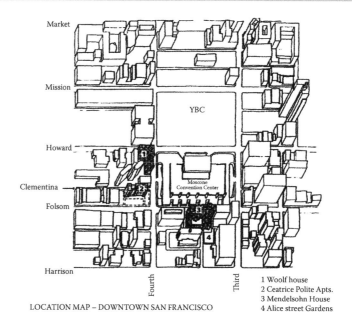

Market

Mission

YBC

Howard

Clementina

Moscone
Convention Center

Folsom

Harrison

Fourth

Third

1 Woolf house
2 Ceatrice Polite Apts.
3 Mendelsohn House
4 Alice street Gardens

LOCATION MAP – DOWNTOWN SAN FRANCISCO

FIGURE 3.10   Neighborhood Downtown San Francisco.

House (182 units) in 1972, Ceatrice Polite Apartments (91 units) in 1982, and Mendelsohn House (189 units) in 1988 (Figure 3.10). As post-occupancy surveys (shown with the Woolf apartment floor plan, Figure 3.11) and experience have fed into a learning curve, each development has become more carefully and more humanely designed than the last.

Robert Herman, in the case study of San Francisco Redevelopment Housing, credited the use of post-occupancy surveys with the increasing success he found in each of the stages of the project

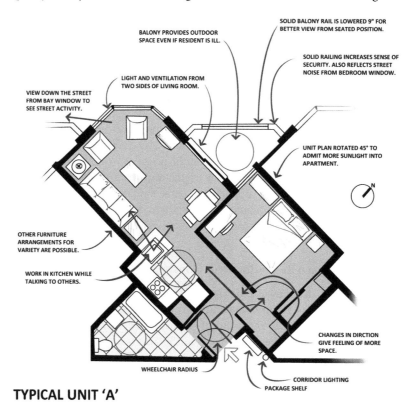

SOLID BALONY RAIL IS LOWERED 9" FOR
BETTER VIEW FROM SEATED POSITION.

BALONY PROVIDES OUTDOOR
SPACE EVEN IF RESIDENT IS ILL.

SOLID RAILING INCREASES SENSE OF
SECURITY. ALSO REFLECTS STREET
NOISE FROM BEDROOM WINDOW.

LIGHT AND VENTILATION FROM
TWO SIDES OF LIVING ROOM.

VIEW DOWN THE STREET
FROM BAY WINDOW TO
SEE STREET ACTIVITY.

UNIT PLAN ROTATED 45° TO
ADMIT MORE SUNLIGHT INTO
APARTMENT.

N

OTHER FURNITURE
ARRANGEMENTS FOR
VARIETY ARE POSSIBLE.

WORK IN KITCHEN WHILE
TALKING TO OTHERS.

CHANGES IN DIRCTION
GIVE FEELING OF MORE
SPACE.

WHEELCHAIR RADIUS

CORRIDOR LIGHTING

PACKAGE SHELF

TYPICAL UNIT 'A'

FIGURE 3.11   Woolf House apartment floor plan.

described by the case study. By combining pre-occupancy surveys/evaluations and research with continuing POEs, he was able to move closer to providing a completely supportive environment for the seniors who were his clientele.

When the Mendelsohn House project (the third phase of the case study described) first began to unfold, the architects and nonprofit housing organization used surveys and interviews to determine the needs of prospective end users, thus involving seniors in the design process from planning to post-occupancy. The planning stage for the entire project took approximately four months to complete, including the time to take the photographs necessary for explaining the choices to end users. Small group sessions were held (4 to 10 people) where slides were presented depicting various design strategies. From these sessions, specific design directives unfolded: specifics on the techniques for establishing design directives from POEs and other information are shown in the teacher's manual that accompanies this book. The floor plan from Woolf House is included here (Figure 3.11).

POE responses that resulted in design directives for all three buildings in the project were as follows:

- Include an active street edge, with some commercial use on the ground floor to connect with the neighborhood.
- Include commercial shops that serve the needs of residents.
- Design a building that looks modern (this was not what the architects had expected to hear since most of the users had previously lived in residential hotels of Victorian vintage).
- Maintain a positive, dignified image through style and details (e.g., carpeting, fine art, use of subtle colors).
- Include a lobby for security and consistency with users' experience in residential hotels.
- Include balconies in the apartments.
- Ensure safety at the entryway and throughout the building.

Through the use of post-occupancy surveys and interviews, observation, and ongoing research into the needs of the elderly, Woolf House, the Ceatrice Polite Apartments, and Mendelsohn House evolved to meet more completely the occupants' needs. These research results elicited the need for the following:

- Place for grandchildren to play
- Automatic opening on main entry doors
- Desk clerk to electrically control the main entry doors, not just residents buzzing people in from their unit intercoms

Mendelsohn House is completely wheelchair accessible, with 19 apartments specifically outfitted for disabled occupants.

Several principles guided the design decisions:

- Establishing continuity with the residents' backgrounds through the use of familiar architectural features
- Avoiding isolation from the outside world by including neighborhood businesses at the ground level and bay windows in residents' apartments or "eyes on the street"
- Encouraging social interaction among residents through a variety of activity areas
- Enhancing a sense of pride and dignity by including more than the "basic daily minimum" of services
- Treating time as an important feature of life measured by natural changes—sunlight, shadows, rainwater, wind, and color, all encouraging a state of alertness

Affordability was maintained by containing unit sizes to US Department of Housing and Urban Development limitations, although amenities such as bay windows and recessed balconies greatly enhance livability for the residents (Davis 1995). The building structure is painted, poured-in-place concrete with infill plaster panels. Added expense was reserved for design features where it counted most: the main entrance canopy, decorative tiles, window patterns, arched openings, and finishes for ground-level shared areas (Figure 3.12).

One enters the building through a lobby reminiscent of a comfortable residential hotel (the type of residence most of the clientele had been living in). Residents' common spaces overlook a central courtyard. At the front door, seating is arranged for inconspicuous people watching. Spaces are organized to encourage mingling and a feeling of independence through the interior architectural layout and choice of furnishings. Handrails are an integral design element, as important for appearance as for utility. As used here, they lend a friendly quality, accentuating the curves of ramps and corridors. A variety of period furniture harmonized by color and fabric selection resemble a typical family's collections gathered over a generation or more. A "history wall" containing photos of the history of the neighborhood, including the residential hotels bulldozed by the redevelopment agency, lines the edges of the public lounge space. Small lamps at the reception desk animate the entry. Structural columns with decorative capitals characterize the ground floor, integrating the lounge, ramp, entry, and elevator foyer. The courtyard, arcaded on two sides, is luxuriantly planted. A fountain and pool, hard-surface exercise area and grassy gathering place for special events, meandering paths for strolling, and a tot lot for visiting grandchildren complete a variety of carefully designed spaces. Two large rooftop vegetable gardens provide an opportunity for productivity and sense of growth for all residents.

A personal sense of security, actual and perceived, was also an essential design goal. Balancing the need for reassuring security devices with their presence as reminders of threatening conditions was a key consideration. The best security originates with the residents themselves, who provided surveillance for the public spaces in the building. The project received a 1991 National Honor Award for Design Excellence from the American Institute of Architects, one of 19 projects selected from over 650 considered. Mendelsohn House also received the first-ever People in Architecture award in 1990 from the California Council/American Institute of Architects. As Mr. Herman pointed out, "Subsidized housing has rarely been the recipient of design awards. Just 20 years ago such housing was a social, political, and aesthetic embarrassment within our cities." Mendelsohn House was selected as a winner not only from among other housing developments but also from a range that included museums, hotels, marketplaces, a symphony hall, and a post office. Woolf House, the Ceatrice Polite Apartments, and Mendelsohn House have continued to serve the needs of low-income elderly persons in the "south of market" area and serve as a successful model for POE and universal design.

Within the residential second to ninth floors, corridors are punctuated by package shelves, deeply inset front doors, decorative apartment numbers, and kitchen windows that borrow light from across single-loaded corridors. Each Mendelsohn apartment has a bay window and most have balconies recessed from the wind, encouraging people who have difficulty getting out to remain visually connected to the rest of the world, much as the traditional front porch on single-family homes connected people to their neighborhood (Figure 3.13).

Kitchens in one-bedroom apartments and in many of the studios are located in the midst of living areas, not isolated, encouraging the person cooking to comfortably talk with

FIGURE 3.12 Artist's rendering of Mendelsohn House.

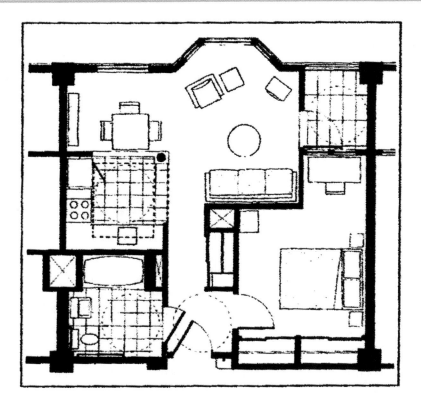

FIGURE 3.13   Detailed floor plan of an apartment in Mendelsohn House.

someone seated nearby. All units and public areas are wheelchair accessible, with 10% of the apartments outfitted for residents with disabilities. All apartments are adaptable; that is, they can be made fully accessible with minor additions. Orientation to the sun was a key factor in the shaping of the building. The seventh- to ninth-story wing casts its shadow over a main street while shielding the garden courtyard from street noise. Lower and narrower building wings complete the sunny and amply sized inner courtyard.

Case Study: San Diego Center for the Blind

The San Diego Center for the Blind (SDCB) is a nonprofit, independent rehabilitation facility founded in 1972 to help people who are blind or otherwise visually disabled reach their highest potential for independence and self-reliance. SDCB accomplishes this through ongoing classes and counseling in such areas as orientation and mobility, activities of daily living, and communication skills. A large segment of the clientele are elderly, with a lifetime of sighted habits, who have a great deal at stake in being able to maintain their independence in their own homes. They need to relearn much that they have taken for granted (Figure 3.14).

SDCB has, since its inception, inhabited a 9000-square-foot building on a busy street near the San Diego State University (SDSU) main campus. The building originally housed a health spa/gym and later a bank. Because of funding concerns and a lack of expertise, the building SDCB inherited in 1972 remained little changed until 1985 when a class of undergraduate design students from the university (as a class project) took on the task of redesigning the existing kitchen facilities that were being used to train people with varying degrees of visual disability. The redesign of the kitchen space prompted SDCB administrators and at least one of the students to ask the professor (this book's main author) if there was anything that could be done to make the students' plans a reality for the center. Dr. Null started making phone calls. Within a few months, over 70 companies had donated over $60,000 worth of equipment and a design team had been formed to oversee a remodeling project—a team that included kitchen designers, Dr. Null, social service personnel, other designers, a low-vision lighting specialist, and the public relations home economist for the local utility

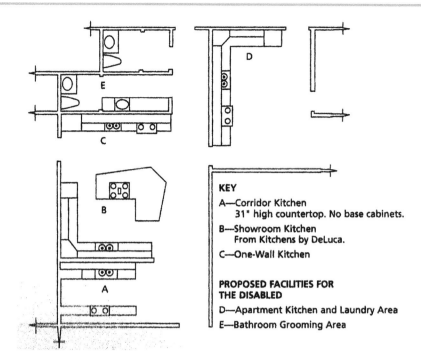

FIGURE 3.14 Floor plan of the kitchen area of the San Diego Center for the Blind.

company who was able to secure volunteer services from labor unions and other community groups to carry out the renovation. The board chairman for the SDCB was a retired contractor who supervised the actual construction (Null 1988; Parrott et al. 2008).

By November of 1985, the one-wall, outdated kitchen (which lacked even basic adaptations for appliances such as large-print controls/instructions) had been transformed into three working models of state-of-the-art kitchen designs, representing the types of kitchen layouts often found in people's homes. The following year, the project was awarded the American Society of Interior Designers' (ASID) prestigious Environmental Design award for the students' part in the remodeling.

As with remodeling projects everywhere, when one thing changes, everything else seems to need changing. A snowball effect took place when the center's staff and design team asked themselves and each other, "Why not?" They recognized that the overall design of the building was not supportive of the workers or the clientele, and they set out to rethink and rebuild the facility with the assistance of an expanded team, including the community service committee of the San Diego chapter of ASID, headed by interior designer Jan Bast; environmental psychologist Ann Gero-Stillwell, who provided a pre-design baseline study (from interviews, surveys, and focus group meetings) and a post-occupancy evaluation (using the same techniques); ASID professional members; and a continuing number of students from several design courses at SDSU taught by Kerry Nelson (who all learned from the work being done and contributed to the final product). Funding for this new project was secured through private donations, government grants, and a community block grant from the city of San Diego, totaling over $500,000. The general goals of the project team were to:

- Improve the feel of the space to create a cheerful and professional image
- Improve wayfinding through special cues and a logical space plan
- Make the building easier to use for the elderly and for people with disabilities by utilizing universal design techniques

Wayfinding concerns covered areas such as space planning to group similar activities; traffic flow to ensure a logical movement from entry to activity and to protect against obstacles;

use of clear, consistent signage; and contrasting color and texture to serve as cues to changing function. Universal design concerns included increasing the overall lighting in the facility, reducing glare, using task lighting and lighting within handrails, and creating contrast in color and shade in doors, doorframes, door handles, and walls, as well as in carpet and tile to mark areas and changing levels. Other universal design features (see Figures 3.14 through 3.22) included the following:

- D-shape or lever handles on doors and cabinets
- Visible and accessible storage using pull-out shelves and lazy Susan corner cabinets
- Storage inserts for pantry and other cabinets
- Adjustable shelves within cabinets
- Sliding doors on cabinets
- Side-by-side refrigerator/freezers with slide-out shelves, water and ice dispensers in doors, and mini-doors for frequently used items
- Large print directions, Braille overlays, and easily grasped controls on appliances
- Arrangement of appliances in order of their use within kitchens
- Ranges with front controls
- Magnetic-induction cooktops that are cool to the touch and sound a warning when a pan is removed or improperly placed
- Pull-out boards below microwave and other side-opening oven doors
- Rheostats for incandescent lighting
- Automatic faucets with preset temperature controls
- Faucets with spray attachments in kitchen
- A variety of heights in tables and counters
- Nonslip grab bars
- Adjustable desks and workspaces

The project also called for the creation of a model apartment that would be used as both a demonstration facility and a training center for adapting one's home to be more fully supportive of activities of daily living.

The team wanted to utilize design features that would maximize the self-sufficiency of all parties concerned and that could easily be applied to most other public and private facilities, not just those serving people with visual impairments. They also understood that the population they were dealing with was not one that needed to have only their visual disability accommodated. Many of the clientele were elderly; many used wheelchairs, walkers, and canes; many had arthritis or some other disability. The design team wanted to create a facility that would be supportive of all people and all their various abilities. While

FIGURE 3.15 Activity room with universally designed Bola Chairs donated by Norm Polsky from Fixtures Furniture.

FIGURE 3.16   Corridor featuring color contrast flooring and doorframe and arrows at room entries.

the project was not covered by ADA regulations (since it was begun before enforcement dates), the design team's decisions and implementations met and in many cases exceeded the requirements of the ADAAG (ADA Accessibility Guidelines).

As often happens when remodeling an older building, unexpected setbacks occurred. Soon after construction had commenced, it was discovered that asbestos insulation had been used in the original building; its removal proved costly and time consuming, forcing a reduced scope of the project.

FIGURE 3.17   Kitchen features: color contrast between cabinets and countertop, sturdy Bola stools, varied height counter/cabinets, D-shaped handles on cabinets.

FIGURE 3.18   Corridor kitchen, D-shaped handles, lowered counter heights, space under counters for wheelchair access, side controls on the gas range.

The three overall goals—more professional image, better wayfinding, and use of universal design features—were met. The POE clearly showed how much had been accomplished and how much remained to be achieved. Thus, it served double duty: not only was it offering a means of evaluating the project, but it could now be used as a baseline study for the next phase of the project. Hence, just as the initial kitchen redesign led to a reconsideration of the facility as a whole, the overall remodeling has led to a further refinement of goals. The award of another community block grant further improved the facility, including an updating of the much-used training kitchens, upgrading the HVAC (heating, ventilation, and air conditioning) systems, and creating a new façade for the building (see Figure 3.22).

The SDCB building was dark, monotonous, cavernous, and confusing (because of haphazard organization). One of the statements made during the pre-design interviews was that the Center looked "hard up," that is, not professional or capable. That image has now been made over. The facility is brighter, with a clear layout and logical plan. Obstacles have been removed, and universal design details have been incorporated to make the building safer and easier to negotiate. Because most training facilities for blind and low-vision persons have been geared to rehabilitation of young, totally blind individuals, the SDCB is

FIGURE 3.19   Model apartment, window to the hall, red handles on cabinets.

FIGURE 3.20   Bathroom signage: protective metal kick plate at bottom of the door, color contrast, correct ADA location (left of the door).

unique in recognizing that most of its clients are elderly and need training in activities of daily living rather than job skills.

The SDCB has gained national and international recognition for its design and programming. The training kitchens and interior design of the Center have been so successful that a new facility was added in the San Diego north county area 15 years ago. Both centers serve approximately 1200 clients per year. In 2010, a major fundraising campaign was launched to expand the comprehensive Assistive Technology Center. Each stage of the project has built on the success of the previous work. It is still difficult to obtain funding for

FIGURE 3.21   Recessed varied height water fountains and dish for seeing-eye dogs.

FIGURE 3.22   New façade side view.

the rehabilitation of older persons, but having a beautiful facility with an impressive success rate has helped generate new sources of funding.

The facility is now gaining recognition throughout the country as a model for other such programs—both for its designs and for the methods used to carry out those designs.

(Additional pictures of the Blind Center are available in the Teacher's Manual and the website: www.RobertaNull.com.)

UNIVERSAL DESIGN TECHNIQUES

ENVIRONMENTAL PROGRAMMING

Identifying user needs is essential to any successful design activity. Environmental programming is a flexible design tool that encourages designers, students, and facility managers (i.e., anyone with an interest in universal design) to emphasize the development of design criteria and the analysis of client needs before approaching a graphic design solution. Environmental programming also provides a design approach that can be used successfully by anyone who does not have high-level graphic design skills, since it instead requires a focus on the social and psychological aspects of an environment. In the case study describing airport design (Chapter 7), the complexity of the projects actually made use of environmental programming in addition to many types of graphic design skills. This process can be used early in the research programming phase.

Environmental programming encourages the designer to emphasize the development of design criteria and analysis of client needs before they approach a design solution. It thus provides a designer with the means to communicate with clients about their needs and desires. Or it could be used by the client to analyze their needs and communicate them to the design professional.

Following a development of profiles for users and the environmental setting, design criteria are formulated that can:

- Be supportive for a chosen user group in a specific physical setting
- Facilitate or support a particular social process
- Meet the needs of a person with a disability

The following example is a complex analysis of a suburban shopping center in relation to how it met the needs of an elderly population. The figures were completed by housing-interior design graduate student Ann Warble-Nienow from an independent study project. In this case, environmental programming was basically used as a post-occupancy evaluation with recommendations for adapting the environment rather than as an initial problem-defining programming effort.

A Coastal Resort Shopping Center Example   The purpose of this environmental assessment project was to analyze a suburban shopping center in relation to how well it met the needs of the permanent residents in a southern resort area. By using behavioral science research techniques to analyze the problem and develop design criteria for architectural adaption, an evaluation model was developed. This model can be used by businesses or community action groups to study shopping facilities in a variety of settings. The techniques may also be expanded to include universal design and ADA compliance evaluations.

The Setting and Clientele—Pineland Mall   Because of the mild climate on the South Carolina Coast, Pineland Mall was designed with a covered walkway connecting the stores and a small lagoon in a park-like setting in the center of the mall. The mall's clientele consisted mainly of permanent residents of this resort area. Many were retired or semi-retired individuals in the upper-middle to upper income level.

Human Traffic Problems   Observation revealed that the front edges of the mall were the most heavily traveled. Close parking spaces were vied for by people just running in for groceries. The space between the grocery and drugstore was one of the busiest in the mall, while the other heavily used stop was the dry cleaners on the front corner of the mall. This site had a buildup of illegally parked cars, resulting in street congestion while people popped in to pick up their cleaning. The fringes of the front edge and the "near" lagoon area in the center of the mall got semi-heavy use.

Environmental Programming Requirements for the Elderly   In order to plan adaptations of a shopping facility to make it a supportive environment for the elderly, recognition had to be made of the diversity and needs of this age group. The environment needed to be free of architectural barriers, which were barriers that discouraged elderly persons from orienting themselves to the environment. Other types of barriers prevented the elderly from passing easily from one area to another and prevented them from manipulating the equipment in the environment. The architectural barriers were identified (see Figure 3.24), and design criteria were established for recommended changes in the shopping environment. Design criteria were based on what was learned about the elderly through observation and other forms of behavioral research reported in the environmental design literature.

Environmental Revisions   Even though the recommended adaptations of the mall were planned to create a supportive shopping environment for the elderly, these adaptations were practical for everyone shopping there. Easy access to stores, adequate restroom facilities, places to sit and rest

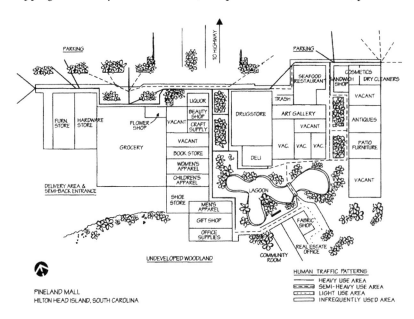

FIGURE 3.23   Human traffic movement, Pineland Mall.

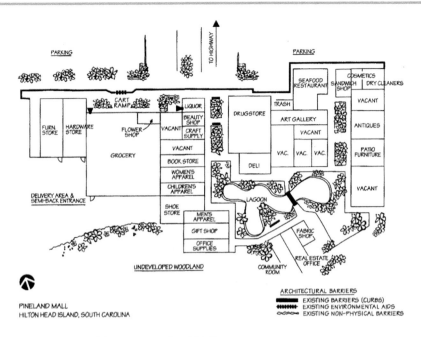

FIGURE 3.24   Arthitectural barriers, Pineland Mall.

with pleasant views, adequate mapping, and identification of store locations are good design for everyone. In fact, many merchants have discovered the hard way that elderly customers don't see themselves as elderly and will shun merchandise and facilities planned for older people. Meeting the varied needs of customers is simply good business. Specific revisions and adaptation for Pineland Mall included the following (Figure 3.25):

- Numerous benches provided for resting, socializing, and waiting
- Public restrooms located to service the whole of the mall were designed to accommodate persons with disabilities
- Mechanical boosters on all entrance and exit doors for ease of use by all
- Ramp-type curbing at various points along the front edge of the mall
- Parking for use by those with disabilities designated according to code
- Directories redesigned to produce signs legible to those with failing eyesight
- Two additional directories added to the site to aid in wayfinding

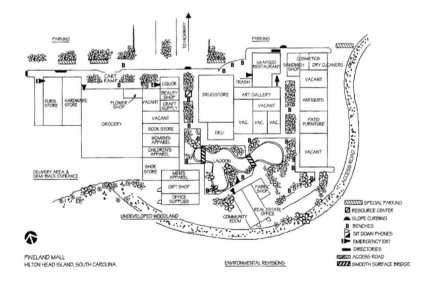

FIGURE 3.25   Environmental revisions, Pineland Mall.

UNIVERSAL DESIGN STRATEGY

Oral history interviews represent a powerful way for people often ignored by society to add their voices and their perspectives to the writing of history that interprets the past, to the understanding of the values that govern the present, and to planning for the decisions that will influence the future. The use of personal oral history research with people with disabilities can serve many purposes, some of which are relevant to designers.

If designers are interested in creating an environment that can accommodate everybody, the needs of people with disabilities must be included from the beginning. The best way of assuring that these needs are met is to involve people with disabilities directly in the process of planning and designing the environment. Oral history interviews constitute an effective way of letting people with different kinds of disabilities contribute to the task of eliminating architectural barriers. Various members of the disabled community have different accessibility needs, and sometimes the removal of a barrier for one group has been done in ways that create problems for another. It is therefore important for designers to include people with all kinds of abilities and disabilities in the design process. While people using wheelchairs need an environment that is free of architectural barriers, people who are blind need a multitude of concrete environmental clues to aid them in their task of orientation and navigation without the use of sight. Thus, curbs represent effective barriers to people in wheelchairs, while they are helpful guides for people who are blind. Oral history interviews with both groups of people could reveal other areas of conflicting needs as well as some creative solutions, such as using textured surfaces on curb cuts (and/or corner placement of curb cuts) that could work for both groups.

In addition to making their needs known to designers, people with disabilities can contribute to the process by becoming designers themselves and participating directly in the task of creating an environment that is responsive to their needs. If design, broadly conceived, consists of the plans and strategies people develop in order to solve their problems and achieve their goals, then people with disabilities get more than their fair share of practice in designing, because they have more than their fair share of obstacles to overcome. A person with a disability needs to plan ahead and anticipate obstacles in most situations where people without disabilities can take for granted that they have easy and quick access to information, resources, and execution of their plans. Viewed from this perspective, people with disabilities represent a tremendous and untapped resource for designers, especially designers who are beginning to think in terms of universal design—the creation of an environment that will be readily adaptable to everyone's needs, including people with disabilities.

Conversations with people with disabilities reveal their capacity for practical problem-solving. While most of them may have a tendency to share this kind of information with each other more readily than with people who are not disabled, they are often as proud of their own resourcefulness as they are frustrated by the barriers. The following story demonstrates some of these points.

One cold and icy winter day when using my crutches, I was slowly and carefully making my way over the ice from my car to the building where I work. I had to pass close by a pick-up truck parked in a "handicapped" parking space. As I came close to the truck, the person in the driver's seat rolled down the window and started telling me about how he deals with ice using his crutches. He had installed "ice picks" into his crutches by drilling holes in the rubber safety tips and inserting barrel-type door latches into them in such a way that he could easily pull the latches in when he was about to walk on a floor, and let them out when he had to go out. This story reveals one of the ways of beginning to take advantage of the life experiences of people unfamiliar with the jargon or technicalities of a specialized discipline such as architecture. One simply needs to ask these individuals with disabilities to talk about how they have solved some of the problems associated with the activities of daily living. Interviews are increasingly being used in many different fields to collect information about life experiences that are crucial to the understanding of different perspectives, especially the perspective of people normally left out of the development process. Oral history interviews are especially well suited to the process of collecting information about life experiences from ordinary people because they are designed to place any information collected in a cultural and historical perspective. This means that the experiences described in individual interviews are being interpreted against a thoroughly researched background of a place at a specific time in history. Therefore, the professionals are not as likely to overgeneralize their findings and develop ideas that are cultural or solutions that are not sufficiently related to perceived needs. Personal oral history interviews can help designers understand both the cultural meaning of disability experiences and what suggestions people with disabilities have for solving some of their accessibility problems.

Collecting oral history interviews among people with disabilities represents a special challenge for social historians. Several studies suggest that people with disabilities respond differently to different interviewers. They are more likely to share certain kinds of information with interviewers who are themselves disabled. If the interviewer is not disabled, the information obtained must be interpreted in a different way than if the interviewer is also a person with a disability. An interviewer who is disabled may be able to elicit more statements about the barriers a disabled person experiences, the anger and frustration at the inaccessible environment, and the solutions that have been developed and put to use by the interviewee.

On the other hand, an interviewer who does not have a disability might be likely to hear more stories portraying coping and acceptance strategies that fulfill the person's need to appear well adjusted. The most widely held idea about how people with disabilities should act in order to get along in society is that they should be cheerful and positive and should make the best effort possible to adjust themselves to society and not ask for anything "extra."

Oral history interviews of and by people with disabilities would greatly enhance and strengthen the work of designers and other people involved in shaping the built environment of the future. The needs of various groups of citizens with disabilities could be worked out so they could complement instead of conflict with each other. People with disabilities would be able to describe existing barriers and rank them in order of the most frequently encountered to the least frequently encountered obstacles.

Through oral history interviews, design ideas from people with disabilities could be recognized and utilized, with credit given for their contributions in the general effort to create an accessible environment. In addition to producing suggestions as part of a design team, people with disabilities should be encouraged to become designers themselves and to work with both the professional knowledge of the field and their life experiences as citizens who happen to have disabilities. Doing so will create the most cost-effective and efficient solutions to some of the practical problems involved in the creation of an environment that is accessible to all.

UNIVERSAL DESIGN EXEMPLARS: TEACHING FACILITIES

Because universal design is invisible, it is important to show good examples of universal design, and then tell *why* they are good examples. The spaces and products in the following case studies show a variety of teaching facilities that have been designed to present UD examples.

Case Study: UD Teaching Facility—Kansas State University

Using space formerly occupied by a household equipment teaching laboratory, Professor of Housing, Dr. Betty Jo White (2004) developed a universal design teaching facility. The demonstration and research laboratory presented state-of-the-art universal design features and products currently on the market.

The purpose of the kitchen, bath, and office components was to:

1. Allow people to experience human-factored design that was accessible, adjustable, adaptable, attractive, and affordable
2. Be usable by people of all ages, sizes, and abilities, and accommodate common age-related changes (e.g., arthritis, heart conditions, sensory limits, and mobility problems)

The state-of-the-art prototype, testing, and demonstration facilities included the following:

1. Accessible/Adaptable Kitchen
   a. Height-adjustable sink and range centers, accessible stock kitchen cabinets, and an adjustable efficiency kitchen unit
   b. Side-by-side refrigerator/freezer with in-door ice/water dispenser
   c. Induction cooktop, side-hinged wall oven, and varied microwave oven placements
   d. Shallow-bowl sink with spray hose, single-lever high-rise faucet, anti-scald device, recessed/flexible drain, and insulated water supply lines; lowered bar sink
2. Bath/Public Restroom Demonstration
   Converted existing laundry area to accessible/adaptable private bathroom and public restroom mock-up with movable wall components and rearrangable fixtures and models:
   a. Barrier-free, side-entry or sit-down bathtub
   b. Raised toilet with grab bars
   c. Roll-in and transfer shower facilities
   d. Adjustable-height lavatory/vanity sink
   e. (Open) framing for installation of assist features
   f. Stacked front-loading washer/dryer with low-vision, easy-grip controls
3. Office/Work Environment Area
   a. Systems furnishings/adaptable work stations with ergonomic seating
   b. Microcomputers with glare-free, indirect lighting
   c. Telecommunication system display units
   d. "Smart" office feature displays/demonstration units
4. Focus Group and Product Testing Demonstration Areas
   The lecture/focus group area was refurnished with round and square tables and chairs in flexible arrangements to demonstrate ADA-compliant restaurant seating for 20–30 persons. An open universal product testing area was used for ADA-compliant store design exercises, and consumer research activities such as use of prototypes.

Case Study: Center for Real Life Kitchen Design

The Center for Real Life Kitchen Design is a teaching and research facility on the campus of Virginia Tech, in Blacksburg, Virginia. The Center includes five kitchen vignettes, a laundry area, home office, and a classroom facility. There are universal design features throughout, and the GE Real Life Design Kitchen, the largest kitchen in the Center, focuses on universal design. The Center, developed through a university grant and the contributions from 22 companies, highlights the products, materials, space planning, and design of residential kitchens. While universal design features were an important part of the concept for the facility, other factors, such as range and variety of appliances and materials, and lifestyle choices, were also part of the goals of the development of the Center. The facility has been the site of a variety of different activities, including university teaching, continuing education programs, and research.

THE REAL LIFE DESIGN KITCHEN

A critical space within the Center is the GE Real Life Design Kitchen. This kitchen was developed by GE Appliances to demonstrate universal design, including appliance selection, to builders and kitchen designers. Mary Jo Peterson, CKD, CBD, was the designer of the kitchen, which was used at national builder and kitchen and bath design shows during the mid-1990s. A video and booklet were developed to highlight the features of the design (GE Appliances, 1995).

UNIVERSAL DESIGN FEATURES

The Real Life Design Kitchen includes many universal features to highlight universal access. While usability by people with various disabilities was a consideration in the design, equal consideration was given to use by people of various ages and sizes.

"Real Life Design is simply good design. It can be appreciated by everybody because it makes so much sense in everybody's life. It takes into account that most people don't fit the stereotypical norm. Baby boomers in huge numbers are finding out that they aren't as spry or sure-sighted as they used to be. Along with the usual problems faced by aging population, Real Life Design also acknowledges a wide range of physical and mental abilities and impairments. It even acknowledges that a great many of our most worthy citizens are children" (GE Appliances, 1995).

Some of the universal features in this kitchen include the following:

Adjustable height sink. A mechanized lift and flexible piping allow the sink to adjust for tall and short or seated people.

Varied counter heights. The kitchen has counters at 30", 36", and 42" high.

Rolling carts. Several different designs of carts help enhance the variety of work surface heights. The carts can be used to transport supplies and ingredients or as a separate work surface. One cart is hidden under the counter, while two are nested beneath the microwave/convection oven.

Raised toe kick. The toe kick has been raised to 9" in certain areas. Cabinets sit on the raised toe kick, which provides extra floor space and raised work surfaces.

Raised dishwasher. One dishwasher has been placed on a 9" toe kick, raising the appliance and minimizing the bending needed to load and unload.

Varied height storage. The varied height of cabinets and shelves provide storage within reach of a range of people.

Drawers and pullout shelves. Deep drawers are located beside the cooktop and sink, and pullout wire baskets are in the food prep area. Multiple pullout cutting boards and work surface provide spaces for seated cooks to work.

Pull-down storage. Uniquely hinged shelves in the wall cabinets beside the sink allow the shelves to be lowered for easy reach and view.

Open shelves. Wall shelves along one wall provide storage that is easy to see and reach. Glass doors on some wall and base cabinets also provide clear views of stored items.

Built-in step stool. Concealed in a corner base cabinet is a step stool that allows short adults and children to reach the higher cabinets.

Knee space. Space beneath counters has been created in several ways. Exposed knee space is located under both sinks and under the cooktop, which has folding doors that hide the space. Removable carts also provide knee areas.

Counter edge contrast. The counters are an off-white color but have a dark blue stripe inserted into the solid surface. The stripe is raised, providing a tactile surface. Counters at the snack area and cooktop and on the carts are white and blue tiles, a heatproof surface that allows for hot pots and pans.

Clipped corner counters. All counter corners have been clipped so that there are no sharp edges to bump into.

Single oven. The single wall oven, placed in a corner, is at a height that puts the lowered door at the snack counter height.

Microwave/convection oven. This smaller oven has a drop-down door. A pullout table with a tile top is located beneath the door to provide a convenient landing area.

Lever-handle faucets. Faucets for the main sink and the salad sink are black, providing a contrast to the stainless steel sinks, and include lever handles and pullout sprays.

Open handles. Most wall and base cabinets have open D-shaped handles.

Magnetic closures. The bank of cabinets on the island use magnetic closures that open and close with pressure.

Under cabinet lighting. Task lighting is placed throughout the kitchen and enhances general recessed lighting.

KITCHEN DESIGN GUIDELINES

For many years, the National Kitchen and Bath Association (NKBA) has produced guidelines to aid designers in planning both kitchen and bath areas. Originally, the guidelines reflected kitchen design standards first established in the 1950s and 1960s. The guidelines were revised in 1996 to include many universal design elements. The GE Real Life Design Kitchen was used to illustrate many of these guidelines in NKBA's manual on Kitchen Planning Standards and Safety Criteria (Cheever, 1996).

The Center for Real Life Kitchen Design has been a successful experiential learning environment that has provided opportunities for hands-on activities related to universal design in the kitchen. The use of appliances and the examination and evaluation of spaces allow students and the public to view and experience design that can be used by everyone. The Center incorporates features and applications that are somewhat unique and new to many of the students. The examples they experience in the Center will influence the way they design and manage residential spaces. The research that has been conducted in the Center will help designers and consumers plan kitchens that meet the variety of needs that all people have (Figures 3.26 through 3.32).

Case Study: Permanent Teaching/Research Facility—The Ohio State University

Involving the community in providing design education and training is beneficial on many levels. The leadership at The Ohio State University succeeded in obtaining the interest and skills of Lowe's Home Improvement to create a Kitchen and Bath laboratory for hands-on, or experiential, learning. This permanent teaching facility (Figures 3.34–3.40) has also been instrumental in influencing professional designers and builders and in evaluating new and proposed standards.

Commercial products in the kitchen and bath facilities can be updated as new models become available, maintaining a site that will always be relevant for the OSU students and the public.

FIGURE 3.26 The GE Real Life Design Kitchen at Virginia Tech features multiple counter heights, raised toe kicks and dishwasher, under-cabinet lighting, open shelves, knee space under the sink and cooktop, carts, and contrasting floors and counter edging.

FIGURE 3.27 An adjustable-height sink with knee space, a single wall oven at counter height, contrasting counter edging and flooring.

FIGURE 3.28 The raised dishwasher and the built-in speed cook oven are placed within the universal reach range. Nested carts under the oven provide a method to move food and supplies around the kitchen.

FIGURE 3.29 Within a raised cabinet in the GE Kitchen in the Center for Real Life Kitchen Design at Virginia Tech, the pull-out board has a cutout that holds a bowl steady for a person with limited dexterity or arm strength.

FIGURE 3.30   The floor space in the Gourmet Kitchen in the Center for Real Life Kitchen Design is large enough for a 60″ turn space. At the sink, a corner wall cabinet includes a lazy Susan for more accessible storage and nearby dishwasher drawers are at a useable height.

FIGURE 3.31   A microwave drawer, bottom drawer refrigerator/freezer, and cabinet drawers provide convenient access to appliances and storage in the Gourmet Kitchen in the Center for Real Life Kitchen Design at Virginia Tech.

FIGURE 3.31 (Continued) A microwave drawer, shown open in the Gourmet Kitchen in the Center for Real Life Kitchen Design at Virginia Tech.

FIGURE 3.32 This small Outpost Kitchen provides a built-in microwave and coffee system within the universal reach range. The drawer under the microwave includes a shelf for a convenient landing area.

FIGURE 3.33 Universal design information table with The Ohio State University student in Lowe's store.

FIGURE 3.34 Broad view of display kitchen at The Ohio State University facility built by Lowe's.

FIGURE 3.35 People using the demonstration kitchen.

FIGURE 3.36   Kitchen cooktop showing the pot filler faucet.

FIGURE 3.37   Kitchen with a raised dishwasher and vertical plate storage.

FIGURE 3.38 Kitchen counter levels for wheelchair and for standing.

FIGURE 3.39 Accessible sink in the cabinet of the UD bathroom.

FIGURE 3.40   Accessible bathtub with opening in The Ohio State University display.

Case Study: The Utah House—Teaching/Research Demonstration Facility

The Utah House's (UH's) mission is to demonstrate, educate, and empower the public about new ways of building homes and creating landscapes that promote the principles of sustainability, energy and water efficiency, universal design principles, and healthy indoor environments. The UH project demonstrates that a healthy, sustainable environment is fundamental to the well-being of any community (Figures 3.41 and 3.42).

UH showcases innovative ideas for saving energy, water, and other resources for reducing waste, using recycled materials, and maintaining a healthy indoor environment. The UH will serve as a forum so that Utah builders, architects, designers, and consumers can utilize the "Learning and Resource Center." Some of the offerings include fact sheets, newsletters, workshops, training, satellite programs, a resource room containing books, videos, Internet connection, and displays where people can see actual examples of new and innovative building materials and techniques. In addition, there are educational programs targeted to consumers; school-age children; professionals working in the design, construction, and landscape industries; public school teachers; vocational teachers; volunteers; and interns.

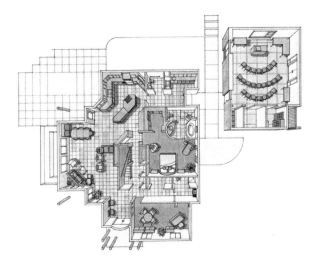

FIGURE 3.41   UH floor plan showing universal design features.

FIGURE 3.42   Exterior of Utah House.

UNIVERSAL DESIGN

- The house is designed for the life cycle. Living space on one floor (kitchen, bath, bedroom)
- All hallway/doorways 36″. Large bedroom with emergency access to the outside
- Lever handles on doors, cabinets, and faucets. Large bedroom on the main floor
- No thresholds into showers
- Light switches at 44″ to 48″
- Electrical receptacles 15″ above the floor 30″ × 40″ parallel or frontal approach appliances
- View windows with 36″ or less sill height
- Crank-operated casement windows. No steps or barriers to the front door
- Level entry to the front door entrance and all other entrances.
- Covered entry leading to front door
- Multi-level work areas to accommodate children as well as seated and standing people
- Placement of task lighting in appropriate work areas
- Knee space under sinks
- Continuous counter for sliding heavy objects
- Front mounted controls on appliances
- Full extensions, pull-out drawers
- Pull-out shelves in base cabinets
- Adjustable-height shelves in wall cabinets
- Some cabinets with glass fronts to easily view contents
- A microwave oven at base cabinet height for easy access (microwave and convection combination)
- Walk-in closet with adjustable height closet rods
- Access to bed on both sides

LARGE BATH

- Large sliding pocket doors for easy access into closet and large bathroom
- Textured tile floors to prevent slips. 30″ × 48″ area approaches in front of all fixtures
- Toilet: 16″ toilet seat height and 18″ from side walls
- A whirlpool tub with edge for sitting and transfer
- Safety rails that have towel bar hooks
- Knee space under bath sink. Minimum 32″ sink counter height
- Grab bar blocking in all walls around toilet, tub, and shower

- Bathtub faucet controls located in the outside rim of tub for easy to reach
- No threshold into a large bath shower
- Lightweight hand shower with a pause control feature, adjustable-height shower head

LAUNDRY ROOM

- Front-loading washer and dryer for water conservation and energy efficiency. Washer and dryer raised 13″ for easy access
- Fold-down, adjustable ironing board
- Adequate storage
- Laundry sink with pull-out faucet

Case Study: Andrus Gerontology Center Bathroom Renovation Project—University of Southern California

This case study highlights the implementation of a proposal for remodeling the two first-floor bathrooms of the Andrus Gerontology Center at the University of Southern California (USC) (Figures 3.43 and 3.44).

PURPOSE

The original purpose for the project was to create two accessible first-floor bathrooms employing the principles of universal design. Funding was to be sought through matching funds and donations of fixtures and services. The Andrus School of Gerontology and the School of Architecture were to collaborate on planning the design of the project. The Architecture linkage did not work so the implementation of the project was completed under the auspices of the School of Gerontology in compliance with local and statewide codes. National Kitchen and Bath Association (U.S.) (2012) Volunteers were also sought to help.

IMPLEMENTATION

The implementation plan had indicated that design for the modification of the bathrooms could come from students of the USC School of Architecture, so as to familiarize students with the principles of universal design and the code requirements of the Americans with Disabilities Act. This did not work, so the project was transferred to the USC Facilities Management department. The project was then expanded to include all six bathrooms in the Andrus building and Mo Hollman, Director of USC Facilities Management, became the director for the renovation. An architectural firm hired by

FIGURE 3.43   Sinks placed at a variety of heights and configurations USC bathroom remodel.

FIGURE 3.44   Each stall has grab bars of various heights, USC bathroom remodel.

the University did a general proposed space plan for all the bathrooms. Then, a team of design consultants, led by Mary Lou D'Auray, a Certified Interior Designer, suggested products and concepts (sketches) for the Architects' consideration as they detailed the spaces. An additional consultant for the project was Universal Kitchen and Bath Designer, Mary Jo Peterson, CKD and CBD. Mary Jo took responsibility for the major bathroom design work on the projects. I, Roberta Null, was a member of the design team and made many of the contacts with fixture companies. The team met several times and also communicated through conference calls.

Some suggestions for the design included the installation of different types of sinks, toilets, and fixtures that show examples available for accessibility and universal design; the use of Braille and foreign languages in signage (a signage, wayfinding expert also joined the design team) exposing, through cutouts in the walls, the use of backing for installation of grab bars in commercial and residential projects; the availability of leaflets in the bathrooms describing the principles of universal design; the use of different types of safety flooring and wall coverings; the implementation of emergency security and safety devices; and the installation of different types of lighting. Not all of these suggestions could be incorporated in the bathroom designs but they helped during the programming phase of the project. Some of the more innovative ideas were included in the second-floor bathrooms.

FUNDING

Funding was initially provided at $10,000, along with a 1:1 matching (cash contributions only) grant of up to an additional $30,000. Because the project had been expanded to include all six bathrooms, the University provided major funding. Donations also came from individuals, foundations, government, and suppliers.

RESULT

The initial phase of the bathroom renovation project was completed, and four of the six bathrooms were showcased at the Morton-Keston Summit meeting at USC on November 10, 2003. All six bathrooms are prototype facilities providing information on universal design to students, staff, and visitors at the Andrus Gerontology Center.

SURVEYING PROFESSIONALS REGARDING UNIVERSAL DESIGN

Nancy Wolford's interest in universal design initially came from her mother's experiences and frustrations with physical space (because of osteoporosis, a broken shoulder, and then broken hips) more than 20 years ago, long before it was popular and most were even aware of it. She has been including these concepts in my teaching of design and space planning nearly as long.

About 1990, Ms. Wolford helped develop a community college course, Special Housing Needs, on this topic, about the same time the Americans with Disabilities Act (ADA) was enacted. Not to be confused with ADA requirements, which apply only to public spaces, universal design concepts apply to all spaces, including residential settings, such as the single-family house. Currently, there are no codes, standards, or consistent requirements for the single-family home that qualify as universal design. There are, however, many suggestions and guidelines, often conflicting, put forth by a variety of organizations for specific populations (i.e., AARP for the senior population, Paralyzed Veteran's for the disabled, as well as ADA and Fair Housing requirements). During the research process, however, she discovered conflicting information, depending on the source.

One goal of her research was to survey the awareness and use of universal design concepts in residential construction by Oregon housing contractors in an attempt to set forth some common universal design guidelines for the single-family residence. Contractors were selected for this study because they are an important link between the architect/interior designer and the client. Their awareness and use of universal design concepts is crucial. They need to be willing and able to implement the design as specified in the architect and or designer's plans that the client wants and has approved. Yet, they are often the most resistant to do so, because it's not the way they've always done it.

This study used a self-administered, mail survey questionnaire developed by the researcher. The Dillman Total Design Method was used as the basis for the survey instrument and its administration. A random sample of housing contractors indicating single-family residential construction as a primary focus of business was taken from the Oregon Construction Contractors' Board list. One hundred sixty-four surveys were returned for use in analysis.

Data analysis included descriptive statistics, mean, and frequency distributions. Paired sample t-tests were used to determine differences between awareness and use of universal design. Multiple regression and Pearson correlations were used to compare universal design use and selected demographic characteristics. Paired sample t-tests determined whether or not added cost to implement universal design affected use. Kendall's tau tests compared viability and mandated use of universal design as part of the building code. The MANOVA (multivariate analysis of variance) test compared current voluntary use and housing contractors' opinions about specified characteristics of universal design.

These analyses found that, of Oregon housing contractors surveyed, there was a greater awareness than use of universal design, which was significant. Barriers and incentives to use were important considerations in the process of adopting universal design. Cost and demand by clients were most often cited. A majority of respondents felt that incorporating universal design standards as part of the building code was a viable idea, even though they disagreed with it. Specified demographic characteristics of housing contractors did not play a significant role in either awareness or use of universal design. Added cost to implement universal design was found to be associated with its use. The more there was an indication of additional cost, the less the use of universal design.

Such findings are important to educators and other housing design professionals, as well as the manufacturers of universally designed products. Education about these features and products, as well as availability and examples of existing applications, need to be pursued and expanded, becoming an integral part of the curriculum.

The advantages of universal design for everyone need to be promoted, as an opportunity for educators. As acceptance, use, and demand for universal design products increase and become more widespread, costs should become more competitive with products currently in use. Again, there is opportunity for manufacturers, developers, builders, and educators to inform and educate not only housing and design professionals, but consumers as well.

The cooperative effort of building model homes and disseminating house plans that are universally designed would help accomplish this. Also, educating professionals and consumers alike about the differences between universal design, handicap accessible design, and the ADA is essential and a challenge for both educators and professional organizations. Government at all levels and other policy makers have an opportunity to encourage universal design use through providing incentives, such as tax credits or other grants for single-family housing that incorporate these features. They also need to understand the long-term and far-reaching implications and advantages of having a universally designed home in terms of the impending increase in size of the aging population. The

AARP has taken steps in this direction with their Home Modification Program; however, more needs to be done. As for viability of universal design standards becoming part of the residential building code, those surveyed felt that it was viable but did not necessarily agree. Code and building officials need to understand this when dealing with code and policy changes.

The advantages and benefits of universal design to all in single-family housing will hopefully be the factors that lead toward the removal of barriers and the eventual adoption of universal design. This would be as a part of not only the building code but also acceptance and use by single-family housing contractors, other housing design–related professionals, and consumers.

The first page of a sample survey is displayed in Figure 3.5.

DEVELOPING AN INCENTIVE PROGRAM FOR UNIVERSAL DESIGN IN NEW SINGLE-FAMILY HOUSING

The purpose of the project conducted by Seattle Pacific Professor, Sandra Hartke, was to make recommendations for an incentive program for universal design (UD) in new, single-family construction, and then to develop each recommendation into a working plan of action. An incentive program might accelerate the adoption of UD features in such housing and could further the goal of transforming the housing market into one in which UD features are the standard for design and construction, rather than the exception.

The project began as a request from the Housing Task Force, Advisory Council on Aging and Disability Services, City of Seattle, to develop an incentive program for UD in residential construction. Members of the task force wanted a program that would encourage builders, developers, architects, and other professionals to include UD features, products, and materials in new, single-family housing. They specifically did not want UD to be required by ordinance or regulation. They therefore suggested that the Leadership in Energy and Environmental Design (LEED) program be examined to determine if its components could be adapted for UD.

Several sources were relied on in developing this incentive program:

- Current federal, state, and local initiatives with language for inclusion of UD features in new residential construction. Such initiatives included four categories of existing state and local policy that address accessibility and visitability features in single-family homes: (1) builder requirements for housing built with a public subsidy, (2) builder requirements or incentives for unsubsidized housing, (3) consumer-based strategies, and (4) consumer awareness campaigns.
- Local and regional incentive programs related to energy efficiency and green design.
- LEED, which is a national program developed by the US Green Building Council, formally implemented in 2000, rates green building applicants according to their degree of compliance with a developed rating system. All commercial buildings are eligible for consideration as a LEED building. The LEED residential program went into effect in 2005.

To accelerate the adoption of UD in new, single-family housing, the researchers' recommendation was to develop an incentive program:

- That is voluntary, consensus based, and market driven
- With a scope broader than the minimum standards of visitability but less complex and costly than the program requirements of LEED

In order to develop such a program, one must:

1. Determine the UD features, products, and materials required at various levels of attainment, ranging from essential (Level 1) to nonessential (Level 2, 3, 4, etc.).
2. Develop criteria by which the UD features are to be evaluated (design criteria and guidelines). Develop a concise UD resource for use by design and building professionals.
3. Develop a coalition of supporters or identify an organization to oversee and implement the program. Implementation includes evaluating the housing plans, inspecting the

projects as they are completed, and awarding certificates to those builders and houses that meet UD criteria.

4. Develop a marketing campaign for the incentive program that is positive and reaches the mainstream consumer-housing market.
5. Evaluate the incentive program.

UD features incorporated into new residential construction would be immediately useful to all residents. The housing stock in the United States would evolve over the next 15–20 years to better meet the needs of all users. A model for an incentive program that includes UD features would bring us one step closer to transforming the housing market into one that better meets the needs of our changing population.

UNIVERSAL DESIGN EXEMPLARS: TEACHING STRATEGIES

Important teaching strategies used to promote universal design (UD) are shown in case studies that will be included in the Teacher's Manual. The Jackalope commercial UD product was designed by students of Pattricia Moore at Arizona State University. (This project actually resulted in development of a Light Rail System in Phoenix, Arizona. See Figures 3.45 through 3.48.) The project included images created by the students, a condensed PowerPoint presentation with 27 images and also a large, complete PowerPoint file that included 88 images.

FIGURE 3.45   Cutaway railcar image from student Jackalope project.

FIGURE 3.46   Accessibility and usability were the primary concerns of the Design Team.

FIGURE 3.47   The METRO Light Rail System premiered on December 2008 and has been a great success with riders and businesses along the route.

FIGURE 3.48   The Interior of the METRO vehicle allows for all riders to determine where and how they will position themselves for their trip. Bicycle riders have dedicated seating areas and hanging bike racks to accommodate their needs.

REFERENCES

Davis, S. (1995). Is affordable housing significant architecture? In *The Architecture of Affordable Housing* (pp. 127–188). Berkeley, CA: University of California Press.

Hartje, S. C. (2004). Developing an incentive program for universal design in new, single-family housing. *Housing and Society, 31*(2), 195–212.

National Kitchen and Bath Association (U.S.). (2012). *Kitchen and Bathroom Planning Guidelines with Access Standards: [based on the 2012 International Residential Code (IIRC) and the ICC A117.1-2009 standard]* (2nd ed.). Hoboken, NJ: John Wiley & Sons.

Null, R. L. (1988). A universal kitchen design for the low-vision elderly: Research applied in practice. *Journal of Interior Design, 14*(2), 45–50.

Parrott, K. R., Beamish, J. O., Emmel, J. M., & Lee, S. (2008). Kitchen remodeling: Exploring the dream kitchen projects. *Housing and Society, 35*(2), 25–42.

Price, C., Zavotka, S., Teaford, M., & Holmes, P. (2004). Universal design: An interdisciplinary partnership to promote ease of living. *Natural Resources and Environmental Issues, 11*(1, Article 8), 6.

White, B. J. (Director). (2004, December 9). Diffusing universal design globally: Learnsites for youth and adult leaders. *Designing for the 21st Century III*. Lecture conducted from Adaptive Environments, Human Centered Design, Rio de Janeiro.

Implications for Facility Managers and ADA Compliance

4

CONTENTS

UNIVERSAL DESIGN IMPLICATIONS FOR FACILITY MANAGEMENT

The Americans with Disabilities Act (ADA) has created a myriad of challenges and opportunities for facility managers. Much of the discussion surrounding the ADA has focused on the physical accommodation aspects of the legislation as expressed in the ADA Accessibility Guidelines (ADAAG). The majority of the professions concerned with devising "solutions"—architects, designers, engineers, accountants, lawyers, and so on—have focused predominantly on physical, financial, and legal issues. They have often acted in a reactive mode, fearing more potential penalties of noncompliance rather than seeking out the potential opportunities of meeting both the letter and the spirit of the law.

Facilities management is one of the most challenging fields to develop recently in this country, affecting a wide range of American business, education, and public facilities. With the passage of the ADA, it is also one of the most demanding, because it is the facility managers' responsibility to oversee the interplay of the physical environment and operations, especially as it affects workers. Facility managers are hired to manage physical property so that it is efficiently and safely used for its intended purpose with both universal design and green design being considered. In the private sector, the facility managers' primary function is probably to ensure a profit for the facility owner. Their secondary function will then be meeting the needs of workers. While these are often concurrent, they are not necessarily so, and many owners and facility managers feel that the ADA demands an unwarranted shift in emphasis from the former (a profit) to the latter. Universal design offers a means of balancing both of these demands along with the recent emphasis on sustainability of any green design.

The opening paragraphs in this chapter present an argument for how the spirit, rather than the letter, of the ADA legislation is the crucial element for true accommodation and show why universal design is both the conceptual and practical answer to the question of accommodation.

This is followed first by a case study that shows how Purdue University created and implemented a comprehensive ADA compliance plan. Several years later, Facility Management played a major role in the renovation of the football stadium at Purdue. The next case study by Mazumdar and Geis (2003) illustrates how noncompliance with ADA (in the building of sports arenas) has resulted in a series of court cases and decisions in the years since the passage of the ADA. Additional case studies related to facility management include a discussion of acoustics standards and virtual meetings; Shoshana Shamberg then discusses how professionals such as occupational therapists can work with facility managers to arrive at viable solutions for accommodating individuals with disabilities. Finally, Nancy and James Canestaro present a case study using a simulation game to discover cultural differences between Japanese and American managers.

The ADA is, purely and simply, civil rights legislation enacted to prohibit discrimination and eliminate it by removing the barriers, physical and otherwise, that deny the "differently abled" full access to public and commercial facilities. To successfully implement goals of the ADA, a detailed knowledge of the technical and physical aspects of the law is not as important as a committed understanding of the intent of the law. It is primarily an attitudinal question rather than one of physical change, accommodation, and associated costs. Changes will have to be made; money will have to be expended. However, the long-term benefits will far outweigh the short-term costs. If the changes are approached in the spirit of "reasonable (and rightful) accommodation" for all, both those who are currently disabled and those who are "temporarily abled" will benefit.

UNITED METHODIST ORGANIZATION MODEL FOR ADA COMPLIANCE

Figures 4.1 through 4.12 show the results of the United Methodist Church's attempts to accommodate people with disabilities, including updated renderings to reflect ADAAG (the drawings are done by Jerry Ellis, AIA). It is important to understand that efforts at accessibility have been going on for more than 20 years.

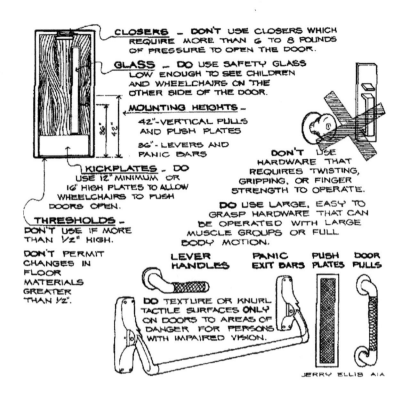

FIGURE 4.1   Door hardware.

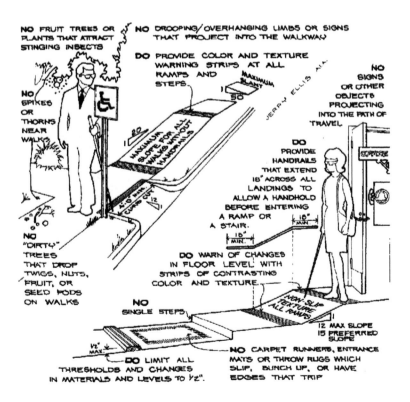

FIGURE 4.2   A clear path.

FIGURE 4.3 Lighting.

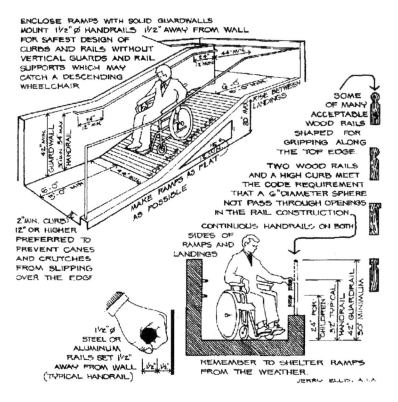

FIGURE 4.4 Ramps.

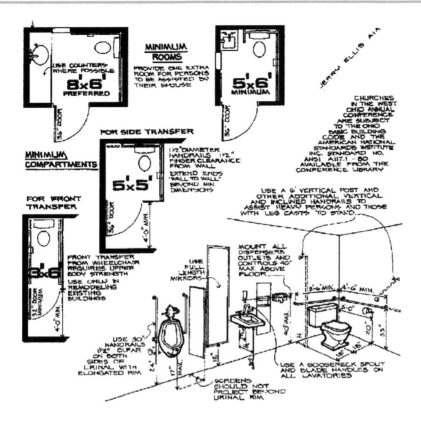

FIGURE 4.5   Toilet facilities.

FIGURE 4.6   Sheltered entrance.

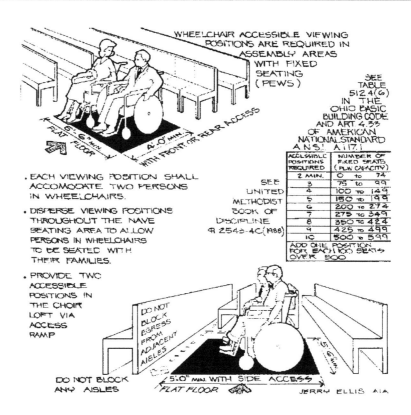

FIGURE 4.7 Seating.

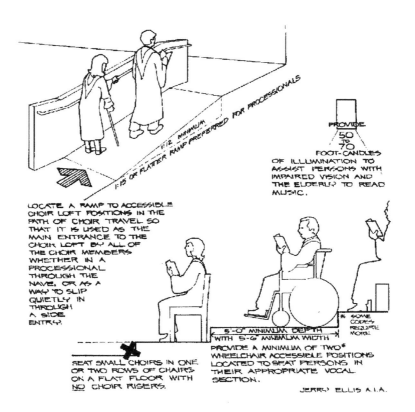

FIGURE 4.8 Choir loft.

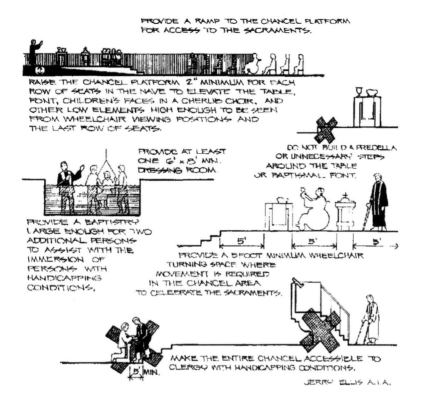

FIGURE 4.9   Chancel platforms.

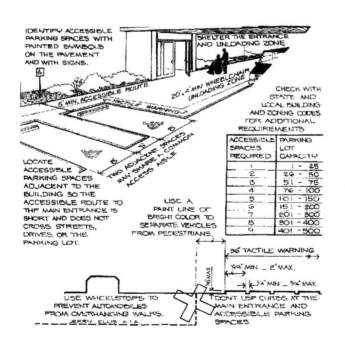

FIGURE 4.10   Parking/Curbs.

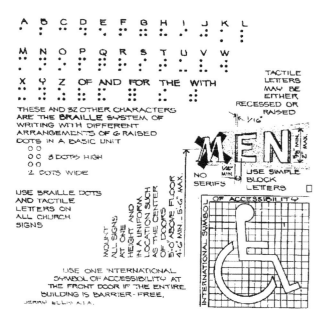

FIGURE 4.11 Signs.

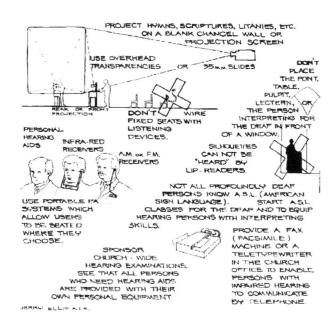

FIGURE 4.12 Hearing and media. Computer technology has made the projected images easier to see and use.

IMPLICATIONS FOR FACILITY MANAGEMENT

Facility management has to do with *facilitation*—how to maximize the effectiveness of the built environment for those who own it and those who use it. Implementing the ADA, and incorporating universal design as a key part of that implementation strategy, is not a question of simple compliance with a set of "minimums" as expressed in the ADAAG. It is not a cut-and-paste exercise of applying template solutions to problems, embodying an afterthought mentality to avoid potential legal hassles. For facility managers, meeting the spirit of the law involves the questions, and the accompanying implementation strategies, of how one treats people. It implies using what is already known about people, their jobs, and the work environment to design the work processes properly.

Universal design is directly related to the ADA. This is understandable since both deal with issues of accommodation and people with disabilities, yet each represents a distinctly different approach to the same issue. The ADA is the most sweeping piece of civil rights legislation in recent history. As law, it is subject to challenge, interpretation, and codification in the courts. It is, by definition, a reactive approach that fosters a compliance mentality to the concepts of accommodation through "inclusive" or "ability-sensitive" design. In contrast, universal design is philosophically different from the ADA because it is both broader and proactive. According to Ronald Mace, who is credited with coining the term, *universal design* means design for people throughout their life. This is an important distinction.

The ADA attempts to deal with certain types of disabilities and recommends accommodations based on common forms of disability. The underlying premise of universal design holds that disability is a contextual state and that everyone at some point is disabled. Disability in this context is an inability to adapt to or adjust to a situation, device, or environment. Since most devices and environments are redesigned, universal design seeks to eliminate elements that force excessive adaptation on the part of the human occupant or user.

Disability is a continuum. We are all disabled to a greater or lesser degree at some point in time. Even Olympic athletes and other champions of physical prowess are disabled in certain situations. Those who are left-handed, are pregnant, are obese, wear eyeglasses, or are children belong to the "universe" of disabled people—depending upon the setting. When considered from this perspective, universal design could be said to advocate design that minimizes contextual disability for the greatest population of users.

Advocates suggest that we "consider those individuals most people think of when they hear the term 'disabled'—the wheelchair users. These people, under most circumstances, are 'handicapped' three times over. Firstly, whatever condition put him/her in the wheelchair, the disabilities concerned will be handicapping in themselves. Secondly, he/she must operate at an eye-level that is some 16 inches lower than that of standing people, which is disadvantageous both physically and psychologically. Thirdly, he/she rolls around in a cumbersome, awkward, space-consuming, distinctive, and inelegant vehicle." It is this group of users that much of the ADA is intended to accommodate. However, under certain circumstances (e.g., long, broad hallways, outdoor walkways, etc.), wheelchair users, especially when using power chairs, are less restricted and encumbered by their environment than are the so-called able-bodied.

Since this approach is very anthropocentric—human focused—there are certain unavoidable parallels with human factors. The discipline of human factors or, more narrowly, ergonomics attempts to define the design requirements of devices, environments, and systems in terms of human capabilities and limitations. The goal is to make the things people use and the ways and places in which they are used as safe, easily comfortable, and productive as possible. Human factors recognize the value of people in human technology and human environment systems and strives to minimize the adverse impact of design on people. Thus, *human factors* is applied universal design.

In this context, universal design goes well beyond the ADA to address the issue of prevention as well as accommodation. Consider the contribution of design (both positive and negative) to current issues such as cumulative trauma disorders and workers' compensation costs. Universal design attempts to identify potential causes of disability as well as barriers to people with disabilities. The goal is to rectify the design environment so that disabilities are *prevented*.

The challenge for facility managers is not necessarily to assess the accommodations required to serve a particular segment of the population—the so-called compliance attitude fostered by many who misinterpret the intent of the ADA. Rather, facility managers should look for ways to maximize efficiency and profitability by removing impediments to all users. Where special accommodations appear necessary, one must question the purpose of the change and the root cause of the inaccessibility.

As a simple example, consider steps leading into buildings. Steps in general are major contributors to slips, trips, and falls. Outdoor steps are particularly hazardous because of varied weather conditions. In northern climates, outdoor steps have associated maintenance costs of snow removal (usually by hand) and the added hazard of ice and snow. A ramp is often added to steps to increase accessibility; however, the ramp really serves only a portion of the population. Because of the angle, surface, and other considerations, it may not be appropriate for certain people (e.g., arthritis sufferers, people with walkers). A ramp also presents additional cost and maintenance concerns such as snow removal, vandalism, and damage. Several demonstration projects have shown that a different approach walkway from the curb to the door not only increases access but also saves money, reduces safety liability, and enhances appearances if well designed and landscaped. Initial costs may be somewhat higher, but the cost of use is considerably lower. The difference between an add-on ramp and a more complete environmental solution highlights the difference between simple ADA compliance and the use of universal design. The former can be efficient: *doing things right.* The latter is effective: *doing the right things.*

The point is that few problems have only one right answer. Inside the facility, one can find many examples where universal design can yield benefits for a wide spectrum of occupants and users. Fundamentally, the focus should be on three areas—safety, communication, and wayfinding. There are many elements that can be used to affect these functions (space, light, sound, color, texture, contrast). For example, signage, an aid to wayfinding, should accommodate all users. Attention not only to the format (e.g., raised letters, Braille) but also to issues that will serve all users (placement, readability, content, and coding of information—color, for example) is important.

Cost is a concern of all businesses when considering universal design or the ADA. If viewed in the context of simple compliance with the ADA, the goal is to minimize costs by doing the minimum necessary to avoid contention. Applying universal design changes the cost consideration to one of investment. As mentioned in the step-versus-grade example, up-front costs can be somewhat higher for doing right things; however, proper understanding and application of universal design principles will yield a greater return. That is not to say that all universal design solutions must necessarily cost a great deal more than those that merely meet the letter rather than the spirit of the law. Many approaches exist that require only appropriate application and use. Spatial arrangements, such as wider corridors, can both accommodate special equipment needs (e.g., wheelchair use) and support organizational requirements (e.g., interaction spaces and informal communication). Systems furniture, for example, can accommodate a wide range of work surface height requirements with relative ease. A variety of accessory, modification, and add-on products is available at relatively low cost. Door handle levers, for example, make sense for nearly all users in most environments and are a modest-price retrofit. Some technological fixes resolve larger design questions and make accommodation issues moot (e.g., energy controls that sense movement require very little interaction on the part of the occupant). The bottom line is that good design—universal design—is cost-effective because it considers all resources (including valuable human ones) and attempts to minimize long-term costs. This is also the goal of the facility manager.

The following are the best criteria by which to plan and judge universal design:

- Does the design present situations to which any people are incapable of adjusting?
- Can environmental elements be changed or removed to eliminate the challenge?
- Does the design offer alternatives—for example, stairs for those who wish to take them as well as free-access transport (e.g., lifts, elevators, etc.); different sink heights for children, wheelchair users, tall people?

There is no one right solution. Universal design as a strategy for achieving the goals of the ADA will help facility managers meet the spirit as well as the letter of the law. It is all a matter of attitude.

Universal design, with its emphasis on being supportive, adaptable, accessible, and safe to the entire range of potential users, goes far in meeting the physical concerns of the ADA. If universal design is to accomplish its intent, it is necessary to focus on removing the nonphysical barriers—intentions and attitudes. These must be supplanted with the good intentions of the organization, its management, and its employees if the spirit of the law is to be achieved. This is the challenge and the opportunity for facility management.

After the passage of the ADA, the President of Purdue decided that it was an opportunity for the University to become a "model" of ADA compliance. Owen Cooks, an interior design graduate of the university, was hired to create and implement the plan. In the following case study, Owen Cooks describes the approach taken by Purdue.

Case Study: Purdue University's ADA Compliance Plan

Title II of the ADA required all public entities that employ 50 or more persons to have developed a transition plan by July 26, 1992, detailing the steps to be undertaken to achieve program accessibility, along with a self-evaluation examining the policies and practices to be completed by January 26, 1993. In keeping with that mandate, Purdue University set as its priority task the removal of barriers that deny individuals with disabilities an equal opportunity to share in and contribute to the vitality of university life. Although the ADA only specifically required program accessibility and did not require that all existing spaces be made accessible, Purdue determined that *all* physical obstacles in *all* facilities should be identified. In doing so, information would become available that gave planners a better overall view of the situation, allowing a more comprehensive and long-term approach to program accessibility.

Several committees were established to oversee the various operations required to comply with the transition plan and self-evaluation. An ADA Steering Committee managed the overall project and subcommittees for the five titles were assigned; focus groups from individual schools, departments, and areas came together to complete the self-evaluations. The Transition Plan was developed through a series of facilities surveys, self-evaluations, and individual and public involvement.

FACILITIES SURVEYS

The first step in Purdue's ADA compliance plan was to bring together a team of university employees to conduct individual building surveys (Figures 4.13 through 4.15) to identify all physical barriers as determined by ADA guidelines. A detailed reference guide was compiled drawing on the ADAAG for structure and standards. Copies of this document— "Facilities Evaluation/Reference Guide to Architectural Barriers to the Disabled"—were used to train and guide the surveyors as they conducted their inspections.

A summary of each identified barrier was prepared as well as a floor plan location (see Figure 4.13). All the surveys were kept on file for reference by estimators and designers working on construction projects so that barriers in spaces being remodeled could be removed or modified to the maximum extent feasible. A computerized summary of costs was assembled from all the surveys to provide information concerning quantity and costs for compliance by both building and barrier type.

SELF-EVALUATIONS

The ADA required that a public entity evaluate its current services, policies, and practices and determine which ones do not meet the requirements of the regulations so that necessary modifications could be made. Purdue University distributed Institutional Self-Evaluation forms to all departments, which established their own focus groups and assigned individuals to complete the form. Recommendations about their findings were then forwarded by the appropriate subcommittee to the ADA Steering Committee for consideration and prioritization.

The evaluation included the following:

Review of information on eligibility, policies and procedures, and physical environment, and comment on the accessibility problems of programs for persons with the following disabilities:

1. Mobility impairments
2. Hearing impairments
3. Vision impairments
4. Learning disabilities
5. Chronic illness

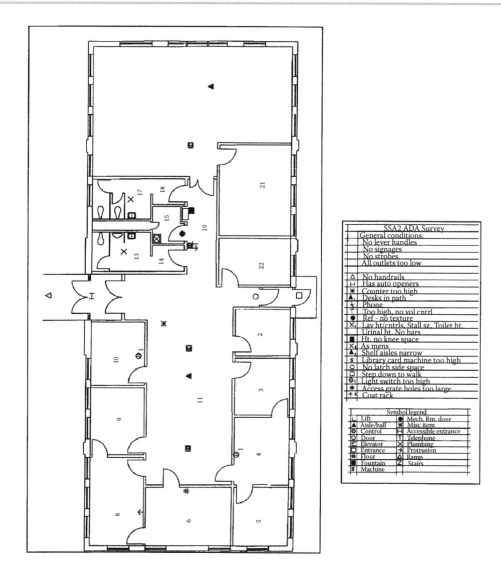

FIGURE 4.13   Sample facility survey floor plan from Purdue's ADA plan.

The planners then described

- Tasks to complete to insure accessibility to programs, services, and activities
- Person responsible
- The estimated cost and completion date

Since these forms were often completed at the department level, summary tables were created for each school.

INDIVIDUAL AND PUBLIC INVOLVEMENT

Purdue students with physical limitations were asked to meet with campus planners and administrators to provide specific information on physical barriers they encountered and their experiences in accessibility. These meetings were grouped around specific disabilities—mobility, hearing, vision, and chronic illness. During the meetings, overall goals of the transition plan for physical barrier removal were discussed and input was requested. Both students and planners left with a better understanding of the problems and possible solutions.

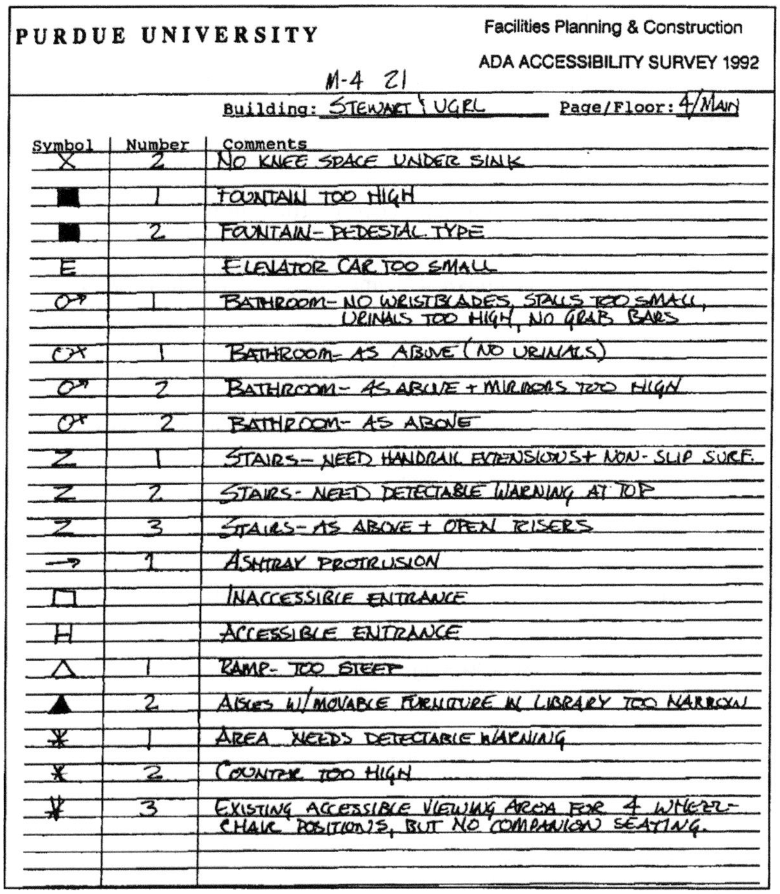

PURDUE UNIVERSITY

Facilities Planning & Construction

ADA ACCESSIBILITY SURVEY 1992

M-4 21

Building: STEWART \ UGRL Page/Floor: 4/MAIN

Symbol	Number	Comments
X	2	NO KNEE SPACE UNDER SINK
■	1	FOUNTAIN TOO HIGH
■	2	FOUNTAIN – PEDESTAL TYPE
E		ELEVATOR CAR TOO SMALL
♂	1	BATHROOM – NO WRISTBLADES, STALLS TOO SMALL, URINALS TOO HIGH, NO GRAB BARS
♀	1	BATHROOM – AS ABOVE (NO URINALS)
♂	2	BATHROOM – AS ABOVE + MIRRORS TOO HIGH
♀	2	BATHROOM – AS ABOVE
Z	1	STAIRS – NEED HANDRAIL EXTENSIONS + NON-SLIP SURF.
Z	2	STAIRS – NEED DETECTABLE WARNING AT TOP
Z	3	STAIRS – AS ABOVE + OPEN RISERS
→	1	ASHTRAY PROTRUSION
□		INACCESSIBLE ENTRANCE
H		ACCESSIBLE ENTRANCE
△	1	RAMP – TOO STEEP
▲	2	AISLES W/MOVABLE FURNITURE IN LIBRARY TOO NARROW
✱	1	AREA NEEDS DETECTABLE WARNING
✱	2	COUNTER TOO HIGH
✱	3	EXISTING ACCESSIBLE VIEWING AREA FOR 4 WHEELCHAIR POSITIONS, BUT NO COMPANION SEATING.

FIGURE 4.14 ADA Accessibility Survey.

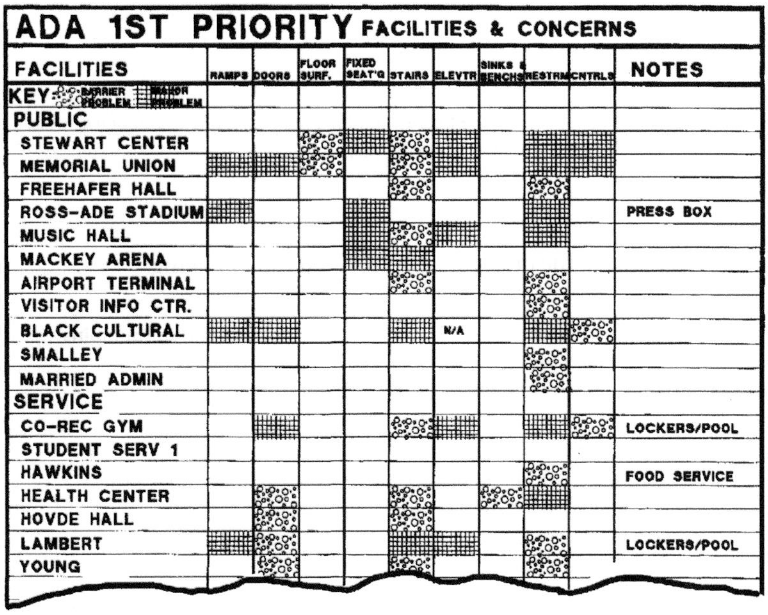

FIGURE 4.15 Community input meeting charts from Purdue's ADA plan.

Questionnaires specific to disability type were prepared and distributed at these meetings. All questionnaires requested the same general information: building identification and use (work, entertainment, study, classes, dining, living, meetings, service) and prioritization of building and area in terms of establishing accessibility. Each specific questionnaire then asked respondents to rate several items by priority and provided for comments and additions.

For hearing disabilities, the questionnaire covered the following:

Telephones

- Accessibility to telecommunications devices for the deaf
- Availability of volume control telephones
- Availability of hearing aid–compatible telephones

Controls and Alarms

- Strobe lights on fire alarms
- Controls in general

Elevators

- Visual controls

Assembly Areas

- Availability of assistive listening devices

Miscellaneous

For visual disabilities, the questionnaire covered the following:

Protrusions (concerns for objects in path of travel)

- Display cases
- Ashtray urns* mailboxes
- Other structural items
- Overhead hazards

Elevators

- Audible controls
- Raised signage and controls
- Braille signage and controls

Detectable Warnings (surface texture changes used as warnings)

- Textured door handles to hazardous rooms
- Exterior curb ramps
- Exterior ramps, stairs, and other drop-offs

Signage

- Raised for interior rooms (classrooms, offices, restrooms, etc.)
- Braille for interior rooms
- Raised for displays (maps, directories, etc.)
- Braille for displays

Stairs

- Handrail height
- Handrail extensions
- Slip-resistant step and landing surfaces
- Color contrast of step and landing surfaces

Floor Surfaces

- Gratings
- Level changes
- Slip resistance
- Color contrast

Miscellaneous

Note: When all campus buildings were made "smoke-free," ashtray urns were removed and attractive metal tiles were devised to cover the holes where the urns were attached to the wall.

For mobility disabilities, the questionnaire covered the following:

Floor Surfaces

- Slip resistance
- Level changes
- Carpets
- Gratings
- Landing size
- Width of ramp

Doors

- Width
- Maneuvering space to open door
- Thresholds
- Hardware
- Opening force
- Automatic openers

Drinking Fountains

- Knee clearance
- Height
- Controls
- Spout location

Telephones

- Height/reach of clearance
- Maneuvering space

Seating and Tables

- Availability of wheelchair locations
- Knee space/height of tables/benches
- Reach over tables/benches

Elevators

- Interior car size
- Time to close door
- Height controls
- Handrails provided

Controls and Alarms (light switches, outlets, and alarms)

- Reach/height of controls
- Reach/height of alarms

Restrooms/Locker Rooms

- Space to maneuver
- Mirror height
- Lavatory height/knee space
- Lavatory controls
- Stall dimensions
- Grab bars
- Toilet height
- Urinal height
- Bathtubs
- Showers
- Lockers

Sinks

- Knee clearance/height
- Reach
- Controls

Signage

- Signs for accessible entrances
- Signs for accessible restrooms
- Directional signs to accessible entrances
- Directional signs to accessible restrooms

Miscellaneous

- Areas of rescue assistance in buildings
- Raised platforms/stages (needing ramps)
- Curb cuts
- Parking near accessible entrances
- Sidewalks

The information from these surveys allowed planners to establish a priority list for barrier removal that was reviewed and revised at a public meeting advertised in all public media. Charts were shown to the participants that identified buildings and barriers by priority (see Figure 4.15). Then, with the group's participation, additional barriers were discussed and priorities were more clearly established.

MAKING FACILITIES ACCESSIBLE: PRIORITIES AND METHODS

On the basis of the facility surveys, the self-evaluations, questionnaires, and meetings, barriers were placed into two categories: nonstructural and structural. Priorities were established first by individual request and second by the priority list created from the above methodology. The number one priority for the university was and continues to be meeting the needs of students and employees wanting access to a specific space.

NONSTRUCTURAL BARRIERS

The plan called for appropriate changes to meet ADAAG standards in the following areas:

- Room signage
- Door hardware
- Telephones
- Seating and tables
- Elevator controls
- Protrusions
- Alarms
- Directional signage
- Assistive listening devices
- Power door openers
- Drinking fountains
- Counter heights
- Parking locations
- Exterior curb cuts
- Sidewalks
- Exterior projections

STRUCTURAL BARRIERS

Barriers that were by their nature difficult and sometimes expensive to eliminate were structural barriers. Alternatives allowed by the ADA included relocating a function to an accessible facility. Purdue University's plan adhered to the following hierarchy: relocation, renovation, and replacement. Among the structural barriers identified were residence hall rooms. Figure 4.16 shows how the renovation was completed without making major structural changes.

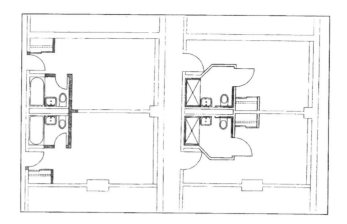

FIGURE 4.16 Residence Hall renovation with areas to be changed in gray on the left; modifications were without major structural changes.

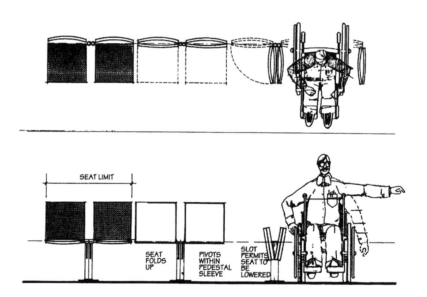

FIGURE 4.17 Accessible seating design for Oriole Park by Kim Beasley.

Widely acclaimed for its architectural beauty and urban placement, the Oriole Park at Camden Yards in Baltimore, Maryland, has also been hailed for its accessible design that meets the requirements of the ADA. Kim Beasley drew the sketch for the accessible seat (Figure 4.17), which was built by the American Seating Company. PVA (Paralyzed Veterans of America) received a patent for the design.

REMODEL OF THE FOOTBALL STADIUM AT PURDUE UNIVERSITY

Several years after the renovations described above, Purdue University needed to renovate the football stadium, and because this was a renovation, the project was required to be ADA compliant. The facility management team utilized the seat design that had been created for Oriole Park in 1993. The Purdue University football stadium renovation highlights the importance of a team approach to compliance. The university was fortunate that Owen Cooks and his staff had already worked together in developing the comprehensive compliance plan. They were successful in creating a renovation plan that met ADA Compliance Plan Guidelines (see photos of the renovated stadium in Figures 4.18 through 4.22).

FIGURE 4.18   Purdue University Stadium.

FIGURE 4.19   Wheelchair platform in gray, discontinuous hand rails, and bold use of contrasting colors. Purdue University Stadium.

FIGURE 4.20   Alternating box seats with wheelchair space; all seats have a good line of sight even when the crowd stands. Notice the blue icons on the side of the seats in both Figure 4.20 and Figure 4.21.

FIGURE 4.21   Removable armrest seat for transferring from a wheelchair; also shown are counters for standing viewers.

FIGURE 4.22   Luxury box seating.

ADA IN ACTION: LEGAL RULINGS IN SPORTS ARENAS

Designed environments should be available for all to have access to, use, and enjoy without prejudice. Yet, historically, this has not been the case, as many buildings are not accessible to persons with disabilities. Since the 1960s, attempts have been made to develop standards and regulations to address this issue. Court decisions on numerous recent lawsuits have brought attention to the issue and changed the landscape of provision of accessible environments, the importance of compliance, the nature of the law, and architects' liability and responsibility. Based on studies by researchers Gilbert Geis and Sanjoy Mazumdar (U C Irvine) of the history of provision of accessible environments, problems of accessible design, regulations including the Americans with Disabilities Act (ADA), and lawsuits, a brief sketch of the current situation regarding the law and its application, as well as the condition of architects' liability and responsibility under the law, as interpreted by the courts in the United States emerges.

Until the mid-1990s, architects had not been held directly responsible, legally or financially, for building mishaps. The Anglo-American legal principle of privity of contract and precedent of court decisions declared that the final decisions about construction lay with the owner.

A number of court rulings on lawsuits based on the somewhat ambiguous wording of the ADA changed that situation in regard to persons covered by the act. The liability of architects was contested in a spate of cases, most of them involving persons using wheelchairs who claimed that they had not been provided equivalent seating in newly built arenas that hosted sports and entertainment events.

The first sports arena case, decided in 1996, concerned seating arrangements in the MCI Sports Arena in downtown Washington, D.C. The Paralyzed Veterans of America sued the firm of Ellerbe Becket Architects & Engineers and others under Title III of the ADA as part of its effort "to eliminate discrimination against its members on the basis of physical disabilities." Judge Thomas F. Hogan, however, dismissed Ellerbe Becket from the case on the ground that a party was legally liable only if, according to the words in the ADA statute, he demonstrated "a failure to design and construct facilities that are readily accessible." This, the judge declared, eliminated the architects since they had not both designed and constructed the facility. Similar interpretations of "design and construct" were put forward by federal judges in Kentucky and California in cases involving branches of the Day's Inn motel chain. The triumph of the architects in the MCI and the two Day's Inn cases proved to be short-lived, however.

In 1997, a Florida judge interpreted the words of the statute very differently. In a case involving the Broward Center, the future home of the Florida Panthers hockey team, the judge preferred the argument that if architects were not liable under the ADA, it was conceivable that nobody would be. To be legally responsible for damages, the court ruled, a party had either to have designed or to have constructed the facility, not both. Essentially, the same conclusion was reached in regard to a Camden, NJ, facility, which had fixed seating for 6000 persons and a lawn area that could hold about 18,000 persons but did not provide satisfactory access for wheelchair patrons.

The courts were echoing an interpretation of the "design and construct" clause in the federal Fair Housing act about which a court had said: "The notion that an architect and a builder working together, one performing the first function and the other performing the second one, are thereby insulated from liability is a frank absurdity."

The architects' problems did not cease with these initial sports arena cases. In October 1999, the US Attorney General's Office in Minneapolis filed a complaint against Ellerbe Becket Inc. alleging that it had "repeatedly designed arenas and stadiums with wheelchair seating locations that do not provide wheelchair viewers with lines of sight to the floor or the field that are comparable with those of other spectators." Two years later, a consent decree brought the Minnesota case to an end. Ellerbe Becket agreed that it would include adequate wheelchair seating and satisfactory lines of sight for all fixed seating that it designed after 1998 and that it would provide an annual report of its compliance for the next three years. It was excused from providing such seating in the upscale luxury suites.

"We are not trying to get them to redesign the stadium," a US Department of Justice spokesperson said. "We're trying to ensure they no longer build stadiums that are inaccessible to the handicapped." For an attorney representing the persons with disabilities, the agreement marked a major change in attitude, though he was not certain that the ruling would be taken to heart by other architects. A law review note observed that if the Ellerbe Becket settlement did not bring stability to this area with the different rulings and interpretations, the US Supreme Court would have to provide a final interpretation.

In the various cases, the courts identified five major issues of contention:

1. Number of accessible and companion seats: The number of seats that must be accessible to persons in wheelchairs. Where must the required companion seats be located? Could they be in the row above the wheelchair patron rather than next to that person?
2. Lines of sight: The number of seats that must provide sightlines "comparable to those of members of the general public" in order to meet the law's specified requirement.
3. Operational means: Whether operational means could be introduced to create satisfactory sight lines. Such measures might include signs requesting spectators seated in front of those in wheelchairs not to stand during the event.
4. Dispersal: The dispersal of seating for wheelchair spectators so that they could make meaningful selections between the views afforded and the cost of particular locations.
5. Suites: Which spaces in the luxury suites could be counted toward the required number of enhanced sightline seats? In the MCI Arena, there were 109 suites, each costing between $100,000 and $175,000 a year, seating up to 24 persons.

The judges' conclusion on these matters, briefly stated, was that the law's formula of "one percent plus one" that applied to sites with more than 500 seats must be followed in regard to the number of seats made available for wheelchair patrons. The MCI Center, with 17,989 seats for basketball games, for example, was required by ADA to have 181 seats accessible to wheelchair patrons.

Regarding the ability to view the event, the judge in one case found unsatisfactory the proposed operational means of posting "no-standing" signs for patrons in front of those in wheelchairs. He believed that this approach was unworkable and that it might create hostility toward persons with disabilities. The judge maintained that the law called for architectural design solutions, not operational ones.

A judge in a subsequent case scoffed at the argument made by the architect's lawyers that the designated wheelchair locations with unobstructable lines of sight (ULOS) provided a special advantage to the disabled. He observed sarcastically that he had not noticed a stampede by persons to be amputated so that they might enjoy such a special privilege. The same judge objected to a proposed policy that would require wheelchair patrons to reserve special seating in advance. This, he believed, was treating them as if they had a noxious disease, such as leprosy.

The judges interpreted the law to imply that all accessible seats must have ULOS of the event even when spectators in front stand up. The MCI case judge granted, however, that it may not be possible for the design to accommodate all the requirements of the law (ULOS, integration, dispersal) and ruled that 78% to 88% compliance with ULOS requirement would be permissible.

In regard to fixed companion seating, the decision in the Rose Garden case in Portland, OR, was that the law's requirement was that companion seats must be adjacent to wheelchair seats, and that seats in the row above would not be acceptable. These companion seats did not need to be bolted into the floor, he ruled, they could be padded folding chairs made by Clarin.

How did architects view the ADA's legal directives? The sports arena and related court cases sometimes showed the architectural profession, as represented by briefs filed with the courts by its attorneys and those filed by the American Institute of Architects, in a rather unflattering light, seeking to avoid professional responsibility. Among other rationales, they claimed that had Congress wanted to hold architects liable, this would have been stated in clear language in the ADA law. They argued that if other parties to the construction rather than architects were held liable, those parties would ensure that the architects would meet the legal standards. In one case, though, precisely the opposite occurred. The architects failed to inform persons building the Rose Garden that their actions might result in court cases against them owing to lack of provision of inadequate wheelchair seating. The architect's defense was that though they were aware of this, they regarded it as confidential information that they could not share with their client.

Conversely, it should be emphasized that the framers and enforcers of the ADA neglected to consider incentives for compliance in order to make the law operate as effectively as it might have. Especially with regard to Title III, the ADA or the US Department of Justice made minimal attempts at best to involve architects, whose task it is to find creative solutions to problems such as access, with efforts to encourage creativity by setting up design challenges and providing guidelines with depictions of end conditions. Instead, the law dictated specific solutions. Rather than bringing out the best in the architects and builders, in some ways the law elicited their not-so-good self-interested side so that they ended up fighting the ADA implementation even though they did not dispute its moral component.

Architects need to take an active role to see that all citizens are adequately served—not only those who hire them but also those who will be accommodated by their products, especially those who need special creative solutions from architects.

ANSI CLASSROOM ACOUSTICS STANDARD: LET THE WORD BE HEARD

ADDENDUM TO THE ADA

ANSI CLASSROOM ACOUSTICS STANDARD

According to audiologist Anne Seltz, in 1990, when the Americans with Disabilities Act (ADA) was signed, it was considered a major civil rights law with more teeth than some prior civil rights laws. In addition to protection against discrimination, this law requires that accommodation be made to remove barriers to full participation by people with physical and mental disabilities. The Access Board is the US Government agency responsible for developing the ADA Accessibility Guidelines to help people meet the law's requirements. The Board has done that and continues to add rules and regulations.

Title III of the ADA, *Public Accommodations and Services Operated by Private Entities*, lists *places of education* as number 10 on its list of 12 categories covered by the Title. Prior to 2002, the United States of America did not have a Classroom Acoustics Standard: one of few developed nations without one. Yet the original Access Board Guidelines did not address acoustical accessibility in schools for students with hearing loss and other disabilities. This was primarily a reflection of society's priorities about accessibility: a major focus had been to remove barriers for persons with mobility disabilities.

A STANDARD TO REMOVE ACOUSTICAL BARRIERS IS DEVELOPED

We would never teach reading in a classroom without lights. Why then do we teach in "acoustical darkness"? Speaking to a class, especially of younger students, in a room with poor acoustics, is akin to "turning out the lights" (John Erderich, PhD, 1999).

HISTORY

The Acoustical Society of America (ASA), The American Speech-Language-Hearing Association, and others had been urging the Access Board to consider research and rulemaking on the acoustical performance of building and facilities, in particular school classrooms and related student facilities. Other active groups included individual acoustics professionals, parents of children with hearing loss, individuals who are hard of hearing, and a consortium of organizations representing people with disabilities. The effect of poor classroom acoustics had been studied for years and the research data overwhelmingly supported the need for all students, but especially young students and those with hearing loss and other disabilities, to have access to classroom acoustics that supported learning and communication.

In 1997, a parent of a child with hearing loss used a formal legal process of petition. She petitioned the Access Board to amend the ADA Accessibility Guidelines. She asked the board to include new provisions for acoustical accessibility in classrooms to remove barriers for children with hearing loss. It was this parental petition that caused the Access Board, in 1998, to publish a request for information to gather public input on the issue of classroom acoustics. The Board subsequently actively supported the ongoing efforts to create a standard.

In mid-1997, the ASA had commissioned a Working Group on Classroom Acoustics in conjunction with the American National Standards Institute (ANSI) to develop a draft standard for approval by the ANSI committee responsible for noise issues (S-12). This Working Group included audiologists, acoustic engineers, building managers, educators, interior designers, persons with hearing loss, architects, acoustical materials manufacturers, parents, professional organizations, consumer organizations, and governmental organizations.

CLASSROOM ACOUSTICS STANDARD APPROVED JUNE 2002

The ANSI standard was submitted by the Working Group to the ANSI Board of Standards Review for approval at the end of May 2002. The ANSI Board approved this standard on June 26, 2002. The Access Board proposed this standard to the International Code Council (ICC) for inclusion in a future International Building Code (in process at time of publication). When ICC approves the standard as Code, future school buildings and renovations will have the opportunity to be acoustically supportive to learning and communicating.

Taken by itself, the new ANSI/ASA standard is voluntary unless referenced by a code, ordinance, or regulation. However, school systems may require compliance with the standard as part of their construction documents for new schools, thus making the design team responsible for addressing the issues. Parents may also find the standard useful as a guide to classroom accommodations under IDEA (the Individuals with Disabilities Act). Some parents might include it as part of their student's IEP (Individual Education Plan).

Advocates for barrier-free acoustic environments now have a powerful tool for influencing future school building and renovation plans. The early decades of the 21st century are expected to see major school building activity in the Unites States. According to the US General Accounting Office, one-third of the nation's schools need major renovation or replacement. Furthermore, census projections indicate that over 400,000 additional students will enter our schools each year for the next 50 years. This growth means we will need about 16,000 new classrooms each year.

GOOD TIMING FOR DESIGNERS

Now is the time for designers to recognize that public awareness of the need for quiet classrooms is growing. Our society is just beginning to develop a priority for quieter listening spaces that support learning and communication. The designing team can facilitate that growth by becoming knowledgeable about the standard, understanding the rationale for the standard, and including it in their plans when approaching clients. Designers can work with local acoustic professionals for technical support as well as with community organizations who advocate for good acoustics.

Even if you do not work with schools, the standard's accompanying support papers provide much information about the rationale for quiet listening and communication spaces. A client need not have a hearing loss or other disability to benefit from quiet environments. Keep in mind that not just classrooms are used for communication and education: hotels have meeting rooms that could benefit from the approach to sound management suggested by this standard. Restaurants, shopping sites, and houses of worship could be more supportive of human communication and interaction if noise levels were reduced. Lifelong learning is the rule now, not the exception. Adult learners will be thankful users of quiet classrooms and places of learning.

THE STANDARD

Its official name is *ANSI S12.60-2002 American National Standards Institute Performance Criteria, Design Requirements and Guidelines for Schools.* It is commonly referred to as The Classroom Acoustics Standard.

The two main components of this standard are as follows:

Noise levels should not exceed 35 DBA in an unoccupied classroom.
Reverberation time should not exceed 0.6 seconds.

This means that in an unoccupied classroom, during regular school hours, with all systems running (including HVAC [heating, ventilation, and air conditioning]), the noise level will not exceed 35 DBA. Also, there will be minimal sounds bouncing off hard surfaces in the room because walls and ceilings will be treated with appropriate tiles and other sound-absorbing materials. When an empty classroom is quiet, the teachers and students can speak at normal levels, without strain, and with good expectations for understanding and being understood (Nelson, n.d.).

HOW TO ACCESS THE ASA/ANSI STANDARD

The ASA sells the standard along with accompanying documents (in 2002, the cost was $35.00). The ASA has a web site, www.asa.aip.org/, or you can email the ASA at asastds@aip.org and ask for information about purchasing the Classroom Acoustics Standard.

SUMMARY

Our world is noisy. Though this standard emerged out of a need to serve students, primarily young students and students with disabilities, its basic principles can be applied in some degree to all living spaces because all human beings communicate in most every space they inhabit. When spaces are designed to be free of acoustic barriers, we create a world that supports good listening for all people. It is a Universal Design Standard.

FACILITATING VIRTUAL MEETINGS: LESSONS LEARNED

About four years ago, designer Daniel Mittleman, along with Robert O. Briggs and Jay F. Nunamaker, Jr., began facilitating virtual meetings, both same time and different time, to support work being done for the US Navy's Third Fleet in San Diego. That group facilitated about 100 virtual meetings, both for the Navy and other related organizations (Ter Bush and Mittleman 2005). These meetings included idea generation, planning, decision-making, issues surfacing, status briefings, environmental scanning, collaborative writing, training, and expert briefings. Some of the lessons this group, specifically Daniel Mittleman, learned from these activities included:

After 9/11, many people were reluctant to fly. Daniel Mittleman, using the list serve for the work environments division of EDRA, shared his experience with Navy video conferencing.

1. Use neutral mid-tone wall colors for background. Many (most?) cameras have an automatically adjusting iris. If the background is too light, the iris will close to darken the shot and faces will be too dark to make out details (a particular problem if some participants are dark skinned). Very dark walls will cause the iris to open and produce too much light on Caucasian faces. A mid-blue/gray works very well.

   Also, avoid busy patterns on walls (and on clothes!) as the camera may pick this up as vibrating noise. A solid wall makes the most sense.

2. What we did with the navy was rather than place whiteboards on the walls, we used a black melamine board and purchased fluorescent pens (the kinds that you might see in restaurant signboards). This worked very well. The darker background worked better for us than a reflecting white background would have (see iris above), and the fluorescent pens showed up well on camera (and went over well with the participants). We did not try mounting black lights above the boards, but that would be an interesting experiment.

   In a room I am currently developing, we are using whiteboards with wood covers that fold back. This creates a very nice look, and we only need to worry about the camera's iris when the whiteboard is open.

3. We used multiple light sources in our room. We had indirect ambient room light (fluorescent), individual task lights (halogen), and wall washers above the black boards (fluorescent). Separate controls for all. In our case, we did not have rheostats, but that (or even better—presets) would be valuable.

   If possible, indirect lighting is preferred as direct lighting will create a glare on participants' faces (and bald participant domes). It is very difficult to aim direct lighting so it will not produce glare on anyone in the room.

4. Camera angles were a challenge for us. We went with one camera in a front corner (front defined as the wall with the public display); let us say this camera was at 11:00. Our other camera was mid-wall, let us call it 3:00. This did a couple of things for us: (1) it let us get a shot of most of the participants looking at a presenter in the front of the room; (2) it let us get a head shot of the presenter from a slight angle but at a close distance; (3) one of the two cameras could shoot the face of every participant.

5. Camera height is an issue. Ideally, you want to shoot from the same height as the head, so you are not shooting down and seeing the top of people's heads. However, the lower the camera, the more likely it is to be blocked. In our Navy room, we placed the cameras at about 75–78″. In my new room (under construction), we may go below 72″.

6. Public display presentation via Video Teleconferencing (VTC) is an issue. Will there be an independent channel for VTC participants to see public display, or does it have to come through the camera shot? If a presenter is going to make gesturing reference to display content, it might well serve to get all this into a shot. Also, I have found that when making a presentation myself from a distance site, it is very helpful to know directly what public display my audience is seeing (it is not always the slide I thought they were seeing). So, to include it in a shot can be helpful.

   If the display information is not key to the meeting process, it could be distracting.

7. Thought needs to be given to the types of meetings to be held via VTC. Meetings that are primarily presentations require video on the presenter but may not require video on participants (a group shot may be sufficient for question-asking periods). Meetings with much give-and-take dialogue may not have a "stage" and may require equal video treatment of all participants. Camera count, camera angles, and camera control decisions will be affected by knowing these objectives.

8. I note that in classroom VTC situations, students at the distant end are less likely to ask questions if they know the camera will zoom in on them. This may also be true of VTC meetings in some organizational cultures. In other cultures, the presence of zoom might actually increase questions (conjecture on my part).

9. There has been some research done on deception and nonverbal cues that found most of the kinetic deception cues come from between the shoulder and the hips. This suggests (at least to me) that VTC headshots lose much valuable nonverbal information (certainly deception cues and possibly other useful cues). However, the trade-off is to shoot a smaller, less detailed picture of the far-end person in order to capture the torso. It is an interesting trade-off and an area that merits research.

THE OCCUPATIONAL THERAPY ACCESSIBILITY SPECIALIST: CONSULTATION FOR IMPLEMENTING UNIVERSAL DESIGN AND ADA COMPLIANCE

Historically, occupational therapists (OTs) have provided training to help individuals with disabilities live as independently as possible. Two historical initiatives have contributed to OTs' ability to successfully help clients work toward that goal: the Independent Living Movement and federal legislation specifically addressing accessibility and discrimination issues. The culmination of decades of regulation to eliminate discriminatory practices is the Americans with Disabilities Act (ADA). Occupational therapy practitioners, who are knowledgeable about these laws, provide a client-centered approach to service delivery that empowers their clients to advocate for their own needs with knowledge and a wide array of resources.

The client and caregivers define what they want and need, and OTs then help them meet their goals. The practitioner may act as consultant, advocate, case manager, and often as a provider of both nonmedical and traditional occupational therapy services in the home or community. The client evolves into an educated consumer who can then bring valuable information to the team of design and building professionals. These individuals then use the information about functional performance, ergonomic, and disability management when designing, constructing, and selecting products for the home, long-term care facility, or jobsite. Public spaces can be designed and built with universal design principles that also address the needs of a specific population for maximum access, safety, independence, and quality of life.

The American Occupational Therapy Association has a referral list of specialists who provide consultation for home modifications, work-related issues, and ADA compliance. An OT who is also an accessibility specialist consults with designers, building contractors, engineers, public and private agencies, and consumers to insure that environments are accessible and safe at home, at work, and in the community. Occupational therapy education emphasizes the importance of the dynamic interplay between the individual and the environment and its impact on functional performance, independence, and safety. The OT practitioner evaluates the needs of the individual by observing and assessing his or her ability to perform daily activities, simultaneously considering the demands of the environment, assistance needed, and requirements for modification and adaptation.

A team approach to accessibility services ensures that the many facets of assessing, designing, constructing, or modifying an environment will be coordinated and consistent with the consumer's needs, financial resources, and priorities. The accessibly consultant's general knowledge of government regulations, universal and adaptable design concepts, architectural products, building construction, specific disabilities and conditions, functional performance, assistive technology, and adaptations is necessary to understand the issues involved.

Specialists work to insure ADA compliance, helping employees, employers, and human resource personnel determine "reasonable accommodations." A reasonable accommodation is "any modification or adjustment to an environment that will enable a qualified applicant or employee with a disability to perform essential job functions" (Equal Employment Opportunity Commission and Department of Justice).

Information on the environment of the client is shared with the client's rehabilitation team for effective and efficient problem solving. Major areas of concern are safety, mobility, balance, coordination, manipulation, grasp, reaching ranges, sensory skills, cognitive skills, endurance, and the ability to use equipment for daily activities. Medical history, precautions, prognosis, progression of the condition, life changes, social/occupational roles, leisure activities, and support system are also considered.

The evaluator walks throughout the exterior and interior of the client's home. Access and use of the following environmental elements are carefully considered: parking, driveway, exterior walkways, steps, handrails, exterior/interior lighting, exterior/interior doorways, interior hallways, and stairs. Space planning in each room (living room, dining room, bathroom, kitchen, etc.), is detailed in the plan. Space planning includes furniture arrangement, floor surfaces, storage, closets, kitchen appliances, laundry facilities, basement access, location of breaker/fuse boxes, personal emergency response systems, fire extinguishers, smoke alarms, and such security features as emergency escape routes.

Accessibility specialists then collaborate with contractors and designers to determine the design, structural feasibility, and costs of the recommended modifications. Oftentimes, modifications are made in phases: once the client is able to move and function in the bedroom, bathroom, and entrance environments, renovations can be phased in until full access is achieved.

For buildings and services funded with federal dollars, mandates require that all points of contact are able to be approached, entered, and used by all people, including those with disabilities. Building contractors need to create the ability to participate and "get in the door." The practitioner can also provide valuable information about the impact of environmental barriers and accessible design.

Legally mandated design guidelines provide a basis from which to formulate solutions to eliminate barriers. In public spaces, specified measurements and parameters are designated. In private homes, the Fair Housing Act Accessibility Guidelines provide recommendations for multi-family and rental housing. If there are offices, businesses, or a daycare center on the premises, these must comply with accessible design guidelines under Uniform Federal Accessibility Standards or ADA Accessibility Guidelines. Local building codes, if stricter, take priority for compliance. In housing, individual accessibility needs must be considered and accommodated if not an "undue hardship" or structurally unfeasible. All guidelines provide recommendations for design of architectural features, installation of equipment, and space planning. The OT provides basic as well as more in-depth information to go beyond the requirements for accessibility so that maximum access to the environment is achieved.

Community spaces must also include accessible parking and routes to approach and enter both buildings and recreational sites such as parks. Curb cuts and sidewalks with a continuous and unobstructed pathway wide enough for wheelchairs provide access throughout the community. Auditory signals on elevators and traffic intersections, varying textures on ground and floor surfaces, contrasting color surfaces and edges for cueing, increased lighting, large print and Braille/raised letter signage, and scrolled handrails for directional cues provide access for persons with visual impairments. Visual/vibrating signals on alarms and electronic devices, telecommunications devices for the deaf, amplification devices, close captioning, and sign language interpreters provide access for persons with hearing impairments. Ramps, lifts, elevators, seating along a route, and handrails provide safe access for persons with mobility impairments. A firm pathway along a continuous route, wide enough for a wheelchair, enables access to a park, playground, beach, golf course, or other recreational sites.

The OT uses universal design as a guide to understanding the issues involved in creating an accessible home. The concept of "Visitability," created by Eleanor Smith of Concrete Change in Atlanta, Georgia, suggests that all environments be designed so that people can visit one another. The features of a visitable environment facilitate inclusion in community settings for people with mobility impairments. Where this concept has been used in the design of housing developments and other residential projects, a person using a wheelchair can visit neighbors, who themselves can bring a baby carriage, bike, or heavy packages into their homes without lifting and climbing up stairs. Visitability suggests minimal standards for access including at least one no-step entrance, access to the main floor of the home, and access to a bathroom for toileting with an accessible toilet, sink, and light switch. Private homes are not required to be accessible; however, it is often recommended that basic features of access be incorporated into the design in case of future need. This eliminates the need for major and often costly modifications, if needed later on.

The benefits of an integrated medical rehabilitation and design/build team approach result in an organized and efficient transition to the community, home, and jobsite. The move to maximum independence of the client depends upon the disability, prognosis, and recovery process of the patient.

The problems are not too difficult to remedy when the client's desires and needs determine the way that service is organized. Networking specially trained professionals, consumers, ongoing communication, and information are the key to successful rehabilitation and maximum independence and quality of life for our clients and family support systems.

Case Study: Facility Management; A Cultural Universal Design Perspective

Universal design is a major issue when cultural differences come into play. For the past five years, a simulation-game, "The Maze: A Facility Management Dilemma," has provided a framework to experience accessibility and issues of communication relating to Japanese and American facility managers.

The "Maze" simulated 10 years in the early life of a company producing puzzles. Time and space were condensed—10 years became an hour and a half, and a 20,000 square-foot office facility is condensed into a 12′ × 18′ space providing potential offices for approximately 60 staff members.

Cultural differences may not seem to relate to universal design. However, simulations were run each summer at the Massachusetts Institute of Technology with a group of Japanese International Facility Management Association members comprising a Japanese puzzle company competing for market share against an American puzzle company. It is obvious from observations of these simulations that the way these two cultures choose to work is a design issue. The following observations of cultural work patterns have remained constant for all simulation games.

Japanese workers in the simulation tried to break down both physical and communication barriers that were limiting the flow of information and material through the assembly line they created (see Figure 4.23). Teams were quickly formed to solve problems rather than individuals striking off to address problems on their own. A decision was made at

FIGURE 4.23   Japanese team working on communication problems.

FIGURE 4.24   Identifying barriers to communication and efficient production.

FIGURE 4.25   Japanese CEO and his management team plan a simulated space layout.

the outset on space layout by the CEO and facility manager and was implemented at once (see Figure 4.24). Very few modifications were made to the space even when problems later became apparent. The Japanese thrived on a group environment where sound and visual stimuli were not considered a distraction to team interaction (see Figure 4.25).

The American team members, by contrast, set up physical and organizational barriers defining their territory and responsibility. This resulted in limited access to critical information and a compartmentalization of the production process. Less teamwork and more individualism were apparent. It was not as critical for the Americans to see each other as they worked, but they had no reservations about yelling to coworkers who were out of sight. The Americans preferred to be adaptable as new situations arose, and they created physical work spaces that were constantly being changed through the duration of the simulation game.

Simulation-game experiments such as this can be used to help designers understand how people currently work in a variety of cultures and environments. It is possible, based on insights about current methods of interaction, to learn to modify the environment to foster new behavior.

REFERENCES

Erderich, J. (1999, June). Classroom acoustics. *CEFPI brief on educational facility issues*. CEFPI The School Building Association. Retrieved October 20, 2004, from www.healthyschools.cefpi.org/issue9.pdf.

Mazumdar, S., & Geis, G. (2003, January 1). Architects, the law, and accessibility: Architect's approaches to the ADA in arenas. *Journal of Architectural and Planning Research*, (Autumn, 20), 199–220.

Nelson, P. B., Soli, S. D., & Seltz, A. (n.d.). Classroom acoustics II-acoustical barriers to learning. *Acoustical barriers to learning*. Retrieved June 11, 2013, from www.centerforgreenschools.org/docs/acoustical-barriers-to-learning.pdf.

Ter Bush, R. F., & Mittleman, D. (2005). Determinants of mutual knowledge on virtual teams with recommendations to virtual environment designers. *EDRA36*, *36*, 137–146.

Enabling Products

5

CONTENTS

One key factor in creating supportive, adaptable, accessible, and safe environments is the products used within them. Barriers to use can exist within a product design as easily as they do in a space. The redesign of the built environment that universal design seeks also to include is the reimagining of the products people use each day.

This chapter, while introducing some products for consideration for their universality, first and foremost seeks to present some guidelines for the creation of such products. The chapter also offers some case studies that reveal the process used by several companies to arrive at new, more universal designs. Obstacles to the adoption or imbedding of inclusive design presented at a workshop at the Helen Hamlyn Research Centre in England were included, as well as a paper by Gregg Vanderheiden that describes research on barriers to adoption of product designs (Vanderheiden and Tobias 1998).

The "Universal Design Strategy" section consists of a short review of a customer usability laboratory operated by Whirlpool Corporation. Case studies present a history of a set of ergonomic tools and a detailed look at the creation of a redesigned school lunch kit and the history of one of the first ergonomic chairs offered in America, the Fixtures Furniture discovery chair. These are followed by a case study from contemporary designer Karim Rachid in which he describes a Designer's New Roles. In this introduction to universal products, the concern has been mainly on design and process, as illustrated in the Lunch Box Case Study. Innovative products and their designs exemplify the excitement that designers for nearly every identified user group have generated. Companies such as Tripod in Japan and designers such as Karim Rachid have produced a vast inventory of universally designed fixtures, furniture, household items, office supplies, lighting, flooring, and so on. Examples of their designs and other universal design products for use within the home are showcased below.

Additional universally designed products are featured throughout this book: Computer and work-related accessories are reviewed in the office chapter, while household appliances are featured extensively in Chapter 8, "Universal Design in the Home." A mixture of interesting products is offered in other chapters on commercial design and marketing.

UNIVERSAL DESIGN STRATEGY

OXO: UNIVERSAL DESIGN INNOVATOR

It all started with an apple tart. In 1990, Sam Farber, a retired housewares executive, noticed his wife's mild arthritis was making it difficult for her to grip the peeler she was using. Hundreds of research hours, brainstorms, and many production models later, Farber, along with help from product developer Davin Stowell, launched OXO, a line of 15 ergonomically sensible kitchen tools—including a new peeler—based on the philosophy of universal design (see Figure 5.1). Since the company was founded, the OXO line has grown to include more than 800 kitchen, garden, cleaning, and organization products. Every product is geared toward making tasks simpler and more efficient, whether it's a salad spinner that can be used with one hand or a liquid measure that can be read from above—no neck craning necessary. "Sometimes our designs are not about invention but about improving existing products," says Alex Lee, OXO's current president. "We believe in always asking 'What's wrong with this product?'—even our own products."

USING CUSTOMER FEEDBACK FOR NEW PRODUCT DESIGN: A STUDY OF APPLIANCE CONTROLS

People who design and develop major home appliances often think differently about their products' use. They may use appliances differently than most consumers do. But it is important that they design appliances for a wide range of users, not just for themselves, whether they term this kind of design "ergonomics," "human engineering," or "universal design."

At Whirlpool Corporation, researchers are trying to increase the usability of appliances for people of various ages and capabilities. They built a one-of-a-kind customer usability laboratory, headed by Sandra Thurlow, PhD in their Research and Engineering Center to learn

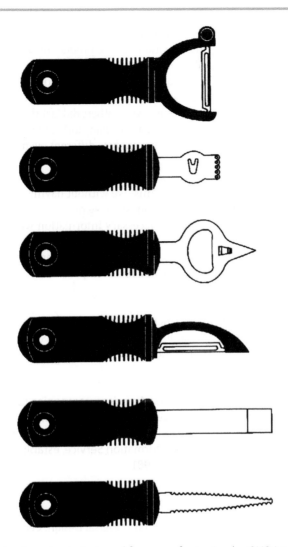

FIGURE 5.1   A sample of products designed for ease of gripping by OXO International.

more about how people interact with appliances and appliance controls. In this laboratory, Whirlpool gave consumers an opportunity to interact with new appliance designs and then recorded and analyzed the product performance and the consumers' thoughts and attitudes to learn more about what they liked and didn't like, what worked well under normal use conditions and what didn't work so well, what they understood and what they didn't understand, and so on. This helped create product designs that meet the needs of a wide range of people while focusing on accessibility and ease-of-use issues (Accessibility Solutions Project 1999). At the laboratory, two rooms were separated by a one-way mirror. A trained researcher took consumers through a series of steps using prototypes of new appliances with computer simulations of the new appliance controls. In the other room, the product design team watched the consumers' interactions with the appliance, and a video recording was kept for future study. The team's goal was to better understand what consumers wanted and needed and to design their new products accordingly.

Whirlpool Corporation donated a microwave oven, a refrigerator, and a washer and dryer to the Blind Center project in San Diego in the mid-1980s.

Whirlpool recently introduced an oven control system adapting research from the usability laboratory where they had looked at many more options before production. They put people at a touch screen on a computer simulating a number of different controls and then had them go through the process of operating the oven. Whirlpool asked questions like "Is the clock on the oven running? If you want potatoes to be done at 6:00 p.m. and it's 3:00 p.m. now, how would you set the controls?" Some users set the timers first and then the temperature, while others did just the opposite. Whirlpool then tried to come up with a design that can accommodate as many different styles as possible.

Some of the questions the design teams used to analyze different oven controls were as follows:

- Will this control allow consumers to use the appliance the way they want to use it?
- Can the user quickly and easily correct a mistake without having to start all over again?
- Can the user change his or her mind without starting over?
- Does the control panel present only the necessary information when it is needed? (Too much information can be confusing.)
- Does it prevent consumers from using the oven in a clearly inappropriate or unsafe manner, such as setting the self-clean option at a broil temperature?
- Is it simple to use?

Whirlpool hoped to make the control panel self-explanatory for all potential users. The oven should do what the consumer wants it to do when he or she wants it done. An optimal design doesn't require a person to look at the instruction manual at every use.

"Whirlpool Corporation was among the first American companies to respond actively to the needs of customers with limitations due to age or disabilities, providing Braille overlays for microwave touchpads and Braille versions of appliance use-and-care manuals free to customers through the company's Appliance Information Service, established by Joy Schrage of Whirlpool's Customer Affairs Department in 1981.*"

James Mueller

*Center for Universal Design (1997). Case Studies on Universal Design. Center for Universal Design, North Carolina State University, Raleigh, NC.

Case Study: Re-Creating the School Lunch Kit— Research in Universal Product Design

Research is perhaps the most important, but, unfortunately, also the most often neglected step in the universal design process. And participatory design—allowing the people actually affected by design decisions to play a significant role in the process—is one of the most neglected methods of doing research. The following case study shows how participatory design can be used by a research team to create a truly innovative product.

The Thermos Company, the market leader in school lunch kits, approached Fitch, Inc., a full-service business and design consultancy, to help design an ecologically sensitive family of lunch kits. The goal was to develop something completely different from what was currently available by first delineating, and then designing for, the as-yet unrecognized needs and wants of children.

It was clear from early discussions with the Thermos Company that the lunch kit design should be based on "what kids need and want today," rather than what design team members remembered wanting when they were children. This focus on creating a product that would respond to children's *current* wants and needs—with a corresponding call to find out what those were—suggested that the design process should include a strong participatory design and research effort. Consequently, a multidisciplinary team of researchers headed by Elizabeth B. N. Sanders and designers worked closely with end users to develop and test increasingly more refined versions of the lunch kit, gradually honing in on the final design.

PRODUCT DESIGN CRITERIA

Fitch, Inc. used the following very simple three-point checklist as a research guide for every product development project to ensure that the needs of consumers were being addressed:

1. Is the product useful?
2. Is the product usable?
3. Is the product desirable?

A useful product is one that consumers need and will use. A usable product is one they can either use immediately or learn to use readily. A desirable product is one they want. A favorite pair of shoes can be useful, usable, and desirable, all at the same time. Fitch believed that for products to be successful, they needed to meet all three of these consumer needs simultaneously. In the past, product success in the marketplace was likely if at least two of these three criteria were met. For example, many home electronic products are useful and desirable, but not very usable (Mueller 2004).

Consider that many of the products targeted toward the aging marketplace today (a population that is growing rapidly and significantly) are useful and usable, but not very desirable—for example, bathroom fixtures with an institutional look. It appears that products developed for the general consumer marketplace have usually and primarily been driven by the need for desirability, while those developed for the aging marketplace have been driven by the need for usefulness. Usefulness, usability, and desirability must all be given equal emphasis and must all be addressed very early in the development process.

The team found that significant roles needed to be played by the following:

- The child—the primary user and the primary decision maker for purchase
- The parent(s)—the primary purchaser, "maintainer," and (most often) "packer"
- The teacher—in whose classroom and lunchroom the kit is used (and abused)

THE APPROACH

The need to combine research and design led to the formation of a strongly multidisciplinary, dynamic project team and a design process that incorporated integrative, tightly interconnected research and design. Researchers and designers worked together to make field observations, analyze data, and create and test designs.

In order to have access to children, parents, and teachers, arrangements were made with several schools and day-care centers in Columbus, Ohio (Fitch, Inc.'s headquarters). The participants were boys and girls from a variety of age groups, and they were involved throughout the development process.

The following examples of research activities describe the methods used to ensure that the lunch kit would be useful, desirable, and usable (the order is different from that listed in the "Product Design Criteria" box because of the level of involvement by the children in the actual design).

IS THE PRODUCT USEFUL?

The first question in the three-point checklist is the most critical. If a product is not useful, that is, if consumers do not need it or will not use it—then why bother to research, design, or develop it in the first place?

PARENT SURVEY

A written survey was sent home with participating children. The survey contained general questions concerning packing lunches, the content of children's lunches, shopping for lunch boxes, and qualities of an ideal lunch kit. The survey showed that, especially with little children, packing lunch helps parents feel connected to their children's lives at school. The emotional importance of this connection was new information for the Thermos Company.

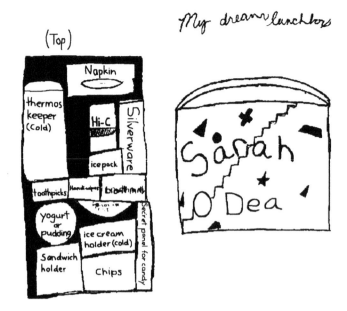

FIGURE 5.2   A child's drawing of an ideal lunch box.

The survey also showed that children's opinions are the largest influence on lunch kit purchase—when it comes to lunch kits, parents buy what their children want.

STUDENT DRAWINGS

All students were asked to draw their ideal lunch kit, a sample of which is shown in Figure 5.2. Many of their drawings were elaborately detailed and annotated. Several needs (that directly influenced the subsequent design of the lunch kit) emerged from the drawings:

- Compartmentalization (i.e., everything should be in its own place)
- Personalization (i.e., ways to mark the lunch kit as their own)

CAFETERIA LUNCH OBSERVATION

The field research teams unobtrusively observed children during lunchtime, paying attention to what and how much children actually ate and how their lunch kits fit into the lunch experience. Written materials were prepared to provide a structure for observation; this ensured that the same kind of information was gathered by all team members while supporting the research activities of less-experienced team members.

Differences between older and younger children were quickly identified. Talking informally with the children, the team learned that the older school-age children thought the traditional hard plastic lunch boxes (that Fitch, Inc.'s client the Thermos Company currently produced) were "for babies" (many carried paper bag lunches, which although leading to "smooshed" food, at least were acceptable by peers).

The teams also found that lunch kits were stored in a variety of spaces ranging from individually assigned shelves within a classroom to laundry baskets outside the room; this helped define size and durability requirements.

IS THE PRODUCT DESIRABLE?

Since the child was the primary purchase decision maker for this product, the desirability of the lunch kit to the child was critical. Fitch created multiple design concepts from the research activity in Step 1. "Desirability" research was then conducted with nine three-dimensional, non-working models.

INTERACTING WITH MODELS

Students interacted with and evaluated the concept models. Researchers briefly described the models, trying not to bias students, and then facilitated conversations about what the students did and didn't like about each concept. At the end of the discussions, students voted on which ones they liked best and which they felt were "kinda weird" (i.e., out of bounds, too extreme, inappropriate). An amazing degree of consistency appeared in their evaluations of the models—two of the concepts were preferred by a majority of the students, and this preference was remarkably similar across age and sex groupings. One of the preferred concepts was the compartmentalized approach in which the compartmentalization was evident on the outside as well as inside the lunch kit.

COLOR CONSTRUCTION KITS

Color is a key characteristic of consumer products, one that can clearly influence the desirability of a product. The researchers asked students to work in small teams with 60 color chips to create color palettes for the lunch box. The colors selected were then tallied and represented on boards, showing the preferences of three groups of children: preschool, second grade, and fourth grade. The preschoolers chose mostly primary colors and very bright contrasting colors; the second-grade palette showed more "fad" colors; the fourth-grade palette showed a mixture of the primary contrasting colors and the more subtle variations chosen by the second graders. On the basis of these results, Fitch suggested that the color direction for the lunch kits be built from the fourth-grade palette, since it represented a blend of the two younger groups. (In an alternate method of gaining information about children's color preferences, the students were asked to choose from a large array of crayon colors to color in sketches of the lunch kits. This provided further information about what kind of color relationships children prefer.)

LICENSING STUDY

The license, or character, on a lunch box is another crucial component of desirability. The researchers created nine categories of licensing ideas and collected many examples for each category: cartoon characters, sports logos, fashion logos, TV stars, popular athletes, and so on. Conversations with the children revealed that most of the current licenses were too "young" for the school-age audience. Fitch recommended to the Thermos Company that they pursue licenses appropriate for older children, as well.

IS THE PRODUCT USABLE?

Usability testing with children and parents was conducted after a full-size, three-dimensional working model (i.e., capable of being opened and packed) of the lunch kit was available.

MINI-FOCUS GROUPS

Mini-focus groups were designed to test usability as well as general likes and dislikes of the lunch kit from the parents' point of view. Each group interacted with and discussed the working model and looked at existing products for comparison. Food was also provided for packing the lunch box.

Generally, parents thought the compartments Fitch had developed were too restricted, which caused some difficulty packing the lunch kit, although they became more comfortable as they worked with it. However, they said that if their children wanted the kit, that would be reason enough to buy it—regardless of their need to change their packing style.

USABILITY TESTING WITH CHILDREN

Researchers began by showing students lunch boxes that were already on the market, encouraging them to think critically about what was available. Next, they were shown the working model and asked questions about it. Researchers then observed students interacting with the model. As with the parent mini-focus groups, the children were given food with which to pack the lunch kit.

Children responded positively to the model, especially after opening it. They were impressed with the fact that the "inside could be seen from the outside," that is, that the outside shape reflected the inner compartments. They enjoyed "playing with" the lunch kit and had no difficulty deciding how to pack it.

CONCLUSION

This lunch kit was first introduced in the fall of 1992, with a "sell through" rate of 95%. Later, the Thermos Company added sports licenses to this line, with very good results. Without the input of actual users (mainly children in this case but also parents and teachers), the striking design of the kit, and its success, would not have been possible. Fitch was fortunate to have the support of its client, the Thermos Company, and the cooperation of end users—kids, parents, and teachers—allowing the team to conduct a rewarding and enjoyable research and development project.

Note: This case study was included in this chapter, even though the backpack has replaced the lunch kit as the means of carrying lunches to school, because it was such a good example of participatory design that could serve as a model for developing a variety of commercial products.

OBSTACLES AND SOLUTIONS TO INCLUSIVE DESIGN

A professional team (Cherrie Lebbon and Ruth Morrow) from the Helen Hamlyn Research Centre in the Royal College of Art in London found that the obstacles and solutions to inclusive design fell into two major categories: those affecting designers and those affecting clients (i.e., those commissioning design). Obstacles to inclusive design fell into three simple categories: external context, professionalism, and relationship. In the user–designer relationship, designers who appeared to have insufficient understanding of and empathy for users and their needs could be brought about within the designers' professional education, by pressure from legislation, and by informed clients and demanding users.

The incomplete and growing definition of inclusive/universal design creates a need for more knowledge and a higher level of detailed understanding for professionals in the field.

Continued Professional Development (CPD) appears to be the solution to the problem of dissemination and explanation of inclusive design approaches—both for designers and their clients. Changes must be supported and guided by an active body of research. The aim is to develop a framework for inclusive design practices and a systematic revision of their practices aided by supportive programs of CPD.

"People who live with some form of disability now number more than one billion worldwide.* An inclusive/universal approach to designing the built environment, from transportation systems to telephones, is fundamental to living, working, and participating in society for these people, their classmates, coworkers, families, and friends, and all the rest of us who simply live long enough."

James Mueller

*World Health Organization (2011). World Report on Disability. Geneva: WHO Press. p. 261.

BARRIERS, INCENTIVES, AND FACILITATORS FOR ADOPTION OF UNIVERSAL DESIGN

As electronic technologies become more integrated into education and employment, the ability to access and use these new technologies becomes critical to people with disabilities (40 million people in the United States alone) if they are to be able to participate in these environments. A key strategy for ensuring access to these emerging technologies is the practice of universal design.

Universal design is a practice that depends on the standard, mass-market product manufacturers for its implementation. Although there are key examples of universal design practices in the electronic and communication industries, they do not represent the majority of current practice. Gregg Vanderheiden, seeking a reason for the lack of utilization of universal design products, explored key factors, motivators, disincentives, and barriers to the practice.

The term *universal design* originated in architecture and was coined by Ron Mace in the 1970s. Since then, its application has broadened considerably, to include the fields of product design, computers, electronics, telecommunication systems, and more. The following is the definition used by our project: Universal design is the process of creating products, devices, environments, systems, and processes that are usable by people with the widest possible range of abilities, operating within the widest possible range of situations (environments, conditions, and circumstances), as is commercially practical.

Universal design has two major components:

- Designing products so that they are flexible enough that they can be directly used (without requiring any assistive technologies or modifications) by people with the widest range of abilities and circumstances as is commercially practical given current materials, technologies, and knowledge.
- Designing products so that they are compatible with the assistive technologies that might be used by those who cannot efficiently access and use the products directly.

In addition, the products are often designed by engineers and product designers under tight deadlines, where insufficient time is available for enough usability testing with the core market, much less those outside of the core 80%. Human factors professionals, with their strong usability emphasis, are often put in the role of defending improvements in usability for people whose performance characteristics are outside that 80%. Contemporary product designer Karim Rachid describes the problems with current production practices in his case study later in this chapter. Human factors and universal design share a perspective on usability that may not be shared by others involved in the product development process.

Advocates of universal design claim that it can increase market share and can increase usability in general. In light of its claimed ability to address typical business motivations, why is it not more widely practiced? What can be done to encourage or support universal design?

UNIVERSAL DESIGN RESEARCH PROJECT

The Universal Design Research Project, a three-year study, led by Greg Vanderheiden of the Trace R & D Center, University of Wisconsin-Madison, was designed to understand why and how companies adopt universal design, and what factors are the most important in making this decision. In addition, factors that discourage or impede the adoption and successful practice of universal design were identified.

A total of 22 companies drawn from telecommunications, media and materials, "edutainment," computer, and built environment industry segments were selected for interviews.

INITIAL RESULTS

Large companies that have succeeded in implementing UD were found to share some characteristics: support from upper management, use of formal product development processes to institutionalize universal design, and the use of cross-functional teams in their product development process. Small companies tended to have universal design champions. They used informal information networks. Their flatter, more "empowered" organization frameworks required less authorization. Size alone did not appear to be predictive of UD adoption. Size did appear to predispose companies to certain styles in their implementation of UD.

COST

COST

Virtually all interviewees mentioned cost as an element in their company's decisions. The tension between quality of design and cost would be familiar to human factors professionals in industry.

RESEARCH AND DEVELOPMENT

Almost all interviewees wanted closer ties to organizations performing research and development in universal design or accessibility. Specific comments were directed toward making research results easier to find, improved market research, and industry participation in the research agenda so that more economically viable products would result. This was an example of evidenced-based design.

CONCLUSION

If universal design is to be adopted across a broad range of mainstream companies, it must graduate from a design approach that is in uneven practice in a relatively small number of companies or by a small unit within a company into a constellation of effective usability practices. For both market and mandatory reasons, universal design is a trend. Human factors professionals are the natural participants (both within companies and externally) in developing and refining strategies and practices that would support universal design. In addition, the core focus of universal design is extending the usability of products to all users and directly reinforces many of the current efforts of human factors professionals. In many cases, the mandatory or regulatory nature also helps bring new focus on the involvement of human factors and usability testing in general with new products.

Case Study: "Discovery" Seating and Fixtures Furniture

In Chapter 4, Timothy J. Springer shows how human factors (or ergonomics) are applied to universal design. Throughout the literature dealing with universal design, human factors, and ergonomics, a common thread appears in each discipline's interest in work environments, especially workstations and most especially seating. This case study looks at one of the first exercises in ergonomic chair design undertaken in America. The discovery® chair is still recognized as one of the leading examples of ergonomic seating design (see Figure 5.3).

Fixtures Furniture, under the chairmanship of founder Norman Polsky, was the first US manufacturer to seek out and create an ergonomic line of seating for offices. The discovery® chair was developed by their German partner firm Froescher and was marketed in Europe in the late 1970s under the trade name "Resumo." The discovery® chair incorporated front-seat tilt and both passive and active ergonomics, along with full adjustability while seated (called dynamic sitting).

Dynamic sitting (see Figure 5.3) allowed body liquids to be assimilated and exchanged within the body's tissue without hindrance, thus reducing slipped disc problems and alleviating strains in the back muscles. Dynamic sitting also had a positive effect on all the muscles of the legs, back, and stomach by preventing the continual pressure in the stomach area that happens with fixed sitting. The blood and lymph circulation in the calves was also improved with dynamic sitting.

A chair that offered variations in adjustments was necessary because the individual measurements of the user must be taken into account and, depending on the type of work, height of working area, position of head to work material, and so on, the seat must be adjusted so that it aided the work and helps bring about greater efficiency.

Fixtures led the industry in promoting ergonomic seating while supporting Dr. Marvin Dainoff's groundbreaking research on ergonomics at Miami University in Oxford, Ohio (Dainoff 2005). From this research, the following 12 elements of ergonomic seating were developed:

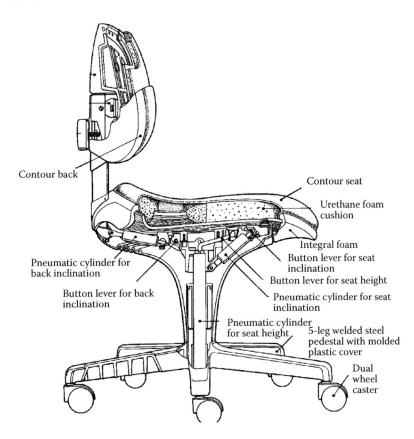

FIGURE 5.3   Dynamic sitting, the discovery® chair.

1. Variety of models for various sizes and statures of people
2. Contoured padded seat and back
3. Seat height adjustment while seated
4. Lumbar support with back height adjustment and full tilt for dorsal flexing
5. Back tilt adjustment while seated
6. Seat from tilt adjustment while seated
7. All adjustments easily made while seated
8. Various static-posture settings or free-flowing dynamic motion
9. Active ergonomic controls versus passive non-controls
10. Mechanical reliability, safety, and ergonomics by passing ANSI-BIFMA and ANSI/ HFS 100 standards
11. User training to assure optimum results
12. Model was field-serviceable and backed by comprehensive manufacturer's warranty including nationwide service capability

The first discovery® chairs were modeled after the Resumo line and featured the following design elements:

- An easily accessible press button for adjusting the overall height of the seat lay beneath the seat on the right-hand side.
- An easily accessible duo button that adjusted the angle of the back between 14° and 35° lies on the right-hand side beneath the armrest support. The duo button worked in conjunction with a sliding control to allow the user to fix the back movement permanently within a chosen range of adjustment.
- Another easily accessible duo button with sliding control lay beneath the seat on the left-hand side in front of the armrest support. This control allowed the seat surface itself to be adjusted for angle of incline and could be set in a fixed or

range-of-motion position. (If both duo buttons are fixed by the sliding controls, the seat could be said to be in a state of permanent motion, supporting the concept of dynamic sitting and even providing a mild form of exercise for the seated user.)

•   The seating area and back were well upholstered with an optimal shape. Central column springs cushioned away extra weight when the stool was in use. In all positions, the user was free of pressure points in either the seating area or back. Seat height could be adjusted to leg length, thus removing unnecessary pressure behind the knee.

•   Body weight was spread over the whole seating area. The lower part of the stool back was slightly convex, the middle part was concave, and the top part was convex again, which supported the natural curves of the spine. Movement of the shoulder joints and shoulder blades was unimpeded, as was sideways movement.

The discovery® line later included five different models, which fit the needs of a variety of work types.

JAPANESE UNIVERSAL DESIGN PRODUCTS

Japanese industrial designers have been leaders in recognizing the needs of an aging population. Satoshi Nakagama and his research and development firm, Tripod Design, has set the standard for universal design in Japan. Tripod Design has showcased many of their products in gallery shows and exhibitions (shown in the examples given in Chapter 9, "Marketing Universal Design"). For each product, the design research process for understanding users includes the following:

■   Web research
■   Field research (recording of actual situations)
■   In-depth interviews

The range of products produced by Tripod Design can be seen in the variety of products produced. These include the following:

■   The HandyWormy (or Handi-tote): A support handle for carrying shopping bags or any bag with heavy items in it. The grip is designed to accommodate ease of use by all people regardless of age, hand size, or grip strength (Figures 5.4 and 5.5).
■   Handi Birdy Series (Mini Birdy pen): a pen designed for both left- and right-handed people. The pen is best suited for a user with small hands or a strong grip (see Figure 5.6).

FIGURE 5.4   Nakagawa "HandyWormy" bag holder.

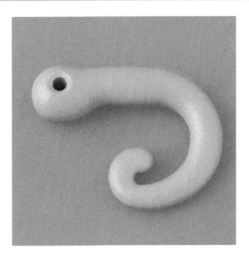

FIGURE 5.5   Enlarged view of HandyWormy bag holder by Tripod Design.

FIGURE 5.6   Easy-grip Mini Birdy pen for the right or the left hand by Tripod Design.

- U-Wing Pen: The unique form accommodates many diverse ways of holding the pen that are not possible with a traditional cylindrical pen. The shape of the pen also enables the pen to be held with the mouth or toes (see Figures 5.7 and 5.8).
- Cutlery: Tableware (KUZO) cutlery in eight variations made from titanium, so it is very safe and lightweight (only 38 g each) and can float in water. The form of the grip enables users to hold it in different ways (see Figure 5.9).

FIGURE 5.7   Wing Pen by Nakagama of Tripod Design.

FIGURE 5.8 Wing Pen by Satoshi Nakagama for the right or the left hand.

FIGURE 5.9 Lightweight Titanium Flatware by Satoshi Nakagama, Tripod Design.

■ Tableware: A series of tableware (mug, plate, bowl, and cup) that can be used in a variety of settings by both children and elderly persons. The stand on the bottom is made of elastomer to help hold the top in good balance and make it difficult to slip or tip over (see Figures 5.10 and 5.11).

Universal design has permeated nearly every aspect of industrial design in Japan in recent years. Ranging from Toto toilets to the less expensive washlet (Figure 5.12), easily adjusted shoes with Velcro closures (Figure 5.13), colorful wavy grab bars (Figure 5.14), adjustable kitchen wall cabinets (Figure 5.15), to a series of flexible, accessible automobiles by Toyota (see Figures 5.16 through 5.20). The wavy grab bar (Figure 5.14) and photos of the adjustable overhead cabinets (Figure 5.15) were submitted by Dai Sogawa, editor of the Japanese Universal Design Magazine and a long-time supporter of Universal Design.

FIGURE 5.10 Dishes.

FIGURE 5.11   Cup with holder.

FIGURE 5.12   Toto washlet toilet with wall-mounted controls.

FIGURE 5.13   Easily adjusted Velcro shoe closure.

FIGURE 5.14   Wavy grab bars.

FIGURE 5.15   Vertically adjustable overhead cabinets.

FIGURE 5.16   Large driver-side front door, Toyota.

FIGURE 5.17   Open sliding driver's door revealing fold-down front seat, Toyota.

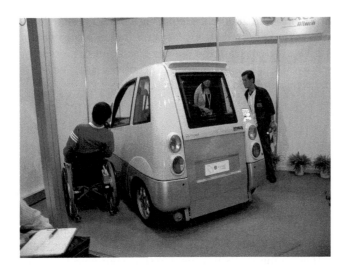

FIGURE 5.18   Rear access door closed.

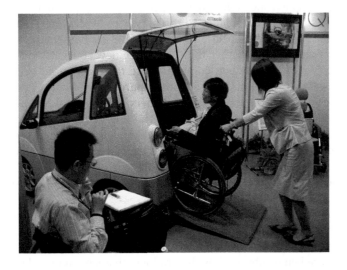

FIGURE 5.19   Rear access door opened.

FIGURE 5.20   Extra wide driver's side door.

Karim Rachid, with more than 2000 objects put into production, reflected on the "new role for designers" in the following case study. His ideas on the problems with getting new designs into production are quite different from the ones expressed in the research reports "Obstacles and Solutions to Inclusive Design" and "Barriers, Incentives, and Facilitators for Adoption of Universal Design" that were described earlier in this chapter.

A NEW ROLE FOR THE DESIGNER

Today, industrial designers approach their work from a different perspective. A poetic design of products has evolved based on a plethora of complex criteria—human experience; social, global, economic and political issues; physical and mental interaction; form; vision; and a rigorous understanding of contemporary culture. However, manufacturing is based on another collective group of criteria: capital investment, market share, production ease, dissemination, growth, distribution, maintenance and service, performance, quality, ecological issues, and sustainability. The combination of all of these components has come to shape our objects and inform our aesthetic, our physical culture, and our human experiences. These issues shape the business—its identity, its brand, and its value.

Even more critical are the inertia and speed with which products are being developed and disseminated. With computer-aided design, the speed of getting products to market has revolutionized the process. All this speed adds to the pressures on designers. And the new global phenomenon of speed has a great impact on our social lives, as it alters the way we absorb and perceive information and objects. Rather than introduce new concepts and conduct intensive research in order to develop a product, companies are more apt to develop products on the basis of existing objects. This is strictly an adaptive approach, and not necessarily reflective of a designer's priorities. While it used to be that designers had time to develop new concepts, the process of product development has robbed them of this liberty. At one time, at least four prototypes were required in order to perfect just one chair. Now, companies expect the first prototype to be perfect in order to bring objects to the marketplace in a timely manner. The demands of the market and of the consumer are immediate, and the condition is irreversible. (The Oh chair, colorful and inexpensive, is a good example of one of Karim Rachid's most successful designs [see Figure 5.21].)

This phenomenon of speed has also had a great effect on progress, communication, the global economy, and our human psyche. We readily accept change as it brings a better standard of living, a more experiential existence, and a savvier, more contemporary consumer. However, our human experiences are being subverted as the products that surround us become more banal. The mass mechanization of the 20th century led to rapid mass production, lack of research and

FIGURE 5.21   Oh chair by Karim Rachid.

development, a decreased interest in cultural significance, and low-capital investments. What have resulted from this phenomenon are few ideas and many variations.

The object as a commodity has lost significant connections with human emotion and artistic experimentation. The postmodernists attempted unsuccessfully to deal with the lack of emotion that accompanied the loss of modernism's hold on popular imagination. The iconoclastic objectivity of the postmodern movement from 1970 to 1990 resulted in a language composed of objects that had very little to do with human life and more to do with an elitist game of history and surface. Although objects were highly visible, little discourse concerning the new ways of informing physical objects took place in theoretical circles. Discussion in the forums of architecture and art in those decades focused on immateriality, ambiguity, hybridization, poststructuralism, representation, simulacra, phenomenology, feminism, semantics, deconstruction, the postmodern condition, tribalism, and so on. The controversial argument among industrial designers revolved around the idea that design was no longer strictly about things or objects. The idea that—in our new synthetic world, that in which the departure is the digital phenomena—objects are meant to cater to our emotional and environmental needs as well as to shape our sensual, intellectual, and ever-changing social behaviors. Are social conditions really changing as digitalia seeps its way into every aspect of our everyday lives? When design accepted the notion that objects really possess meaning, was this a point toward a new direction in design: a new pedagogy?

Industrial design today is changing, and the role of the designer is plagued by the pressure to conform to the knockoff marketplace, not being necessary in a knockoff marketplace, or by only being able to bring a level of involvement to the surface of objects. Instead of having an instrumental role, their involvement is often limited to the visual aspects of design. The political and social role of design has become apparent in the hyperexplosive period of the 21st century, the advent of a critical stance on and celebration of the object as an "immaterial" infinity. Traditional design methodology and a globalization of the consumer product have brought an end to the modern language of the object. Yet the object has become an endless appropriation of its predecessor, a cloning of clones as we trade spiritual values for material values and originality for simulation. The design process is marking time with works that become the foundation for other works via the continuum of concept from subject.

Designers must be involved in the initial stage of development. It is necessary that they play a rigorous role in the conception of ideas and in the definition of the object's value, performance, and intelligible existence. In turn, industrial design must become more visible in the media so that design may evolve from a marginal subject into a pedestrian interest and desire. Greater awareness creates the need for more design participation, and the result is a richer and more relevant commodity landscape. Here, design engages the three most important roles of successful business—quality, service, and loyalty.

Are companies really interested in people and in culture? Do manufacturers discuss personal rituals, the depths of private relationships, the warmth of family, the codes of love, the signs of human emotions, the regard for happiness, freedom, personal expression, and the well-being of our human existence? And do they address these questions through the product they sell? Businesses can only be holistic and comprehensive if they are able to address these issues. A brand would then come naturally, and it would evolve from the product rather than be created by a marketing myth. Remember that brand allegiance today is based on a product living up to or surpassing the expectations of the user. (Artemide is a brand of lighting fixtures that has achieved a high level of brand allegiance from the design community. See Figure 5.22, a lamp designed by Karim Rachid.)

There is no brand allegiance anymore—if a brand lets people down, it is instantly dropped in this hypercompetitive, ever-growing marketplace.

Can design transcend its past and become a subject or a methodology—a tool for self-expression? Can we expect from product design an actual commentary on cultural and social issues? Or is design relegated to serving purpose—to simply fulfilling needs and to creating a more comfortable, convenient situation or condition? We realize that the proliferation and hyper-consumption of commodities is dangerous. At the same time, our everyday lives are shaped by the objects with which we constantly interact. Three objects should be removed for every object that is placed on the market today. That is good design.

The banal discussion around the notion that form follows function is completely irrelevant when considering sophisticated products with complex subjects, digital components, and nearly mythological hyperobjects that engage little interaction, if any. So how do we shape these objects? We must return to the subject as the origin of discussion—a place to study meaning and myth. Inanimate objects do not have meaning—we project meaning onto objects. Design is the study of the subject that informs the objects. Form follows subject.

Subject is not always driven by necessity or function. Subject may be a philosophical position on contemporary culture or a tenet on a changing condition. It may be human experience, behavior, movement, cultural phenomena, medial flexibility, sustainability, variation, reconfigurability, immateriality, materiality, consumption, transparency, or digital production. The subject may be erroneously banal, for example, in the case of tooling modification. These issues constitute a

FIGURE 5.22   Artemide Doride Floor Lamp by Karim Rachid.

language that deals with the subject rather than the object. This is the First Order of good design. To quote Immanuel Kant, "In a product of beautiful art, we must become conscious that it is art and not nature; but yet the purposiveness in its form must seem to be as free from all constraint of arbitrary rules as if it were a product of mere nature."

We are amidst a postindustrial age—one in which autonomy, diversity, and vicissitude exist harmoniously with technological manifestation. The past is pointless, yet there is a great deal to learn from it. One must learn everything about history and then forget it all. In order to make a change in the world, one must understand where the world is going. Change must be created within an existing contemporary condition, not within a condition that is misanthropic and historic. The Old School spoke of a state of industrial design in which industry and technological production worked hand in hand and in which the machine dictated the limitations of the production process. In 1982, Sottsass noted that we were entering a time during which we would control the machine, and during which the machine would not control the outcome. While the machine still has great limitations, we should focus instead on possibilities and opportunities. Few designers embrace technological process. Often designers act like naive artisans, disregarding manufacturing in the name of self-expression. If designers and manufacturers continue to bifurcate as such, industrial design will depreciate as a relative profession. Industrial design must mature in the new digital transformation. The process of design involves using novel ideas to develop intelligent solutions to manifest the objects of our milieu. If we embrace the concerns and demands of production and industry, we are able to produce responsible objects. By responding to the current subject at hand and by developing highly experiential, communicative, and interactive yet poetic works, good design can exist respectfully, visibly, and effectively in our commodity landscape. Designers will become cultural editors, or cultural engineers—cultural purveyors and business strategists of culture.

In my opinion, the utopian condition exists in a world in which every object is so perfectly cyclic that we can consume without guilt—that we can perpetually have new objects in our lives, experience heightened pleasure, and enjoy a dynamic and forever-changing commodity landscape. We will not own anything, but rather borrow or lease the components of our physical world. Our lives will be layered with evanescent phenomena—each object will yield an experience, and each new thing will afford us an even greater experience. As the digital age becomes more seductive, our physical world will demand perpetual newness of stimulation, phenomenological unexpected behaviors and events, and a seamless reformation of beauty and change.

INDUSTRIAL DESIGN CURRICULUM AT SAN FRANCISCO STATE UNIVERSITY

One of the leading training centers for industrial designers in the Unites States is the program located at San Francisco State University (SFSU) in California. The program is especially focused on designing products for an aging population. SFSU design or industrial students are developing new ways to make products more senior friendly. With innovative concepts ranging from a power dish scrubber to an electronic medication dispenser, students ended each semester by presenting their drawings, models, and prototypes to a team of outside design and marketing professionals. To better understand real-world needs, many students did their own usability research. Tuyet Tran, a design senior who created a multiuse laundry bag, observed a laundromat to see how people carried their laundry (Figure 5.23). Other products showcased included a food container system to keep delivered meals fresh (see Figure 5.24).

Brian Donnelly, the class instructor, said that "The challenge for the designer is to create products that incorporate a high level of functionality with sensitivity to quality, aesthetic appeal, and convenience that boomers have grown to expect." Donnelly, a furniture designer in his own right, has created ergonomically successful chairs for hospitals and senior center use (see Figures 5.25 and 5.26).

To get first-hand accounts from older people, the upper-level product design class collaborated with a group of SFSU's Sixty-Plus Club. At several points during the semester, the seniors offered comments on the students' work.

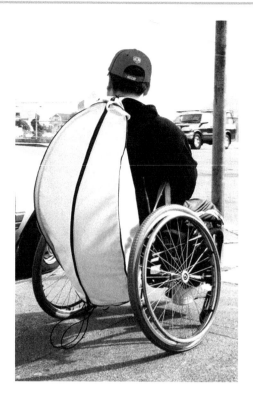

FIGURE 5.23   Student-designed laundry bag.

FIGURE 5.24   Student-designed food container.

SHOWCASE: PRODUCTS

Universally designed products for personal residences now appear everywhere, from specialty catalogs aimed at specific audiences to major department/hardware stores (lever handles, touch-control lighting, large-handled appliances, and grab bars, for example, are carried and displayed by many national chains). These products all stand out not because of any "institutional" look (a common detriment to including accessible products in a residence in the past), but rather because they are strikingly appealing. Examples of some of these products are shown in the Product Showcase in this chapter. In addition, a wide variety of these products are featured in Chapter 8, "Universal Design in the Home." Schools help create a demand for products that are universally designed.

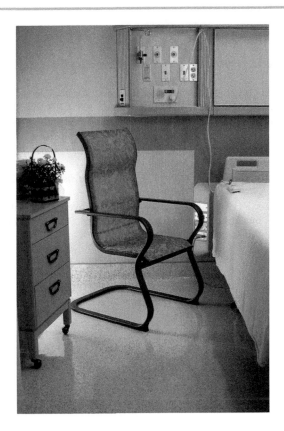

FIGURE 5.25   Spring Chair in a hospital room by Brian Donnelly.

FIGURE 5.26   Spring Chair by Brian Donnelly.

REFERENCES

Accessibility Solutions Project. (1999). *Meeting the needs of a diverse global marketplace.* Irving, TX: Nokia Corporation.

Center for Universal Design. (1997). *Case studies on universal design.* Raleigh, NC: Center for Universal Design, North Carolina State University.

Dainoff, M. J. (2005). *The effect of ergonomic worktools on productivity in today's automated workstation design.* Oxford, OH: Center For Ergonomic Research, Miami University.

Mueller, J. (2004, October 28). *Design for inclusion: Creating a new marketplace industry, white paper.* Washington, DC: National Council on Disability.

Mueller, J. L., & World Health Organization. (2011). *World report on disability* (p. 261). Geneva, Switzerland: WHO Press.

Vanderheiden, G. C., & Tobias, J. (1998, January 1). Barriers, incentives, and facilitators for adoption of universal design practices by consumer product manufacturers. *Proceedings of the Human Factors and Ergonomics Society Annual Meeting, 1,* 584–588.

Universal Design in the Office

6

CONTENTS

Of all the environments affected by the Americans with Disabilities Act (ADA), the one most universally impacted is the office. Typically, the spaces we work in are at least in part open to the public and thus covered under the public accommodation section. And, of course, the employment provisions are directly related to the office. Flexible workstation design is needed to meet the demands of a multigenerational and diverse workforce. Employers need to be able to manipulate work environments across the age groups.

This chapter opens with a detailed illustration on ADA compliance in an open office plan. James Mueller has identified universal design elements in an open-office design (Figure 6.1). He highlighted efforts by the manufacturer, Herman Miller, to integrate workers with disabilities into the workforce through universal design (Herman Miller Inc. 1995).

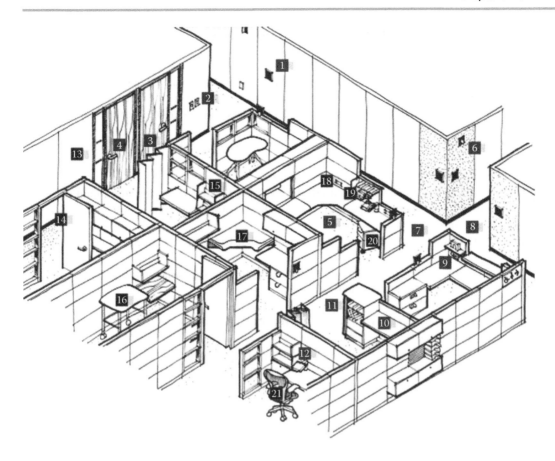

FIGURE 6.1   Open Office Plan showing Universal Design features.

ADA GUIDELINES

The lists presented here are intended as a starting point for consideration of disability issues in the design of business environments. They are not intended as a complete and comprehensive guide to ADA compliance.

Some ADA Guidelines for Interior Design

1. High-contrast, tactile signage 60″ above floor
2. Controls and operating mechanisms 36–48″ above floor
3. At least 32″ passage width through doorways
4. Lever, push-bar, or U-shaped hardware on doors; minimize door-closer resistance (no more than 5 lb.)
5. At least 60″ diameter space for wheelchair turnaround
6. Visual and auditory alarms no greater than 50′ from any worker; mark accessible evacuation route
7. Stable, firm, slip-resistant floor surface; thresholds no higher than 1/2″; 1/2″ maximum pile thickness for carpeting
8. At least 36″ corridor width
9. Electrical and phone outlets no less than 15″ above floor
10. Work surfaces 28–34″ above floor with knee clearance at least 30″ wide, 27″ high, 19″ deep
11. At least 30″ × 48″ wheelchair approach space
12. Storage shelves 9–48″ above floor

Some Additional "Designing for Accessibility" Guidelines

13. Surface textures that reduce glare
14. Contrast in color, brightness, and texture between walls, floors, and doorways
15. Open storage wherever possible; if doors are necessary, select side-hinged rather than top- or bottom-hinged doors; provide hardware that can be operated with a single closed fist
16. Height-adjustable work surfaces to suit workers and their tasks: 25–34″ for seated work and 33–45″ for standing work
17. Surfaces sturdy enough to lean on for balance
18. Workspaces that can be reconfigured—for example, to suit left- or right-dominant workers, or to enable hearing-impaired workers to face visitors and customers; avoid isolating workstations
19. Additional work surface space and electrical outlets to accommodate electronic assistive technology
20. Mobile, ancillary work surfaces and wheeled storage for transporting materials and locating them within easy reach
21. Work chairs with
    - Five-star swivel base with casters appropriate to floor surface
    - Firmly padded or breathable upholstery
    - Full backrest with adjustability for leaning forward and reclining
    - Adjustable lumbar support
    - Adjustability for height and angle from a seated position
    - Padded, height-adjustable armrests

The majority of the areas focused on in this book are relevant to accommodation within the office. Chapters on facilities management, enabling products, and public and commercial environments all connect to office-based universal design. And the techniques and strategies described in this book can be utilized throughout the work world.

In product design, Chapter 5 researchers at the Helen Hamlyn Centre, Royal College of Art

> "Job accommodations usually benefit coworkers without disabilities as well as the coworker requesting accommodation.... Accommodations developed with this in mind bring employers the double benefit of accommodation as well as preventing disability."
>
> "Accommodation of workers with disabilities through job accommodation and workplace design is here to stay. By instilling a universal design approach among those responsible for the development of work environments and products, the incidence of work disabilities can be reduced. And those accommodations that are required for workers with disabilities will be much more likely to be reasonable accommodations."
>
> Mueller, J. (2010). "Office and Workplace Design." In W.F.E. Preiser and K.H. Smith (Eds.), *Universal Design Handbook, Second Edition.* McGraw-Hill, New York.

in London and Greg Vanderheiden from the Trace R & D Center identified problems in getting manufacturers to adopt universal design products, and in addition, prolific contemporary designer Karim Rachid called time pressures from manufacturers limiting the time needed to develop prototype designs a major concern. Computer-aided design techniques have facilitated a quick turnover and modeling (less risk). Technology has improved our ability to react quickly to changing work environment needs and allowed custom design for a smaller group of users (see Figure 6.2)—office chair design for "little people"; another manufacturer "created a line of office chairs for workers that were small in stature" (Asians).

In recent years, we have seen more innovation in design of office furniture because of evidence-based research findings. In addition, heating/ventilation/air conditioning, lighting, and noise levels influence furniture and equipment choices within individual and group work areas. The

FIGURE 6.2   Note the thickness of the back of this Sitmatic chair custom designed for little people's ease of access to a desk or table.

importance of creating supportive work environments has been based on the landmark research by Michal Brill and Sue Weideman from BOSTI Associates.

BOSTI Associates divided the primary costs of doing work over 10 years into four areas:

- People account for 82%
- Technology for 10%
- Workplace for 5%
- Operations for 3%

They went on to show that investment in the workplace and operations would have minimal effect on overall costs but would have a major effect on productivity and people's job satisfaction (Kimball International, "Disproving Widespread Myths about Workplace Design").

This good example of evidence-based research and the time-tested results of the BOSTI research have had a major impact on design of office furnishings because they have demonstrated that relationship between the workplace and job performance, job satisfaction, and productivity (important for all organizations). Research also suggested that the dollar value of appropriately designed offices was substantial.

With reinforcement from the ongoing research in work environments, manufacturers of office furnishings have been able to put the research results into production.

The publication *Interiors & Sources* presents a NeoCON Preview each spring (NeoCon—the Contract Furniture show held in Chicago in June). The magazine also features award-winning office furnishings. NeoCON, regional exhibits, and the Ergonomics Expo in Las Vegas held in November have provided manufacturers the opportunity to showcase their new products. An office chair with adjustable armrests, a standing desk, and adjustable-height tables were all examples of new designs introduced at NeoCON over the last few years.

FIGURE 6.3   The Sitmatic Bariatric chair with its dual pedestal base will support up to 600 lb.

A recent "white paper" on obesity and other research on this emerging problem provided incentives for manufacturers (even in a downturned economy) to introduce a line of bariatric chairs (Figure 6.3).

The bariatric issue has created major challenges for manufacturers—research has shown that even though obese users need furniture that is safe and supportive of their weight, they will reject furniture that looks very different from that of their coworkers (even when they are given so-called "Executive Chairs"). People don't want anything that will set them apart because of their size. Years ago, President Taft, who was known for his large size, was scheduled to stay at the famous Mission Inn in Riverside, California when he was visiting Los Angeles. The Inn very proudly had a magnificent chair created for him. (It still sits in their lobby.) What the hotel personnel usually don't mention to visitors is that he refused to sit in his "special" chair. People don't want to be reminded of their size.

Consequently, this chapter then moves into a global, theoretical discussion of the features that make up a universally designed office. We have sought to define "quality" as the term relates to the interplay between worker, work, and environment. In the process, we arrived at an "ecology" that fulfills our description of universal design.

A case study by Steelcase continues with a picture of the changing profile of office workers today and ways to meet their needs. Virginia Kupritz (2001) reported on her research related to design of office environments. Lighting is the focus of a case study by Eunice Noell-Waggoner, titled: "Light—A Universal Need." Another case study highlights the work of lighting designer Doug Walter and his report on current views of LED lighting. Finally, Franklin Becker and his colleagues at Cornell University conducted research on office design and management around the world. The report ends with a comparison of open and closed offices and how balance can be achieved. Their case study included post-occupancy evaluations and other assessment tools and offered some exciting possibilities for change in the American work environment. Charles Schwab (2009) then describes how universal design techniques can be incorporated into the design of the home office.

OPEN OFFICE DESIGN TO MEET USER NEEDS

Historically, little attention has been paid to providing a work environment that can serve *all* the needs of a company, including maintaining corporate identity and increasing worker safety, satisfaction, and productivity while facing financial realities. This is now changing. Universal design presents a critical approach to rethinking conventional solutions for office design by drawing on knowledge of ergonomics and human factors, as well as information on special needs groups, changing technology, and changing work styles (Figure 6.4).

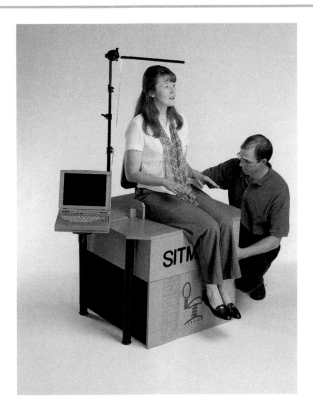

FIGURE 6.4   Sitmatic measuring device for custom chair design.

If a person is unable to perform a task owing to physical constraints in the workplace, is unable to interact with others because of barriers or physical limitations, or experiences a loss of self-worth owing to a lack of control over the immediate environment, then there is a design problem that needs to be addressed.

Two interrelated theories—ergonomics and *proxemics*—are useful to universal design since they can help the designer understand how an environment that adapts to the individual can be better created. Ergonomics is the study of people's reactions and behaviors in relation to their work. The general aim of such study is adapting work conditions to the physiological and psychological nature of human beings and optimizing the fit between people and machines relative to their needs and activities. This results in the most important principle of ergonomics: fitting the task to the human being (Figure 6.4). From this concept, furniture systems are developed, new equipment and machines for the office are created, and new workstation configurations are conceived (see Figure 6.5).

Proxemics is the study of the personal, cultural, and spatial needs of humans and their interactions with the environment—the spatial relationships between people and things relative to activity, purpose, culture, and time. Design activities that affect proxemics include lighting strategies, spatial organizations, and workstation layouts. These components directly affect personal interaction, communication, and territorial definition within the workplace.

The physical office environment is an integration of building systems (structural, electrical, spatial) and furniture systems (desk and work surfaces, seating, storage, and accessory products), equipment and machines (computer, monitor, keyboard, mouse and pad, telephone, typewriter, pencil sharpener, calculator, etc.), lighting specifications, indoor air quality, climate control, acoustics, and material, chemical, and finish quality.

It is important to consider less tangible, or tacit, contributions to office design: management, personal response, degree of self-control over one's microenvironment, and environmental stress. Although these conceptual aspects receive academic and scholarly attention, the general public (including many professionals) is most attuned to the visual and tangible. But both aspects are necessary for a complete understanding of universal design in the office environment.

FIGURE 6.5   Canvas Office Landscape; multiple flexible cubicles by Herman Miller.

THE TACIT OFFICE ENVIRONMENT

HEALTH, SAFETY, AND WELFARE ISSUES

The health, safety, and welfare of workers are crucial concerns within an office design. This area encompasses a wide range of factors—from short-term physical concerns to long-term psychological effects. Definitions of health, safety, and welfare are currently expanding to include a greater emphasis on long-term issues of health maintenance, well-being, and personal growth. The ADA and the American National Standards Institute/Human Factors Society 100 Act of 1988 (see ANSI/HFS 100-1988 1988) are examples of legislative action that considers long-term issues.

Industrial designer, Jim Mueller, worked with designers at Herman Miller to create a list of ADA Guidelines for offices shown at the beginning of this chapter (Figure 6.1). These guidelines were intended as a starting point for consideration of disability issues in the design of business environments. The floor plan illustrates 12 ADA Guidelines for Interior Design, plus another set of additional "Designing for Accessibility" Guidelines. Although not intended as a complete and comprehensive guide to ADA compliance, it is a good resource for designers of office environments.

Creating a healthy, safe office environment can be facilitated by asking the following questions:

- Does the office environment (at minimum) meet existing federal, state, and local health and safety standards?
- Do the design and spatial layout of the furniture and equipment systems promote ease of use for as many people as possible (Figure 6.5)?
- Are access and egress designed with safety in mind?
- Are the work areas and surrounding environments safe for all existing employees at all times?
- Are fatigue, discomfort, and stress evident in the users (or are they expressed as problems on surveys or questionnaires)?
- Are there excessive repetitive or functional activities that might lead to repetitive motion strain injuries?
- Does the environment provide individual climate, acoustic, light, and air quality controls?

Figures 6.6 through 6.8 are examples of workspace furnishings that comply with the ADA Guidelines.

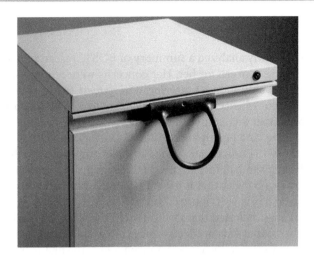

FIGURE 6.6   Meridian's Enhanced Access pull designed specifically to meet or exceed ADA requirements.

FIGURE 6.7   Flexible panel dividers by Herman Miller in Ethospace Reception Area.

FIGURE 6.8   Meridian Filing and Storage in Abak Workspace with Aeron chairs.

AN UPDATE FROM THE BOSTI GROUP

Audiologist Anne Seltz recently analyzed a summary of BOSTI research. Acoustics in the workplace has been a career-long concern for her. Her concerns were reinforced by the discovery of The Kimball International publication, *Disproving Widespread Myths about Workplace Design*. It reported on the research of Michael Brill. He found "that over a 10-year period the cost of hiring and maintaining personnel was estimated to be more than 13 times the cost of constructing, enclosing, operating, wiring, and equipping an office." During the next 15 years, Michael Brill, Sue Weidemann, and the BOSTI Associates added to their original research by studying emerging business trends. The fine Kimball International publication, discussed their research findings. It is difficult to summarize a booklet that is itself a summary but it focuses on the specific areas of acoustics, distraction, and privacy (Figure 6.7).

The BOSTI group findings indicated that investment in the workplace and operations will have minimal effect on overall costs but a major effect on productivity and people satisfaction. They further list the strongest effects on workplace qualities in ranked order. The first is ability to do distraction-free work.

Various philosophies have influenced office design: some emerging from perceived need; others from theoretical perspectives. Acoustics has seldom been mentioned as a priority but this BOSTI report cites noise or acoustic distraction as a major cause of reduced workplace quality.

Their analysis indicates that the four work groups spend the majority of their work time in their own workspace:

- Managers, 78%
- Professionals, 82%
- Engineers and technical persons, 80%
- Administrators, 86%

Thus, one's own workspace is still the *primary spatial tool for work*, even in highly interactive, team-based organizations. Therefore, the workplace needs to support the ability to do distraction-free work. For those in open offices, only 19% report they are relatively undistracted, while 65% of workers report being often distracted. For those in private offices, 48% report they are relatively undistracted, while 29% are often distracted. Of interest are the types of workspaces people have (among the entire database of 10,000 workers): 25% have private offices; 4% share private office; and 71% are in an open office.

As BOSTI summarizes, A PUZZLEMENT. Of all their time, people in all job types spend by far the most doing quiet work. Supporting quiet work is one of the two top productivity enhancers and job satisfiers. However, two-thirds of those in open offices (the most prevalent workspace type) are "often distracted by others' conversations" and can't do undistracted work. Then why are open offices so prevalent? The BOSTI group explores the six myths prevalent in workplace design:

- We can't have both distraction-free work and easy interactions. They're opposites.
- We can provide a distraction-free *open* office.
- We are moving toward being a more *open organization*, one with better communications.
- We *learn* more in the open from overhearing others' conversations.
- We can't have all enclosed workspaces… our *space utilization rate* will skyrocket.
- We *can't afford the cost* associated with providing enclosed workspaces.

Various solutions are offered with primary focus on workspace design and reduction of noise. These research-based findings and solutions can influence the design of workplaces and workspaces so that distraction-free work can occur.

ABC'S OF EFFECTIVE ACOUSTICS

While several environmental factors, including light and temperature, are generally recognized to affect workplace performance, research conducted by the Center for the Built Environment and others indicates that proper acoustics must also be a top design priority.

The workplace should provide occupants with speech privacy, comfort, and freedom from distracting noises, and enable them to work without disrupting others. The formula many acoustic professionals use to achieve these results is the ABC Rule, meaning Absorb, Block, and Cover Up.

ABSORB NOISE

Absorptive materials reduce the volume of noises reflected back in to the space, the length of time they last, and the distance over which they will travel.

The most significant source of absorption in a facility is typically the ceiling. Ideally, open spaces should feature a tile with at least a 0.75 Noise Reduction Coefficient. Tiles used in closed spaces should have a high Ceiling Attenuation Class, because they will be better at containing sounds. It is also important to ensure consistent coverage throughout the facility.

In order to limit the lighting system's impact on the absorptive performance of the ceiling, select a system that incorporates a minimum number of ceiling fixtures while still meeting the specified lighting requirements. When it is not possible to install an indirect system, use a deep parabolic lens instead of a solid plastic lens.

Minimize the size and number of reflective materials, such as glass and metal, used in the space because these will reflect noise and conversation, causing them to be heard over greater distances.

Soft flooring should be used to reduce footfall noise.

BLOCK NOISE

Another method of controlling noise is to block sound transmission. Closed plan designs use walls and plenum barriers to block sound, but blocking is also a relevant technique for open concept offices.

Locate noisy office machines and high-activity areas, such as call centers, in remote or isolated areas.

Maximize the distance between employees and minimize direct paths of sound transmission from one person to another by seating employees facing away from each other on either side of partitions.

Partitions that are 60 to 65 inches (150 to 165 centimeters) are effective because they extend beyond seated head height, though using taller partitions in high traffic areas can be beneficial. If they are shorter, they achieve little more than holding up the desk.

COVER NOISE

Many people believe they achieved good acoustics after applying the preceding strategies and the sound level in their facility is very low. But, just as with lighting and temperature, the comfort zone for the volume of sound is actually not zero. If it is too low, conversations and noises are easily heard and more disruptive.

Ensuring a comfortable and sufficient background sound level is the final requirement of the ABC Rule, which is to cover up conversations and any noises that remain in the space.

Sound masking systems provide the only way to truly replenish and keep the background sound level of an appropriate volume, which is typically between 42 and 48 decibels.

A sound masking system consists of loudspeakers installed in a grid-like pattern in an open ceiling or above the suspended tiles. The system distributes a comfortable background sound at controlled levels across the facility.

Though the sound is often compared to that of softly blowing air, it has been specifically engineered to increase speech privacy. It also improves general acoustical comfort by covering up a lot of intermittent noises or reducing the amount of disruption they cause by decreasing the magnitude of change between baseline and peak volumes. In addition, if the sound masking technology features small adjustment zones with fine control over both volume and frequency, and is installed throughout the space, it will also ensure acoustic consistency across the facility.

REDUCE NOISE

The ABC Rule overlooks one valid and frequently used method of addressing office noise: reducing noise at the source. This task can be accomplished by modifying employees' noise-producing behaviors and replacing noisy office equipment with quieter technologies.

An environment conducive to speech privacy, concentration, and productivity can only be created through the balanced application of all four methods of noise control.

Manufacturers of office furniture are continually researching the changes in demographics and societal needs and ways in which they impact office design. The following case study is an example of research conducted by Steelcase.

Case Study: Wellness in the Workplace

CHANGES THAT IMPACT WELLNESS IN THE WORKPLACE

Differences in job, gender, physique, age, ethnicity, and culture and personal taste all place unique demands on the office environment.

- There is a mix of older and younger generations, each with his own needs and characteristics.
- Workers are getting bigger: as much as two-thirds of today's workforce is considered overweight and one-third is obese (see Chapter 7, Obesity Research).
- Work has changed. Technology has given us freedom and independence to work anywhere and anytime, and we're collaborating with others now, more than ever (Figure 6.9).
- More workers are now experiencing back pain as office work has increased.

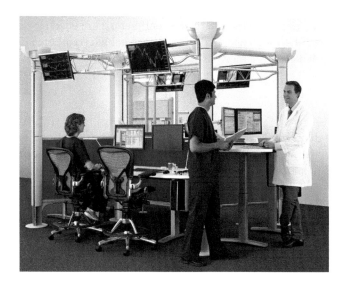

FIGURE 6.9   Meridian Stackable Storage in a collaborative area by Herman Miller.

FIGURE 6.10   Standing desk from Biomorph by industrial designer Stephen Barlow-Lawson.

Mobile technology, new kinds of jobs, and increased collaboration force people to work in spaces that don't support, ergonomically or otherwise, the way people now work. Physical needs include movement throughout the day and supportive furniture for those who prefer to stand while working or to change positions for different tasks (see Figure 6.10). Workers need the ability to connect and collaborate effectively with others in the office. Spaces need to be designed to meet the unique needs of the work group.

Individual workers need to feel they have control over the conditions of their work site. Without workspaces giving workers the control they need, stress is inevitable.

Today, businesses are taking a more proactive approach to employee wellness. Some provide on-site fitness centers or gym memberships, and others offer incentives for healthy living such as lower rates for health insurance for workers who quit smoking, lose weight, and commit to a regular exercise program. The employer benefits from increased worker productivity, higher employee retention, and fewer employee absences.

MODERN OFFICE ERGONOMICS: ENCOURAGING HEALTHY MOVEMENT FROM THE SEATED POSITION

No one sits still. In the real world, there is no such thing as the office worker depicted in some ergonomic guidelines—head up, eyes focused on the upper third of the monitor, spine erect, feet supported, elbows angled at 90 degrees, wrists straight with hands poised just above the keyboard. At least, this is seldom seen. Yet, textbooks and ergonomic guidelines habitually use such illustrations to represent the "ideal" seated position. Is it?

Scientific (and casual) observation of what office workers do at their desks reveals a different picture. People are always moving. They change positions constantly: they recline while taking a phone call, bend forward to resume work, turn around to look at someone, reach up for a report, slide their chairs forward, backward, and sideways to access items within their work areas, and so forth. The variations are endless.

This is good. Movement stimulates circulation, which supplies nutrients to tissues and removes wastes, resulting in greater alertness and less fatigue throughout the day. Current ergonomic thinking calls for acknowledgment of this natural behavior, and for solutions that work with it, rather than strategies aimed at modifying or overriding it. Freedom of movement is the goal.

DYNAMIC SEATED WORK: THREE ESSENTIALS

Encouraging safe, comfortable movement for the seated worker requires attention to three essentials: stability, clearance, and position.

STABILITY

Stability is the basis of controlled, comfortable, and efficient movement from a sitting position. It is achieved by supporting feet, buttocks, back, and forearms.

- *Supporting the Feet.* Feet should rest squarely and firmly on the floor or on a footrest, with knees bent somewhere between the angles of 85° and 110°. This provides "ground reference," that is, connection to the ground. If the chair's seat is too high and the user's feet don't reach the floor, a primary source of balance and support for the upper body is missing. In contrast, a seat too low compromises ground reference. Though allowing contact with the floor, it elevates the knees, weakening the optimal structural relationship between floor, feet, legs, and the rest of the body. Thus, the first step toward stability is adjusting seat height to an effective level.
- *Supporting the Buttocks.* In seated position, the *ischial tuberosities,* or "sit bones" of the pelvis, along with the surrounding musculature of the buttocks and upper legs, bear the weight of the upper body. This is the body's principal anchor, its main source of stability. For optimum support and comfort, chair seat pans are sized and shaped to distribute this weight over as much area as possible, with padding that will compress and conform to the user's body.
- *Supporting the Back.* A normal, healthy spinal column has a series of curves that support the weight and movements of the upper body. For most people, this natural S-shape is the back's "default position" when standing. However, when sitting, the inward lumbar curve of the lower back (sometimes referred to as the "lumbar lordosis") tends to flatten, causing the shoulders to droop forward and the pelvis to rotate backward so that the tail bone curves down and under. The slumped position introduces stress in the thoracic and lumbar regions of the back, in the pelvis, and in the shoulders and neck. To counteract this and allow the spine to maintain its naturally supportive curvature, modern office chairs all offer some level of lumbar support. These usually come in the form of passive supports (backrests that passively adjust to maintain the curve in the small of the back), and active supports (manually adjustable backrests that "push" against the lower back). Innovative designs are now providing dynamic, adaptive lumbar technology that provides an appropriate level of support—from sitting upright to reclined—as the worker moves and changes position spontaneously throughout the day.
- *Supporting the Forearms.* Lifting and extending the arms repeatedly and holding them in extension introduce stress to the rotator cuffs and bursae of the shoulders. Thus, prolonged use of a keyboard and mouse can lead to repetitive strain injuries in shoulders if arm supports are not present or used. Likewise, carpal tunnel syndrome and tendonitis may result from mousing and keyboarding with wrists out of a comfortable neutral position. Both problems can be avoided by supporting the forearms. Properly adjusted, these supports take the load off shoulders and neck, and eliminate awkward wrist angles, allowing easy efficient use of arms and fingers.

CLEARANCE

Freedom of movement requires elimination of barriers. Barriers in the seated worker's immediate surroundings mean inefficient movement, and sometimes strain, to avoid or accommodate them. Not all such maneuvers are harmful, but some can throw a stable posture out of balance, putting stress on the musculoskeletal system and introducing the risk of injury. Two common clearance problems relate to seat depth and work surface height.

■ *Seat Depth.* Seat depth, or the front-to-back length of the seat pan, poses two potential problems if not adjusted: First, the lip of the seat may contact the *popliteal fossa*, the area behind the knee through which the primary blood vessels and nerve pathways lead to the lower legs and feet. Compression here may lead to pain, numbness, or swelling. Second, if the user moves forward to allow clearance, the lower back will also move forward, taking away the backrest's support of the lumbar curve. This will lead to the drooping, round-shouldered posture described above. Correct seat depth adjustment will allow firm contact between the chair's lumbar support and the lower back, leaving 2–3 finger widths between the front edge of the seat pan and the back of the knees.

■ *Work Surface Height.* Any work surface too high in relation to the forearms and hands creates a clearance barrier. (A visual indicator of the problem is elbows below and hands above the level of the work surface.) In such a case, where wrists cross the edge of the surface, direct contact and wrist flexion compress tendons. Repetitive stress of this kind usually leads to carpal tunnel inflammation. The ideal position is elbows and forearms even with or slightly above the height of the work surface. Proper clearance can be achieved via one or more adjustments: (a) chair height, (b) armrest height, or (c) work surface height. (Since height-adjustable workstations are still uncommon in most offices, usually it is chair and armrest heights that are adjusted for an effective relationship to the work surface.)

REAL AND IDEAL

Considering the principles of stability, clearance, and position, it should be clear that this actually is an ideal seated posture for desk and keyboard work. Ideally, feet, back, buttocks, and forearms are well supported, providing stability. It is important that there are no barriers to movement or circulation. And the head, neck, forearms, wrists, hands—in fact, the entire body—be in the green zone, the desirable neutral position. At the same time, this idealized picture does not imply that a worker should stay put. (In fact, for prolonged-focus tasks like data entry, workers are encouraged to get up from the chair occasionally to stretch, bend, and move.) As we have said, workers do not sit still. With modern office furniture, most of their natural, spontaneous movement is encouraged and supported. Healthy freedom of movement with comfort and safety is today's performance standard. It is sound ergonomics and good sense. Ergonomics at Allsteel: our ergonomics team studies workers: who they are, the way they work, and what they need to be comfortable and healthy. These insights are built into every product we make. Commitment to ergonomics is clearly reflected in advanced chair designs with dynamic adaptive back support and height-adjustable work surfaces. For more information, visit www.allsteeloffice.com/ergo or e-mail ergonomics@allsteeloffice.com

CREDITS

Scott Openshaw, M.S., heads Allsteel's Ergonomics Group. With an academic background in Human Biology and Biomedical Engineering, Scott applies human factors and ergonomics principles to the design of office furniture at Allsteel.

Drew Bossen, P.T., is founder of Atlas Ergonomics, an ergonomics consultancy with expertise in multidisciplinary prevention-oriented programs for minimizing occupational injuries in office settings, healthcare facilities, the industrial workplace, and transportation fleets.

ERGONOMICS AND YOUR BOTTOM LINE

Your guide to a return on your investment and ergonomics. When it comes to office productivity and health, few decisions you make will be as critical as your purchase of seating and workstations. This educational guide can help you understand the costs and benefits of ergonomic furniture, choose the correct product options, and ensure you are using them correctly.

INJURIES AND COSTS

Furniture choices affect health, safety, and productivity, as well as occupancy and move-related costs.

Adjustable height furniture may decrease injuries because it can eliminate poor fit and awkward, uncomfortable postures.

Static postures associated with fixed-height furniture can contribute to decreased productivity, weight gain, diabetes, and other chronic illnesses (Mayo Clinic/Duke University).

Some facts about injuries at work

- Back pain is second only to the common cold as a cause of work loss (University of Missouri study).
- Repetitive motion injuries account for more than $2 billion in workers' compensation costs annually (Liberty Mutual Workplace Safety Index).
- Non-adjustable desks generated twice as many ergo issues and higher facilities and workers' compensation costs compared to user-adjustable ergonomic furniture (Fortune 500 Financial Firm, internal study).

ERGONOMICS—AN INVESTMENT IN THE FUTURE

Ergonomics means fitting the job to the worker, because one size may not always fit all.

Ergonomic seating can provide user adjustability so that the seat, backrest, and armrests better accommodate the user.

Since fixed, non-adjustable work surfaces or desks may not provide as much accommodation and customization as adjustable furniture, providing height-adjustable keyboard trays and appropriate ergonomic seating as options for adjustability can improve accommodation.

If using facilities-adjustable work surfaces, adjust the height to suit the user. The work surface should be near the resting elbow height.

User-adjustable work surfaces within the sit range (24″ to 32″) can accommodate most users and enables desk sharing for multiple shifts or telecommuters.

User-adjustable furniture within the sit-to-stand range (24″ to 40″) offers the best ergonomic benefits and enables desk sharing for multiple shifts or telecommuters.

Basic ergonomics training and posture evaluation can also significantly improve user comfort and reduce injuries.

ERGONOMICS AT ALLSTEEL

Finding new ways to make offices more productive and people more comfortable is leading to great things at Allsteel. We're not only conducting intensive ergonomics research, we're also tracking the latest in workplace trends and helping companies understand how our innovative products facilitate a healthier, happier workforce.

For example, we pioneered chairs that move with you from sitting upright to reclined positions, while maintaining your line of sight to your monitor and expanding your torso for better circulation. Our ergonomics team also helped redefine systems furniture by placing storage within workstation walls at desk level, so there is no uncomfortable reaching above the shoulders and below the waist for frequently used items.

Whether it's the Stride height-adjustable work surface bringing cost-effective ergonomics to the masses, or our seating portfolio that promotes healthy movement, we are constantly finding new ways to improve the office.

OFFICE WORK ENVIRONMENTS

Enacted in 1988 and overturned by the Superior Court of California after only a year in legislation, the American National Standards Institute/Human Factors Society 100 Act (ANSI/HFS

100—limited to the city of San Francisco) was written for individuals working in a video display terminal (VDT) work environment. The ordinance required companies to provide their employees working at VDT workstations more than four hours a day with the following:

SPECIFIC MANDATES OF THE 1988–1989 LEGISLATION

- Pneumatic seating—height adjustment ranging between 16 and 20.5 inches
- Armrest, wristrest, and footrests when requested by any operator who routinely performs repetitive keystrokes
- Angle and height-of-back adjustment and seat-depth adjustment on chairs
- Document holders with angle and height adjustments upon request as appropriate
- Seat rake (angle between seat back and seat pan) fixed back: angle shall be 90° or more; adjustable back: range shall include some part of the range from 90° to 105°
- Minimal seat width of seat cushions at least 18.2 inches measured at the spindle center and measured to the edge of the seat, to the edge of the fabric
- Footrest of at least 2 inches in height for those using a keyboard if they are unable to place their feet flat on the floor
- Special lighting—a combination of adjustable task and ambient lighting is needed
- Rest breaks negotiated by workers and related to specific tasks
- Noise reduction—appropriate use of acoustic materials to minimize problems with noise
- Appropriate lighting for VDT use (proper window treatments, antiglare screens, and other measures to ensure visual comfort)

While this ordinance was in effect, it was applicable to all businesses, private or public. Even though the ordinance has expired, the guidelines remain valuable for office designers today.

PSYCHOLOGICAL AND BEHAVIORAL ASPECTS OF OFFICE DESIGN

Maintaining psychological and behavioral health among workers in an office environment is obviously important. Unfortunately, acceptable levels of privacy for individuals, aesthetic preferences, and successful interaction strategies are difficult to measure or define. Nonetheless, people must be able to interact with their environment and with others in that environment.

The designer must thoughtfully address the following:

- Do the overall qualitative aspects of the workplace (character and ambiance) support the image and personality of the users?
- Do workspaces promote privacy, allowing one to concentrate while also providing the opportunity for occasional distraction?
- Is there flexibility in occupying individual workstations?
- Are there facilities for small and large group gatherings?
- Do conference spaces promote interaction, allowing one to be both part of a group and an individual?
- Does workspace configuration support the changing needs of office workers?

(See Figure 6.10 for a good example of flexible office workstation.)

NON-TERRITORIAL OFFICE: CORNELL UNIVERSITY'S INTERNATIONAL FACILITY MANAGEMENT PROGRAM

The Becker/Sims case study described their research on facility management and design (Hedge et al. 1989). The non-territorial office was one of the plans included in their research. In this office footprint, the intent of management was to stimulate informal communication and more teamwork by removing panels and walls, thereby creating the likelihood that people would both sit

FIGURE 6.11   Hub space informal learning environment.

next to and talk to different people as they occupied different unassigned work areas (Figure 6.11). Thus, the non-territorial aspect of this case evolved from the goal of stimulating informal communication and teamwork, not from cost-cutting.

PHYSICAL DESIGN

The floor plan (see Figure 6.12) shows the original free-address installation. The headquarters building was designed for all users who spent 60%–70% of their time out of the office. The system's salient physical characteristics were a complete open plan (with the exception of a glass-enclosed conference room); a wide range of work settings and furniture including residential style

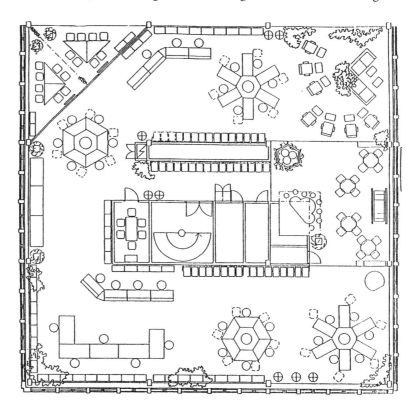

FIGURE 6.12   Lounge-type work area by Becker and Simms.

sofas, reclining easy chairs, and an outdoor swing set; large, colorful wall murals; and a small, centrally located but visible kitchen with stand-up break areas.

OFFICES THAT WORK: BALANCING COST, FLEXIBILITY, AND COMMUNICATION—NEW YORK: CORNELL UNIVERSITY INTERNATIONAL WORKPLACE STUDIES PROGRAM (IWSP)

THE OPEN VERSUS CLOSED OFFICE DEBATE

Fifty years after the introduction of panel-based open plan office systems, we still vigorously debate the value of open versus closed offices. Phrased this way, the question makes no sense. Both open and closed offices serve useful purposes. The meaningful question is, "What's the right balance between open and closed offices?" Close on its heels is, "What do we mean by 'open' offices, anyway? Is it a high-paneled cubicle (one cannot see over a panel when seated)? Is it a low-paneled cubicle (one can see over a panel when seated)? Is it a cluster of 'pod' of low-paneled workstations separated from another pod by higher panels? Is it a shared enclosed office (2–12 people in an enclosed space)? Is it a team-oriented bullpen, with a small group of desks in a completely open area?" (See Figure 6.13 for one cubicle example.)

It is, of course, all of the above. Which is precisely why describing a work environment as "open" serves so little purpose. It is like using "meat" to describe everything from hotdogs to filet mignon, or "car" for everything from Ford Escort to Rolls Royce. It is correct, but learning that some people hated "meat" or "cars," and others loved them, would not be terribly edifying without knowing which kind of meat or car each group had experienced. The same holds true of understanding people's reactions to "open office environments."

Finding the right balance of open and closed offices required understanding the purpose of the office and, even more so, the nature of the work being done. We don't buy a Porsche to haul pianos. We buy (or rent or borrow) a pickup truck. We consider the purpose, the intended activities. One vehicle cannot serve every purpose equally well, but it can serve several purposes, hopefully the primary one, to the highest level. Why not look at the office environment in the same way? The office's primary (not only) value, we believe, is as a place for face-to-face interaction: a place to meet coworkers and managers, to inspire, coach, be motivated, share information, debate goals and objectives, socialize, make friends, and so on. It is as much or more a social setting as it is a refuge or technical or information center.

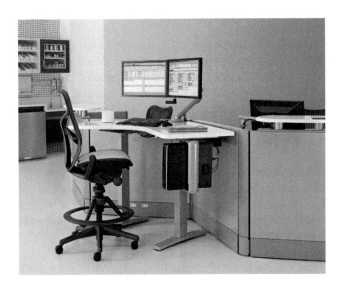

FIGURE 6.13   PACS Application in Healthcare Area, Chicago showroom, NeoCON 2005, Herman Miller.

Given this way of thinking about what an "office" is, we need to understand, first, how different forms of office design, from closed offices to a variety of forms of open plan offices, affect communication and interaction. Secondly, we need to understand how communication and interaction affect valued organizational outcomes such as decision speed, organizational learning, and employee job satisfaction and commitment.

THE CLOSED OFFICE

Our research, done with employees in job functions ranging from software development to marketing and business development, indicates that the more open the "open" plan office environment, the more conducive it is to overall work effectiveness, when communication and interaction are critical elements of the work process. Few jobs or professions don't qualify. Yet for most employees, the closed, cellular office is the preferred office type, for well-known reasons. With few exceptions, it is easier to control unwanted distractions and interruptions, and noise is typically less of an issue. Comments from several software engineers capture this view of the closed office.

Software Engineer: For me personally, if I'm focused on the given task at hand, peripheral noise is really distracting. It takes away from the thought process. There's sort of a momentum that builds up that, with constant interruptions, it's like stop-and-go traffic. You never quite get up to that cruising speed where you feel like you're being productive.

Software Developer: I find I work better as well when I'm alone. I find I code better; I think better when I'm in a closed office, especially in the design stage when I'm thinking about the program. Before I start programming, I need some quiet time, like a white board or something, and (a place to) just go and think on my own. If there are people continuously interrupting you, the process is much slower.

Few would argue that most employees, whether they are software developers, web designers, business strategists, or human resource professionals, need time to think, concentrate, and reflect, as well as to communicate, share information, and interact socially. But as Kellner argues, the reasons for feeling more effective in a private, closed office reflect deeply held values as much as simple utility: "Several forces conspire to keep software work an individual activity, including: desire for autonomy; a culture that rewards individual efforts far more than team efforts; concentration of crucial application knowledge by a few individuals; desire for privacy regarding individual development efforts; the Not Invented Here syndrome and its more personal form (not invented by me); large productivity differences between individuals; political considerations of powerful individuals and of managers." "The Holy Grail" is finding the right balance.

What is surprising about our data is that the more open type office environment, what we are calling team-oriented bullpens and pods and shared closed offices, may come closer to achieving this balance than either closed offices or high-paneled cubicles. In part, this comes from the recognition, even from employees in closed offices, that while communication certainly occurs in closed offices, the pace, frequency, and nature of that communication are significantly different from what occurs in team-oriented bullpens and workstation pods and shared closed offices.

Software Engineer: Email and phone… we're really collaborative that way.

Software Engineer: Communication with colleagues—I think my door is open all the time if I don't have meetings. So, anybody could stop in for a chat or whatever.

Software Engineer: I think it is important in an enclosed office that you are able to do conference calls more easily with the speaker phone on.

Conference calls, email, and scheduled meetings were viewed by those in closed offices as providing sufficient communication. The same types of communications occur in open plan environments. The difference is the value attributed by those in closed versus more open environments to short, frequent, and fast communication. In more open environments, such communication patterns were viewed as contributing to one's own and the team's (project's) productivity—both quality and speed. In closed offices, respondents focused on the perceived benefit of not having to interact with others serendipitously and the benefit to their own concentration. Yet some of those in closed offices recognized that their privacy came with a price, in terms of reduced communication with colleagues that weakened the project or team's performance.

Software Engineer: We suffer because we don't have much of a sense of team. And I don't think people understand the relationship of their work to others and how if they don't execute well it will impact others. Or if they're negligent and not getting their job done right, it has impact and consequences to others. Because they don't have this fabric of a team, they don't understand when they're not performing well that they're impacting somebody else. I don't think people are as committed to getting things done and helping their teammates, because they often may not even know who they are.

COMMUNICATIONS SUBTLETIES

More surprising than the fact that more and different kinds of communication occurred in more open environments was that such offices also could reduce unwanted interactions and disruptions. Had we relied only on our survey data, we might well have concluded that there were no differences among open plan offices of various sorts in terms of how they influenced employees' communication and interaction patterns or their ability to concentrate. That is because almost all respondents, regardless of the type of office they were in, reported in the survey that they had high levels of communication and interaction, and that privacy and the ability to concentrate were more difficult in more open types of offices. The focused interviews and systematic observation data painted a different, subtler picture.

When we looked at transcribed interviews, we found that respondents in closed offices said something to the effect that "Yes, I communicate a lot. I often email or talk with coworkers by phone, and whenever I need to see someone, I can easily drop in on them or arrange a meeting." In effect, "frequent" communication for those in closed offices meant interacting several times a week in a scheduled meeting, but not often on the fly. For those in workstation "pods" and small scale, team-oriented bullpens, "frequent" communication and interaction meant literally dozens of quite short communications throughout the day. Without understanding the employees' underlying internal metric, relying on the survey data alone is a bit like asking an obese and an anorexic person whether they ate a lot, and if they both said "yes," assuming the amount of food consumed was comparable.

The interviews revealed more than just differences in frequency of communication. They revealed subtleties in the communication process itself. Respondents in high-paneled cubes described what they called "pseudo-privacy." The high panel supposedly created privacy, yet one could overhear all of a neighbor's conversations.

Design Manager: When we were in [high-paneled] cubicles, we still sat very close to each other and I would overhear bits and pieces and I would either have to interrupt by standing on my desk and peering over a cubicle wall or running around the corner. And it felt invasive. I felt like I was stepping into a conversation that I wasn't invited into. And my choice was I could either be obnoxious or I could sit there and pretend I didn't hear it. And that was problematic. I was really afraid that I was going to have to … let people kind of flounder.

What do you do when you overhear a telephone conversation and you realize you have information that could help resolve a problem, but you don't want to admit that you overheard the conversation? Is this eavesdropping or just the unfortunate by-product of minimal acoustic separation? What does being "civilized" mean in this kind of situation? If you cannot see over a panel, how do you know if your neighbor is there or not, and therefore whether you need to modulate your speaking voice, or just not have certain kinds of conversations? High-paneled cubicles exacerbate these kinds of problems, while more open team-oriented bullpens and pods, with their unobstructed visibility from a seated position, provide useful cues that govern interaction, build trust, and reduce unwanted interruptions.

BUILDING TRUST

Trust and comfort among team members is a theme that came up repeatedly in interviews with employees in team-oriented bullpens and pods. It mirrors the recent interest in the concept of "social capital," which argues that social and emotional relationships affect everything from work output to organizational commitment. From this perspective, effectiveness cannot be defined solely by individual talent, effort, or output, despite the fact that for most of the last 100 years or longer, most firms in this country have conceived, evaluated, and rewarded performance on an individual basis. In a world dependent on the constant flow of information, and the need to attract and retain the best workers possible in a sellers market, the connection between social relationships and performance takes on a new meaning. Recent research shows that people are about five times more likely to turn to friends or colleagues for answers than to other sources of information such as databases or policy and procedure manuals.

Becker's (1990) examples of post-occupancy evaluation (POE) forms are pictured in Figures 6.14 and 6.15. These questionnaires contain rating scales for

- Common workplace requirements
- Personal workplace requirements

FIGURE 6.14   An example of a post-occupancy evaluation form. Adapted from *The Total Workplace* by Frank Becker.

PERSONAL WORKPLACE REQUIREMENT

Please rate your personal workplace on each of the characteristics below. First indicate your satisfaction with each one, and fill the number corresponding to your answer in the left side box. Then rate how important each one of the characteristics is to you, and fill your answer in the right side box. If you have some specific reason or explanation about your rating, please put your comments on the last column of each question.

1. Dissatisfied
2. Somewhat Dissatisfied
3. Neutral
4. Somewhat Satisfied
5. Satisfied

1. Not Important
2. Hardly Important
3. Neutral
4. Somewhat Important
5. Very Important

Satisfaction   Importance   Comments

☐   1. Location of your workspace   ☐   _____
☐   2. Arrangement of furniture   ☐   _____
☐   3. Amount of work surface   ☐   _____
☐   4. Function of furniture   ☐   _____
☐   5. Amount of storage for work materials   ☐   _____
☐   6. Function of storage   ☐   _____
☐   7. Display area for graphic materials   ☐   _____
☐   8. Style of furniture   ☐   _____
☐   9. Color of furniture   ☐   _____
☐ 10. Comfort of chair   ☐   _____
☐ 11. Degree of privacy   ☐   _____
☐ 12. Suitability to your work   ☐   _____
☐ 13. Opportunity to personalize workplace   ☐   _____
☐ 14. Image of workplace   ☐   _____
☐ 15. Overall satisfaction with workplace   ☐   _____

16. What do you like most about your current personal workplace?
_____

17. What do you like least about your currrent personal workplace?
_____

FIGURE 6.15   An example of a post-occupancy evaluation form. Adapted from *The Total Workplace* by Frank Becker.

Such forms can be specific to a building and its occupants, but these illustrate the questions and format of standardized POEs of the voluntary organization, American Society for Testing and Materials.

LIGHT—A UNIVERSAL NEED

For sighted individuals, nothing could be more universal than the need for light. And, while light for vision is easily understood, there is a growing body of research regarding the non-visual or photobiological effects of light, which impacts our health and wellness.

According to lighting specialist Eunice Noel, as we age, the quantity of light that we need increases for both vision and photobiological effects; at the same time, our access to daylight becomes greatly reduced owing to decreased mobility.

LIGHT AND SUPPORTIVE ENVIRONMENTS FOR OLDER ADULTS

In his 1992 keynote address to the American Hospital Association, Dr. C. Everett Koop stated that in order to control health care cost, the one thing we cannot have too much of is prevention. A "well lighted," supportive environment with access to daylight indoors and physical access to the outdoors are proactive approaches to "wheelchair avoidance" and must be viewed as preventive medicine and included in design criteria for universal design.

Falls are the leading cause of accidental death in the elderly population. Hip fractures alone are currently estimated to cost $10 billion a year in direct cost, resulting in death for one-third of the victims, with another one-third never regaining their prior level of activity. With the growth of the older adult population, the cost may increase to $240 billion in 2040. Environmental factors are linked to almost all falling accidents, with inadequate light at the top of the list.

LIGHT AND VISION

Vision: Decline in both visual acuity and visual performance is a common fact of life for older adults. Normal changes of the eye prior to age 65 typically are pre-retinal with retinal or cell loss occurring after that time. Age-related eye diseases further negatively impact the remaining visual abilities of older people (see Figures 6.16 through 6.19).

Recent research indicates that the loss of contrast sensitivity limits mobility and independence of older adults equal to arthritis. Arthritis is the #1 chronic condition affecting older people. Older people are also much more sensitive to glare than younger people. Older eyes adapt much more slowly to changes in light (brightness) levels. The adaptation from bright light to dim light takes longer than the adaptation of dim to bright. These changes continue throughout life with greater reductions in light reaching the retina, slower adaptation, loss of contrast sensitivity, and greater sensitivity to glare.

Environmental interventions that help minimize the effects of normal aging vision and maximize the remaining abilities of older adults include both appropriate lighting—quality and higher quantities (as defined below)—and value contrast between objects and their background. Since color perception is altered by the yellowing of the lens of the eye, value contrasts (white–gray–black) are more distinguishable than color contrasts. Value contrast enables one to see better in both low light and bright light conditions, whether indoors or outdoors, during the day or at night.

In addition to normal age-related vision changes, the eye diseases affecting older adults, that is, cataracts, glaucoma, and macular degeneration, limit or distort vision and require higher light levels and elimination of glare to enhance their visual function (Figures 6.16 through 6.19 of the same scene with normal and eye diseases).

The Illuminating Engineering Society of North America (IESNA), the recognized authority to establish lighting standards in the United States, Canada, and Mexico, has published a recommended lighting practice for older adults entitled *RP-28-98: Lighting and the Visual Environment*

FIGURE 6.16 Normal vision.

FIGURE 6.17   Sight through a cataract.

FIGURE 6.18   Scene as viewed by someone with glaucoma.

FIGURE 6.19   Loss of central vision with macular degeneration.

for Senior Living, which is now an ANSI standard. This document addresses both the quantity and quality of light and should be used as a reference guide for all spaces that serve older adults. Lighting in private homes should be specific to the residents. (These environmental interventions are illustrated in the case study of the San Diego Center for the Blind in Chapter 3.)

UNDERSTANDING QUALITY AND QUANTITY OF LIGHT

Before we talk about how much light is needed for vision, it is important to discuss what constitutes "quality" lighting, because, if the glare is too great, older individuals will close their eyes, so it won't matter how much light is available.

QUALITY

Glare—Enemy #1: Glare must be avoided. Glare can be *direct* from the light source (unshielded light bulbs and tubes, or daylight) or *reflected* when a bright directional light bounces off a light value or smooth reflective surface. Glare is classified as either *discomfort glare* or *disability glare*. *Discomfort glare* is defined as having discomfort, but not necessarily interfering with visual performance or visibility. *Disability glare* results in reduced visual performance and visibility, and is often accompanied by discomfort as well.

Uniformity in Ambient (General) Lighting: Older eyes adjust much more slowly to changes in brightness, especially adapting from bright daylight to low interior light levels. Within a space, the lighting should be uniform on walls and floors, keeping areas of interest (highlighting artwork, etc.) not dramatically different from the surrounding area. The IESNA *RP-28-98* recommends a ratio not greater than 3:1.

Balance and Diffuse Daylight: To avoid glare from daylight, brought about by the high contrast ratio between the intensity of daylight and dim interior electric lighting, it is important to balance the light by bringing daylight into the space from more than one direct, or high up on the wall or from skylights. If this is not possible, increase the ambient electric light to balance the light in the room. If you have visited the homes of older people, you might find the blinds closed completely, or open only 12″ above the sill. The problem is that the daylight is so bright in contrast to the room's ambient light; this extreme contrast is perceived as *glare*. To avoid experiencing the pain from the glare, they will close out the light. To eliminate the glare, daylight needs to be diffused within a space.

Flicker-Free Lighting: The magnetic ballast used in the "old-style" fluorescent fixtures gave a perceptible flicker, which could cause a range of problems from mild annoyance to headaches. The constant hum is also annoying.

Orientation of Light to the Visual Task: Light should be directed toward the visual task and not toward the eyes of the viewer: A common application of this concept is having the light come over your shoulder, shining onto a book when reading. Another example is when light is directed to a step, the light will call attention to the change of level, whereas if the light is aimed toward the eyes of the person, causing disability glare, he or she will be temporarily blinded, resulting in not seeing the step and falling down.

Color Rendition: The color rendering index (CRI) measures how true colors will appear under a given light source. The closer to 100, the truer colors appear. Daylight and incandescent lamps are rated at 100. Since color perception of older adults is altered by the yellowing of the lens, it is important to use a light source with a high CRI.

Color Temperature: The color temperature of a light source is identified in degrees Kelvin (K). Kelvin temperatures range from warm to cool with lower numbers representing warmer light and higher numbers for cooler light. For example, a candle flame is about 2000 K, 100-watt incandescent lamps are 2700 K, an average blend of sunlight and daylight is about 5000 K, and daylight alone is 7500 K.

Reduction of visual busyness: Since older people have difficulty processing multiple stimuli simultaneously, it is important to remove visual clutter. Stairs or level changes should be free of competing visual distractions.

QUANTITY

Challenge yourself and others to think in terms of "optimum" rather than "minimum" lighting levels. By including daylight in design solutions, we can allow ourselves to think in terms of the optimum solution. In the words of the late Meredith Morgan, OD, PhD, Professor and Dean Emeritus of the School of Optometry at the University of California, Berkeley, at the age of 83, "As long as glare can be avoided, it is almost impossible to have too much light." Because of the normal age-related changes, plus eye diseases, older adults require between 3 and 5 times more light than young people. When the photobiological effects of light are taken into consideration, requiring roughly 10 times more light to synchronize circadian rhythm than for vision, we must think beyond the use of electric lighting and include daylight. As a safeguard, the IESNA *RP-28-98* does list the *minimum* that light levels should be for older adults.

LIGHTING APPROACH

How can we increase light levels without creating glare? Provide the ambient (general) light in the space indirectly using both daylight and energy-efficient fluorescent luminaries (fixtures) and provide additional task lighting where it is needed. An indirect lighting system conceals the bright source of light (bulbs, tubes, or sunlight) and directs the light to the ceiling and walls; the light is then reflected and becomes diffused, providing general ambient light in the space, plus, lighting up the ceilings and walls to eliminate the "cave" effect.

DAYLIGHTING STRATEGIES

Lighting expert and architect, Doug Walters says one size or solution does not fit all! It is best to consult a daylighting designer or utility company in your area to determine the appropriate percentage of daylight openings for the space, that is, a combination of windows, clerestory windows, skylights, or roof monitors, and the appropriate glazing, direction, and location of these openings. Because of the intensity of sunlight and daylight, windows need diffusers. Sheer draperies or mesh shades diffuse the light and reduce the brightness. Shades are available in a variety of opacities, combinations of opacities, and control options.

ELECTRIC LIGHTING

The Energy Act of 1992 changed the course of fluorescent lighting in the United States from the "industry standard" 40-watt, T-12 lamp with magnetic ballast to the more energy-efficient fluorescent T-8 lamp, and now T-5 lamp, with electronic ballast. The "old-style" T-12 lamp used 40 watts of energy to generate 3050 lumens. The new T-8 lamps use 32 watts to generate 2900 lumens. In addition, these lamps are available in a choice of CRI (which determines how "true" the color will appear) and a choice of color temperatures, ranging from warmer (3000 K), midpoint (3500 K), and cooler (4100 K). The electronic ballast does not flicker or hum and uses less energy than the magnetic ballast.

COMMON INDIRECT LIGHTING SOLUTIONS

Pendant mounted luminaries direct the greatest percentage of light up to the ceiling; the light bounces off the ceiling, diffusing the light into the space below. The minimum distance allowed between the floor and the bottom of a luminaire differs state by state (7′-0″ and 7′-5″ are the most typical), so be sure to check with your local code officials.

Ceiling surface mounted luminaries distribute the light out onto the ceiling, using the ceiling as a reflector to further distribute the light to the walls and into the space.

Built-in cove lights direct light continuously to the ceiling or upper wall area.

Wall-mounted valances wash light down the wall and up onto the upper wall and ceiling.

Above-cabinet lighting in kitchens or bathrooms provides an ideal space to hide a strip fluorescent fixture that will provide indirect ambient light to the space.

Closet light is many times overlooked. A built-in closet with an opening above the header provides both high light levels inside the closet and ambient light for the room. Light with a high CRI is important for matching colors and checking for spots and soil.

COLORS AND VALUES MAKE A DIFFERENCE

An indirect system requires high light reflectance values (LRVs) on the ceilings' and walls' surfaces and medium reflectance on the floor. Dark colors absorb light, and light colors reflect light. The rating of the LRV gives a higher number for greater light reflectance, starting at 100. Most paint manufacturers list the LRV either in the index or on the paint chip. I recommend the following values: ceiling—LRV of 80 or above, walls—LRV of 60 or above, and floors—LRV of 30–40.

Too often, appropriate quality and quantity of light for older adults is sacrificed for decorative style, where the lighting fixtures are selected solely as decoration, rather than an important building system that affects the health and wellness of seniors. The following are the most common *inappropriate* solutions currently being used: (1) chandeliers with exposed light sources that produce glare and very little usable light, (2) wall sconce lights, without any other ambient lighting, and (3) recessed down lights in low ceiling areas that produce shadows and glare.

LIGHTING CONTROLS

All fixtures with the exception of day-use-only activity spaces should be able to be dimmed or configured for step-level switching to lower the overall lighting evenly during the evening and nighttime hours. A dimming ballast with a light sensor (photocell) makes it possible to reduce the electric light when daylight is present. This is termed *daylight harvesting* and is another way to save energy. Your local utility company may assist in the purchase of this equipment.

TASK LIGHTING

The definition of "task lighting" for activities of daily living extends beyond the typical reading activities associated with an office environment, to include shaving, applying makeup, matching colors of clothing, and so on. Well-shielded light sources that direct light to the visual task may be built-in, that is, under-cabinet lights or adjustable table and floor lamps. The design and the location of the controlling device (switch, dimmer) are important. A high percentage of older adults have arthritis in the fingers, making it difficult to reach around or turn a tiny little knob that is located in a hard-to-reach area, that is, up under the lampshade. For adjustable task lights, the control should be within easy reach, not on the cord. A lever switch located on the head of an adjustable task light or on the base of a table lamp is preferable.

OUTDOOR ENVIRONMENTS

The design of transition areas between the outdoors and building interiors needs to have special attention. As we have discussed previously, older eyes adjust more slowly to changes in brightness (light levels). Windows, clerestory windows, and skylights (roof monitors) allow the light inside to be balanced with the light outside, that is, bright during the day and dark at night. Electric light levels should be higher during the day and lower at night. Seating should be provided in two locations, immediately inside and outside the door, so that there is a place to sit safely out of the way of traffic, while eyes adjust to the change in brightness.

The principles of lighting the outdoors at nighttime are similar to the principles of lighting interior spaces, that is, avoid spotlights that direct light into the eyes of the viewer, diffuse the light, conceal the brightness of the light source, and direct light on the path or visual task. Avoid "over-lighting" an area, for example, service station canopies, owing to the slow adaptation of the eyes of older drivers. During the day, some shade should be provided over walkways and seating areas by vine-covered trellises, leafy tree canopies, or an umbrella to protect from the hot sun and provide soft diffused light.

Value contrast between objects and their backgrounds and between surfaces where a change of level occurs is very important in the outdoor environment because of the extreme brightness of sunlight and extreme darkness at night. Sidewalks or other walking surfaces should not be white or light in value, but rather a medium value. The extreme brightness of daylight or sunshine reflected off of a white/light surface bounces up into the eyes of the person, creating either discomfort or disability glare, rendering the person temporarily blind. Reflective vertical surfaces, that is, smooth white plastic fencing or even white patio furniture, may be problematic for older adults when bright daylight is present.

Case Study: The Effects of Workplace Design Features on Performance for Different Age Groups

In a recent research study, Virginia Kupritz found that aging office workers are changing the face of the workplace.

Today's multigenerational workforce has created new demands for workplace designs that support the physiological needs of workers for all ages. Meeting these demands is more important than ever because of the increasing age diversity in the workplace. We have four distinct generations working side by side in the workplace for the first time in history (Millennials, born 1981–2000; Generation Xers, born 1965–1980; Baby Boomers, born 1946–1964; and Traditionalists, born 1922–1945). While Traditionalists are being phased out of the workplace in the near future, traditional notions of retirement for the older set of Baby Boomers who are now 65 is being replaced by a trend toward longer periods of employment.

Health professionals have cautioned for some time that design changes may be necessary to support age-related changes that occur as we age (e.g., decreased hearing, eyesight, and stress tolerance and an increased risk of injury, such as musculoskeletal disorders). Studies are needed to determine which design changes will provide older workers with the same opportunity as their younger counterparts to perform efficiently in the workplace. The present case study compared workplace design features that different age groups perceived as supporting performance.

The research findings suggest that completely different workstations may not be necessary to accommodate workers as they age, but flexible workstation designs that offer varied solutions at different stages of aging may be the key to meeting the demands of our multigenerational workforce. The best scenario for today's workplace is one in which organizations provide diverse environments that accommodate worker flexibility and control. This means that workplace design features should support the workers' ability to manipulate their work environments across age groups, including the heating/ventilation/air conditioning, lighting, noise levels, and furniture and equipment within individual and group work areas. This flexibility enhances the workers' sense of control as well.

Lack of control can evoke stress. That is, people not only respond to threats, they are equally affected by the expectations of these events and what they have previously experienced. The adverse effects of job-related stress are well documented. Surrounding stressors (e.g., office noise) interacting with psychological variables, such as the perceived lack of control, can negatively affect employee health and well-being. In a broader context, the present study offers additional insight about a broad range of design features perceived as important to perform work. Organizations need this type of information to alert and direct them to where they should channel their finite resources and to evaluate alternative design solutions that support age diversity and facilitate work practices (Preiser et al. 2001).

> "Given the ever-increasing rate of technological advances, it seems that just about anything that can be imagined may very well be technologically possible. Among these advances, broadband and wireless technologies greatly enhance opportunities for employers to accommodate flexible work styles, schedules, and workplaces for workers with and without disabilities."
>
> Mueller, J. (2010). "Office and Workplace Design." In W.F.E. Preiser and K.H. Smith (Eds.), *Universal Design Handbook, Second Edition.* McGraw-Hill, New York.

COMMUNICATION AND PRODUCTIVITY

Planning strategies throughout the late 1960s led to radical departures from the norm in the physical layout of office interiors in order to optimize communication and productivity. Such departures are evident in the work of the German Quickborner Team, a design firm responsible for many innovative office space plans. It may well have been the absence of formal training and involvement with interior design and architectural professionals that made it possible to conceive such fresh proposals: "Let us play in a green field," they said, meaning a nonrestrictive open space with no set boundaries or limitations. Personnel would continue to be placed at workstations strictly grouped according to working needs for easy communication; rarely was space planning executed on a grid. Their planning schemes looked simple but were actually highly organized around a series of rules and concepts related to openness, lighting, communication, interaction, participation, and revision. The solutions they offered were arrived at by raising and answering these questions:

- Does the design facilitate personal communication at a variety of psychological and physical levels and through a variety of means and techniques?
- Are important patterns of daily communication reinforced by the spatial layout of the office and design of the objects, furnishings, and equipment within the work environment?
- Is there space available for confidential conversation? For public interaction?

A more recent trend in office planning is seen in "shared offices"—a space allocation practice in which single individuals are not assigned a desk, workstation, or office for their exclusive use. Shared offices can reduce costs, improve flexibility, and maintain effectiveness for people who often are out of the office. This is a concept adopted by the Shimuzu Institute of Technology's Planning Department and the San Francisco offices of Anderson Consulting. Typically, shared offices include an initial investment of portable phones and laptop computers, but they often pay for themselves within the first year from space savings. This is an example of universal design wherein the same size "footprint" (a plan for an individual workstation) is capable of housing different furniture and equipment components depending on specific job functions and work styles. This plan may include flexible furnishing abilities.

THE UNIVERSAL DESIGNED SMART HOME OFFICE

Technology has created many opportunities to work from a home office or new flexible combinations. There is little information available to home office users, but in his case study, "The Universal Designed Smart Home Office," architect Charles Schwab addresses the problem of planning for the home.

Now, more than ever, owing to information technology and rising fuel costs, more Americans than ever are moving their offices into their homes. According to the International Data Corp., in 1999, 49 million Americans worked from their homes at least part time, a figure that grows by 8% each year. Because of recent gasoline price increases and increasing improvements in technology, that figure is also on the rise.

It is my opinion, and I often hear the same from clients, that for the person with special needs, the home office is the third most important room in the house besides the kitchen and bathroom. The home office has obvious benefits for those with special needs. Internet technology has leveled the playing field for everyone. When working from home, it is no longer necessary to deal with physical barriers in the public built environment, and who wants to spend the money for increasing fuel costs and lost time in traffic anyway? While working from home, let's face it, you can also be free from some people's prejudices against people who use wheelchairs and other people with special needs.

So allow me to share some of the design features that I include in my *Universal Designed "Smart" Homes for the 21st Century* (Home plan book available from *Special Living Magazine* or www.universaldesignonline.com/).

The first thing you need to do, even before beginning to plan or build, is to check with local municipal zoning officials. With such considerations such as traffic and taxes, there may be limitations as to what you can do. Some cities may not even allow you to work from home. Most cities require a business license to practice within its city limits. Remember that a portion of your home office may also be tax deductible if used for genuine business, excluding hobbies, and so on. That is another topic altogether. See my article in *Special Living Magazine*, Spring 2004, and consult your accountant.

The location of the home office needs to be in a portion of the home that is free from distractions, has good lighting, and is accessible to everyone. It will be the environment where you will need to be productive, efficient, and professional. It is for this reason that I always locate the home office at the front of the house.

The (rarely used) 20th Century Living Room/19th Century Parlor has been replaced with the 21st Century Universal Designed "Smart" Home Office. I recommend the home office have a separate entry with a sidelight, a covered porch, and a package shelf. Many of our designs also feature a thru-wall mail and package drop detail, eliminating the need to go outside to get mail. All have zero-step entries with a maximum 1/2" threshold. We specify sound batt insulation with resilient clips in the wall between the office and the rest of the home to insure privacy.

Most home designs feature a fully accessible powder room or full bath immediately accessible or adjacent the office. This will meet all your needs without going into the private portion of your home, enhancing professionalism and privacy. Some of the home office designs that feature upper or lower levels also have elevators with double doors (one from the private side, one from the office side). This allows your office to extend to the lower (or upper) level for a conference room or additional storage space.

A 10 × 10 sq. ft. room can be large enough for a one-person office, excluding a visitor's area. A 150 sq. ft. room may be large enough for a two-person office. The home office should be designed with a 10 ft., 12 ft., or 14 ft. length room for maximum space efficiency. This will vary depending on your storage and individual business needs. Remember to consider whether or not you will meet with customers in your home office.

A separate "comfort zone" such as a couch, reclining chair, and table can also be a nice break for you if you want to change your visual point of view or posture, kick back, and do a little research or break time. Often the home office also serves as a guest room. The old-fashioned Murphy bed may be a smart alternative to a futon or hide-a-bed. There are mechanically operated Murphy beds now available and they have shelf space while in the upright position.

Speaking of the "comfort zone," do not underestimate the power of bringing the outdoors inside or being accessible to your home office. A reasonable amount of natural light is necessary for maximum productivity and encourages positive and energetic thinking. It is also proven to be motivational. It is for this reason that Wal-Mart now includes skylights in most new stores. If you do "daylighting" research, you will find interesting studies with schoolchildren that show increased learning and test scores in classrooms with increased natural light. Store sales also improve in outlets that have abundant natural light.

If you have the luxury, consider including a courtyard garden as an extension of the home office. Landscaping, a water fountain, or just a spot in the sunshine can be a nice stress reliever and provide a nice productive break from the telephones. Some of your best ideas will come to you when you are in a more relaxed state of mind.

Make sure that your desktop workspace faces opposite the windows to avoid glare. Varying height counters will make your office work for people of all heights and abilities. There are several adjustable countertop mechanical systems on the market that raise and lower per your needs (see resources listed below). We also provide details that allow for less expensive manual adjustments to the custom workstation.

Cabinets should include "pull-down" and "pop-up" hardware so that higher and lower shelves may be reachable by everyone. I specify quality hardware products made by Hafele of the Americas (see Resource Section). The same company now provides Accuride full extension hardware for drawers. This is extremely useful in offices with limited storage space and for serving people with limited mobility.

Adequate telephone lines and data lines are a necessity. Consider both your current and future needs. In a new home, I recommend category 5 wiring throughout. This will allow you to have outlets that serve existing and future electrical devices and systems everywhere in the home.

Provide a dedicated separate 15–20 amps of service. A separate ground is another precaution that will isolate your office equipment from other home appliances. It is nonetheless essential to have a battery-operated full-time power surge protector for all of your computer equipment.

According to the American Optometric Association, glare is one of the leading causes of eye-strain, especially if you work on a computer for more than two hours at a time. Fortunately, there are some great products on the market that can help you burn the midnight oil and cut your energy costs at the same time.

I mentioned natural lighting as being very important to your productivity and general health. In addition to natural light, there are generally two types of indoor lighting known as ambient and task lighting.

Ambient lighting typically comes from a ceiling fixture that evenly distributes light throughout an entire room. I recommend 4 ft. T-8 lamps with electric ballasts in the home office. A good color choice would be 3500 K. The long life, great color qualities, and high efficiency make these a logical choice. Also, the large size of these fluorescents provides a large radiant surface that helps minimize shadows. For ambient lighting, we sometimes specify surface mounted fixtures, indirect pendants, wall-mounted shelves, and sconces, which can add a sophisticated touch. Don't forget wall washers on your favorite artwork.

Task lighting is located on or near your work surface, and switched on to illuminate specific areas. If there's little room on your desk, consider a space-saving model like 3M Ergonomics glare-cutting polarizing task light. The lamp employs a special filter technology to provide ample lighting while reducing glare levels and hot spots on your work surface. Do not forget under-cabinet and shelving lighting, also a task lighting source. Most of you are probably aware of the new compact fluorescent lamps that use only about one-quarter of the energy of incandescent bulbs and last 10 times as long.

Last but absolutely not least is the important issue of you and your own home office safety. Keep the wires out of the way and make sure you do not have too many things plugged into the same socket. For this reason, I like to design outlets every 6′-0″ on center (one-half the distance required per code). Always remember fire safety. If you do not have a separate door into the office from the outside, make sure at least one of the office windows is an egress window. If you use a wheelchair, make sure the sill is no higher than 30″ above the floor for easy escape in an emergency. This is (often 12″) lower than what is required by codes. Verify that your office also has a smoke detector hardwired in series with the rest of the home. Be sure to make regular backups of your computer data and keep that in a separate location outside the home, maybe in a safety deposit box. Keep very important documents in one place so that you can quickly grab them if you need to get out in a hurry. Look into home office insurance. This may be included as a rider to your homeowners insurance. Do not assume your homeowners insurance will cover your office equipment and other materials.

Also remember office interior finish materials that encourage and improve indoor air quality (see the article I wrote for *Special Living Magazine*, Summer 2005). Interior finish materials with low or no volatile organic compounds and formaldehyde are just as important as proper lighting and ergonomics to avoid sick building syndrome, thus keeping you feeling well and productive.

There is much more we could discuss about your home office. Most of that will be special and unique to your own needs. I hope this has been helpful. I wish you all the best and many productive years in your Universal Designed "Smart" Home Office. Thanks for asking.

REFERENCES

ANSI/HFS 100-1988 (1988). *American national standard for human factors engineering of visual terminal workstations.* Santa Monica, CA: The Human Factors Society.

Becker, F. D. (1990). *The total workplace.* New York: Van Nostrand Reinhold.

Hedge, A., Sims, W., & Becker, F. (1989, October 1). Lighting the computerized office: a comparative study of parabolic and lensed-indirect office lighting systems. *Proceedings of the Human Factors and Ergonomics Society Annual Meeting, 33*(8), 521–525.

Herman Miller, Inc. (1995). *Designing for accessibility: Beyond the ADA applications guide and video.* Zeeland, MI: Herman Miller, Inc.

Kupritz, V. W. (2001, March 8). The role of the physical environment in maximizing opportunities for the aging workforce. *Journal of Industrial Teacher Education, 37*(2), 66–88.

Preiser, W. F., Smith, K. H., & Mueller, J. L. (2001). Office and work place design. *Universal design handbook* (2nd. ed., pp. 23–39). New York: McGraw-Hill.

Schwab, C. M. (2009, July 1). America's first universal design home. *Exceptional Parent, 39*(7), 24–28.

Steelcase, Inc. (2009). The movement toward wellness in the workplace. *Steelcase ergonomics.* Retrieved June 11, 2013, from web.steelcase.com/wellness/index.html#/physical/.

Design in Public and Commercial Environments

CONTENTS

Public and commercial environments have been heavily affected by the passage of the Americans with Disabilities Act (ADA). As noted in Chapter 2, virtually every place of business and every public building must be made accessible to individuals with disabilities. For example, the entrance to the Cooper-Hewitt Art Museum is accessed by beautiful stairs (Figure 7.26) and an alternative entrance ramp (Figure 7.27). The cost for meeting this requirement can often seem daunting, especially for places with limited resources. However, as pointed out throughout this book, there are methods for making accommodations that are not unduly expensive, and the returns on investment are high in terms of increased service to a neglected and changing population.

The case studies and strategies offered in this chapter cover a wide spectrum of public and commercial environments. Each is an example of how research can be applied to ADA compliance and universal design through comprehensive team activity. Anne Seltz discusses the need for increased focus on the acoustic environment and includes numerous examples of accommodation from her years of experience as an audiologist interested in the designed world. This is the theme of Kerry Nelson's theoretical essay "Defining and Designing Public Spaces." Henry Sanoff's case study of a child development facility uses a team approach in a very complex environment and shows the importance of end-user feedback throughout the design process. Another child care center is described in the case study about CHAMP House in Japan. Bradley Knopp and Joseph Maxwell describe a universal design approach taken in the creation of an air terminal, including wayfinding mechanisms and the redesign of seating elements to better serve the entire population of air travelers. This is followed by observations on the success of public bus design in Kumamoto, Japan. Environments that benefit everyone precede studies on spaces specifically designed for those with impairments. Dr. Janetta McCoy's design of a summer camp for children with disabilities serves as an introduction to the last part of Chapter 7. It is followed by the case study on signage and wayfinding by Sharon Toji. The last part of the chapter features health care environments including adaptations required for the growing obese population, retirement facilities, healing gardens, and other specialized universally designed environments for aging. Design decisions for health care facilities have been influenced by a growing number of evidence-based research projects.

Other considerations in the design process for public and commercial environments include the obesity crisis facing us today. We have already seen one technique for addressing this in Chapter 6 ("Universal Design in the Office") where many manufacturers developed a line of bariatric chairs to accommodate obesity in the workforce (see Figures 6.8 and 6.9).

DESIGN FOR ACOUSTIC ENVIRONMENTS: LET THE WORD BE HEARD

by Audiologist, Anne Seltz

> In those days the world teemed; the people multiplied,
> the world bellowed like a wild bull, and the great god
> was aroused by the clamour. Enlil heard the clamour
> and he said to the gods in council, "The uproar of mankind
> is intolerable and sleep is no longer possible by reason of the babel."
> So the gods agreed to exterminate mankind.

From "The Epic of Gilgamesh"
Translated by N.K. Sandars
(Penguin Press, 1972)*

Perhaps there are less dramatic solutions to our noisy world.

"Speech produced in one place in a room should be clear and intelligible everywhere in the room" (Nabelek and Nabelek, 1985).

Having practiced audiology for over 40 years, I've learned from my students, clients, and patients, as well as from personal experience, about the great need for good acoustics. A quieter world would enable all of us to communicate more accurately and comfortably. But is that what our society wants? Has our world been noisy for so long we don't know the value of well-managed acoustic environments?

My greater metropolitan area developed a downtown high school in the late 1990s to bring students into the inner city. Interaction with the lively arts and business community was the goal. The building received prestigious architectural awards. During its first year, students withdrew because of the bad acoustics. As one teacher said: "Teaching here is like working with a vacuum cleaner running in your room." The open architecture and untreated surfaces created unusable teaching spaces: retrofitting cost hundreds of thousands of dollars. Why did neither the clients nor the designers consider acoustics for good teaching, listening, and learning?

An architect friend was working on a school design. Because of my strong interest in Classroom Acoustics, I urged him to create good acoustics in this new school. "I don't have control over that." he said. "This school superintendent is in charge and his main interests are a large gymnasium and attractive landscaping." Could my friend have said something to change that superintendent's mind?

UNIVERSAL DESIGN STRATEGY FOR ACOUSTICS

Whose responsibility is it to advocate for good acoustics in our built spaces? If the client doesn't ask for good acoustics, should you encourage it anyway? How do you engage a client in conversation about acoustics during the planning stage? In other words, how can you teach your clients to value acoustics, to request good acoustics, to be willing to pay for it, and then be able to recognize good acoustics when they occupy the space? The challenge to the design/building professions is how to advocate for something that is invisible.

Though acoustics is invisible, we all experience acoustics every moment of our days and nights. We may not always be conscious or aware of good acoustics in a space, but each of us can remember poor acoustic environments that caused us to

- Miss information
- Turn off a program
- Be fatigued from difficult listening
- Have trouble understanding a person
- Leave a session because it was too much work to stay
- Experience a poor telephone connection made worse by a noisy airport

---

* That quotation came from *QUIETING: A PRACTICAL GUIDE TO NOISE CONTROL*, published by the US Government, Department of Commerce in 1976. This booklet, still available, is a classic primer on practical acoustics, reads well, and, despite its age, is of great value. It can be purchased for about $15.00 from the web site www.soundproofing.org.

- Stop talking in a noisy car because listening was so unpleasant and unsuccessful
- Be annoyed by hearing transmitted sounds from above, below, or next door in an apartment or hotel
- Watch as a relative with poor hearing laughs at the wrong time because he couldn't understand the jokes

Add your own experiences; it's important that you do. When you relate your experiences to clients, you make them aware of acoustics as an important priority, and they will start remembering their own difficult listening experiences. Personalize listening experiences, remember them, talk about them. Then you and your clients will be ready for the acoustical consultant.

Respect good acoustics. Value good communication in all built spaces: both indoors and outdoors. Create opportunities for your clients to value good acoustics. Include acoustics in your designs. Your clients will benefit and that's good business.

As an introduction to the general topic of defining and designing public places, the universal design strategy for the design of the acoustic environment has been included. Like good universal design, good acoustic design is invisible. Because it is invisible, it is also one of the first things cut in budget-crunching efforts. Again, like good universal design, good acoustic design needs to permeate all design decisions.

UNIVERSAL DESIGN STRATEGY

DEFINING AND DESIGNING PUBLIC PLACES

When one initially thinks about the term *public place*, parks and plazas, post offices and courtrooms, and libraries and schools may come to mind. But public places, as defined in the Americans with Disabilities Act (ADA), also encompass entities of public accommodation such as social service agencies, retail stores, banks, professional businesses, hotels, motels, restaurants, and theaters. All have one thing in common—service to the public—yet each functions differently, depending on the type and means of service provided.

Additionally, each public space possesses macro, middle, and micro environments—separate scales that interlock to form the whole. The *macro* environment includes the largest scale considerations, such as site and building access, the architectural envelope, horizontal and vertical circulation, and spaces for the general public's use. The *micro* environment involves the smallest scale considerations, private or individual spaces, and also specific areas such as storage rooms, restrooms, an employee workstation, or a barrier-free parking space. The middle environment, as the name implies, falls in between the largest pubic and smallest private spaces. It might include secondary access corridors, an employees' lounge, an interactive holiday display in a retail store, or a private conference facility within a hotel.

The boundaries among these three sub-environments are difficult to define, since they often overlap. However, an understanding of the interdependency of the three is fundamental to the design of public places. The success of a design project is determined by the cohesive interaction of all of its parts. A failure of the design within any of these sub-environments can result in an unsuccessful public place. The macro, middle, and micro environments must combine to form the entity and must unify to support its intended purpose of public accommodation, as any design's content must support its intent.

THE DESIGN PROCESS FOR THE BUILT ENVIRONMENT

Any design process, including one that strives for a more universal design, can be divided into the following three states of activity:

- Defining and analyzing the problem
- Solving the problem and communicating the solution
- Actualizing and evaluating the solution

Each of these stages can be further subdivided into phases, which, when followed sequentially with intuition and infused creativity, should result in a valid and appropriate design:

Define Problem	Design statement: who, what, where?
	Research and gather information
	Write up/diagram the program
Solve Problem	Conceptualization
	Schematic design
	Design development
Actualization Solution	Documentation
	Implementation
	Post-occupancy evaluation (POE)

The universal design process outlined in Chapter 3 shows that of these phases, attention must be devoted to improved definition of the problem through thorough research and user input. The process also requires actualizing a universal solution through the use of POEs. Finally, the entire process can be facilitated by using a team approach.

THE TEAM APPROACH TO UNIVERSAL DESIGN

Interior Designer, Kerry Nelson, introduced the Section on commercial design by describing the team approach needed for a project on the scale of a shopping center. Typically, the creative design of a commercial or public place project will involve many specialists: engineers, architects, landscape architects, interior designers, consultants, and contractors. Each specialist represents a discipline that often overlaps in scope with another. The territorial boundaries of specializations can often be hazy, resulting in the creation of incompatible solutions or areas of concern left unattended. Errors and gaps allow for design failures to occur in the interactive flow of the macro, middle, and micro environments. To maintain fluidity among the sub-environments, it is important that, in the design of public places, a team approach be employed with a high level of communication and coordination among the players.

To illustrate the importance of a cooperative effort, consider an icon of American culture: the shopping center or mall. In the broadest sense of the macro environment, potential users must be alerted to its existence. Advertising agencies and graphic designers might be retained to mount campaigns and create signage to lure and alert customers to the site. Site and building access is required for both vehicular and pedestrian traffic to get customers safely inside, which falls within the purview of the urban planners, engineers, architects, and environmental and landscape designers. Within the building, users must be provided with all the stimuli possible to promote them to circulate, shop, and spend both time and money; this becomes the creative responsibilities of engineers, architects, lighting and interior designers, visual display artists, and an array of other consultants.

The middle environment in this case might be considered the department and specialty stores, food courts, public areas, public toilet facilities, and telephone and water fountain areas. Kiosks that sell everything from jewelry to cell phones have added to the flexibility of shopping centers. They can easily change function and locations. It might also be considered the circulation within the selling floor of a retail shop and the cash/wrap counter where customers make their purchases. Hence, the middle environment might be within the realm of responsibilities for just about everyone on the project, from engineers to store owners, architects to salespeople, and interior designers to subcontractors. Similarly, the micro environment could encompass the contents of atrium planters, dressing rooms, managers' offices, individual toilet stalls, or the other side of the above-mentioned cash/wrap counter from the salesperson's perspective, and again could be influenced by any of the disciplines.

If a cooperative effort among design team members is employed, it is highly likely that a good design, providing smooth transitions from the macro through to the micro environment, will be created.

Case Study: Redesigning a Child Development Center

The successful design of a complex public space such as a child development facility requires the integration of research and programming strategies. In the following case study, Professor Sanoff described a collaborative design process that combined the following components: design research, design participation, and design development (Sanoff 1994).

This case study describes how research findings can be integrated into a design process. The techniques described were based on the results of personal experiences in designing and programming child development facilities. Earlier experience suggested the need for a new approach that could engage the architect and the client/user group in a process that linked children's developmental needs to facility requirements. Strategies are described that engaged parents, teachers, and administrators in collaboration with the architect during the initial stages of design. This process has produced teachers with new capabilities in playroom planning and organization, as well as an understanding of the way in which architects make decisions. Although the example described is a campus child development facility, the techniques can be generalized for other types of facilities as well. The concept of the nonpaying client is integral to the process. For programming purposes, people who use the building are the clients of the architect, whether or not they pay for services. Reference to the user as the nonpaying client then attaches greater significance to the importance of user contributions and to a more binding relationship between the paying and nonpaying client.

Programming and design consultation were requested by the planning group of a proposed 75-child facility and training center for Wake Community College's Child Development Program in North Carolina. They contacted Henry Sanoff, Director, Community Development Group, North Carolina State University, for design assistance. Since this facility was intended as a demonstration site for the country, the department head and client representative, the teaching staff, and the educational consultants to the program were anxious to follow a planning process in which research findings, their expertise, and education philosophy would be linked to design decisions.

THE DESIGN PROCESS

Three major components of the collaborative design process are *design research, design participation*, and *design development*. These components precede production, construction, and evaluation. In this model, programming represents the synthesis between design research and design participation. This collaborative design process is a departure from traditional programming approaches since the client, the nonpaying client, and the architect are directly involved in all decision-making stages. Furthermore, the stages described as design research and design participation subsume what is normally referred to as facility programming. This distinction enables the identification of discrete activities for each stage and clarifies the difference between information received from secondary sources, such as surveys and databases, and from primary sources, such as direct, face-to-face involvement.

Typically, institutional client groups planning a child development center initiate a formal needs assessment that includes the following steps:

1. Campus survey of student childcare needs
2. Surveys of campus childcare center
3. Site visits to childcare facilities
4. Consultation with childcare experts
5. Department planning

The above steps constitute the *design research* phase of the collaborative design process. The design research phase included a needs assessment, visits to other child development centers, and the establishment of educational goals, which included desired staff–child ratios and other factors inherent in a high-quality center. Although typically initiated by the client, the architect can often provide guidelines for more systematic fact-finding procedures. Surveys and visits to existing facilities, if properly organized, can reveal valuable insights into their functions, since casual visits frequently reveal only obvious results.

DESIGN PARTICIPATION

During this stage of the process, background research findings are integrated into the activity analysis. Accompanying the area requirement for usable activity space for each child is the need for well-defined areas limited to one learning activity, with clear boundaries from circulation space and from other activity areas. Well-defined activity areas or center may be created with surrounding partitions, storage cabinets, changes in floor levels and surface materials, or other visual elements that suggest boundaries. Spatially well-defined areas support social interaction, cooperative behavior, and exploratory behavior, and they also prevent ongoing play from being disrupted by intruders. Running and chasing activities are common in classrooms where boundaries are not well defined. Conversely, well-defined activity centers, with clear boundaries from circulation space and from other activity areas, and with some visual or acoustic separation, decrease classroom interruptions and contribute to an increase in attention spans of the children. This implies that activity centers within the classroom require the development of a building program that can spatially respond to the developmental goals of the teachers of young children, as well as to goals identified in child development literature.

MODELING THE PLAYROOM

Since the playroom is the basic spatial unit of a children's center, prior familiarity with its organization can enable teachers to enter into a productive dialogue with the architect. Playroom modeling is an activity developed for a teacher's workshop that allows participants to manipulate fixed and movable playroom elements in order to achieve the desired developmental objectives (see Figure 7.1) (Sanoff 1979). Working in teams of three people each, teachers were asked to design a playroom for a specific age group, such as infants or toddlers. Materials including cardboard, wood blocks, Styrofoam, construction paper, and

Kitchen   Locker/wrapping   Entry, reception lobby, locker   Reading

Toilet diaper changing   Eating   Blocks dramatic play

Art sand water play manipulative   Climbing indoor active   Protected outdoor   Sleeping music

FIGURE 7.1   Block tool icons used for spatial arrangement planning; participants can easily move blocks, or areas, as they seek to optimize the use of their space. (Courtesy of Henry Sanoff, from his book *Design Games.*)

plastic were provided along with instructions to the teachers for measuring and cutting the materials needed to construct a three-dimensional model.

The model making was preceded by an exercise where developmental objectives and corresponding activity areas for specific age groups were discussed and agreed upon by each team. Model results were discussed by participants; playrooms were then joined together to resemble a building for different age groups. At this juncture, issues of playroom adjacencies, building flow, and locations of services were discussed by participants in an exercise that lasted for 4 hours.

Playroom modeling is an effective method for preparing the client group to actively and constructively participate in planning a child development center.

RECORDING ACTIVITY DATA

Planning the center began with focusing on the child as the basic unit of development. Next, the design participation phase involved the collection of behavioral data relating to each activity in which infants, toddlers, and preschoolers would be engaged. The activity center was the conceptual framework used for the design of the facility. The teaching staff of the child development training program identified the developmental objectives for each activity center by age group and the specific (or "molecular") activities that would occur in the activity center (see Figure 7.2 for one example).

The water play area, for example—the objectives of which were sensory and perceptual acuity, concept formation, and eye–hand coordination—would include such molecular activities as pouring, measuring, mixing, and floating objects, all of which are related to the primary activity. Activity data sheets were prepared to record the relevant activity

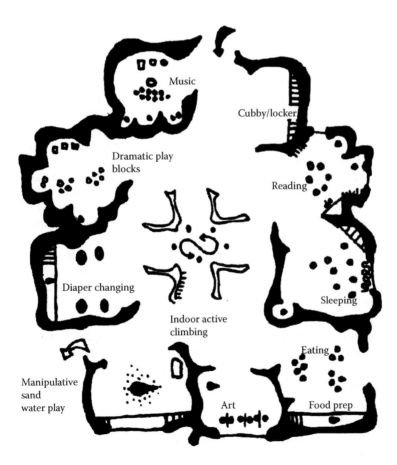

FIGURE 7.2   A teacher's arrangement of a toddler playroom developed by teachers. (Courtesy of Henry Sanoff, from his book *Design Games*.)

information that served as a program and resource for future decisions. The data sheets provided a format where specific equipment needs could also be identified for future purchasing.

SPATIAL PLANNING

Since the planning of a child development facility also reflects a particular ideology about child development, a space planning exercise was developed to engage the teaching staff in decisions related to playroom layout. Since a planning guide of 50 square feet of usable space per child limited the number of activity centers that could be included in a playroom, scenarios were written by teachers about a typical child's day. The constraints encouraged the teaching staff to use trade-offs effectively since they were required to decide which activity centers were most important for various age groups. Graphic symbols corresponding to each activity center (as shown in Figure 7.1) enabled the manipulation of children's movement patterns in the playroom and was the first step in providing environmental information to foster the development of mental images. Spatially organizing activity centers on a "game board" corresponding to a playroom permitted the determination of which centers were to be fixed and which were flexible. The spatial layout process required teachers to consider planning concepts, adjacency requirements, circulation, and visual and acoustic privacy between activity centers. Most of all, the process reinforced the concept of activity centers.

The teachers worked through a playroom layout by manipulating activity symbols for each age group. They outlined the flow process from entering the facility, greeting the staff, removing coats in the cubby area, and moving to various activity centers. When planning the infant room, the teachers identified the diaper change area as the focal point, with surveillance to all other activity areas. To avoid the clustering of unsightly cribs, the teachers proposed decentralizing the sleeping activity into several crib alcoves. This process entailed small group discussions that required consensus in all decisions. When agreement was reached, the symbols were fastened to the base to constitute a record of the group's decisions.

Cardboard scale models of each playroom, with movable walls and furniture, were then constructed by the designer that corresponded with the flow patterns in the diagrams developed by the teachers. This stage of the process permitted the teaching staff to visualize the three-dimensional implications of their decisions. Schematic models of the playrooms limited the amount of information presented at one time, conveying only the most significant issues in order to minimize information overload. Teachers could reconsider earlier decisions, particularly when they saw conflicts arise that were not easily predicted in the two-dimensional diagrams. Although circulation between activity centers was considered in the development of the activity symbol diagrams, the scale model conveyed the need to establish clear boundaries between centers to prevent distraction, while permitting the teacher an unobstructed view of all children's play areas. The scale models included information not shown on the activity diagrams, such as furniture and equipment, but the movable pieces were easily manipulated by the teachers as they referred to the activity data sheets.

When the teachers reached agreement about the best playroom arrangement form, diagrams were developed by the designer, elaborating on their spatial decisions (see Figure 7.2). These diagrams combined activity centers into playrooms for different age groups. The diagrams allowed teachers to gain an understanding of "conceptual relationships." Teachers were thus better able to visualize how education objectives could be enhanced through the design of playrooms.

DESIGN CRITERIA

The results of the participatory exercises helped to generate design criteria and modify the requirements of the building program. Several statements were developed to describe the fundamental environmental characteristics of an effective child development center.

The environment must be comfortable and inviting for children and adults. It should reflect an atmosphere conducive to children's growth. Materials and equipment should be easily accessible to children in order to encourage independence and self-esteem. An effective means of organizing the environment is to develop interest centers where the playroom is divided into areas that focus on specific activities. It is advisable that quieter activity areas be placed in proximity in order to promote a quiet atmosphere. Activity areas demand visual clarity and well-defined limits if children are expected to interpret cues on appropriate areas for certain types of play. A quality playroom would include the following activity areas:

- Creative expression/art
- Literature/language art
- Dramatic play/housekeeping
- Block building
- Self-image, personal hygiene
- Science and exploration
- Cooking
- Water play
- Carpentry
- Manipulative play
- Music and movement

More specific guidelines that influenced the final solution included the following:

- Protected outdoor play area adjacent to each playroom
- Southern orientation for playroom and adjacent outdoor area

FACILITY DESIGN

The teaching staff was involved in organizing the building components into a facility design using graphic symbols that corresponded to the major areas, such as playrooms, kitchen, offices, corridors, and lobby area. Age group adjacencies were considered, with opportunities for different age groups to have visual contact with each other. This was achieved in many ways, including low windows in each playroom for children to look into an adjacent room. The planning concept that emerged from the discussions was that of a "central spine" from which playrooms would be connected. The spine would be more than a corridor, yet similar to a street, where parents, teachers, and visitors could look into the playrooms and observe children's activities (see Figures 7.3 and 7.4). To emphasize the street concept, the area was filled with daylight through the use of overhead skylights. Each of the playrooms would have a central spine leading to a covered outdoor play area. Spatially well-defined activity centers were located on either side of the playroom spine. These playrooms included fixed areas for art and water play and centers that could change their focus at the discretion of the teacher. The phrase "spatially well-defined centers" implies the need to be distinctly different from adjacent centers. This differentiation was characterized by physical features such as partially surrounding dividers or storage units, implied boundaries through the use of columns, changes in floor level or ceiling height, changes in floor covering, and changes in light levels. Learning materials, furniture, and equipment also contributed to the distinctiveness of the activity centers.

TEACHERS' RESPONSE TO THE PROCESS

The diagrams and scale models provided a clear sequential procedure to which all decisions could be traced and subsequently modified. The teachers, however, had some difficulty comprehending the consequences of many spatial decisions. While they were able to follow the process of playroom organization, they had difficulty visualizing the implications of alternative playroom arrangements; continual reference to scale models and

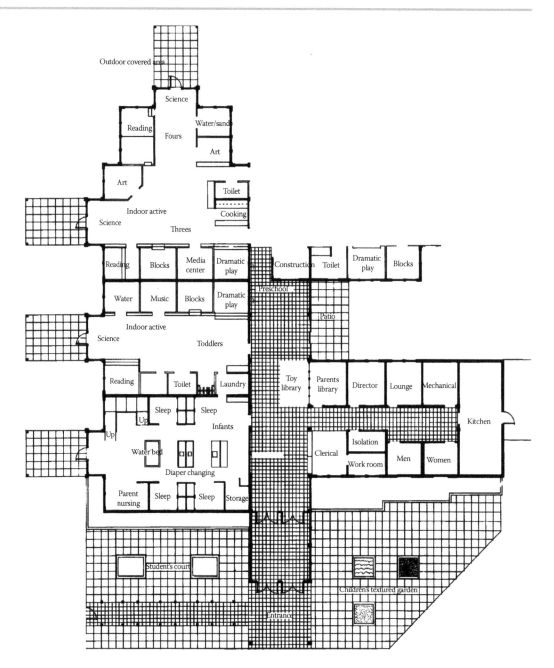

FIGURE 7.3 Interior view of "street" connecting all playrooms of the children's center, Wake Community College's Child Development Program in North Carolina.

perspective drawings helped the teachers substantially in contributing their expertise to the design of the building. Teachers remarked that this process provided them with a better understanding of the principles of spatial planning and the role of the architect. They experienced the "ripple effect," where minor changes in adjacency relationships manifested themselves into major revisions in the spatial layout of the playroom, or the building. This diagnostic procedure of examining flow processes and linking objectives to activity centers enabled teachers to develop a conceptual understanding of playroom and building, layout principles.

CONCLUSION

The interaction between the teachers and the designer described in this project was clearly a departure from the traditional approach to facility development. Conventional practice

FIGURE 7.4 Interior view of "street" connecting all playrooms of the children's center, Wake Community College's Child Development Program in North Carolina.

usually denies the expertise of the users (the nonpaying clients) and their involvement in design decision-making. Traditional designers also focus on the formal and visual issues and give less attention to the behavioral factors that may equally influence the form of the building. Facility designers typically consider defining relationships between playrooms and other areas, disadvantaging the teaching staff because of their inability to comprehend floor plans and the consequences of spatial decisions. The teachers' expertise occurs at the level of the children's behavioral interactions within the playroom, a factor that is usually considered only after occupancy of the facility.

A structured process was provided to enable child development professionals to lend their expertise to the initial programming stages of the design process. Use of activity data sheets, activity symbols, and form diagrams permitted the designer to integrate knowledge about children's behavior and their requirements into a format that was conducive to making space planning decisions. Integrating the expertise of the staff in this structured process established clear linkages between child development goals and the types of places where these goals could be fulfilled. The teaching staff's continual involvement in the building design process encouraged the exchange of ideas and concepts with the designer, which increased the staff's ability to act as effective design team members. The active part of the process usually terminates with the schematic design of the children's center, which has resulted from the team's involvement.

The effectiveness of a collaborative process is contingent upon the involvement of the architect from the inception of the project. When the architect is an integral part of the process, the building design proposals are clearly understood by the user/client group of teaching staff, parents, and administrators. On those occasions when the program is completed prior to the architect being commissioned for the project, significant communication

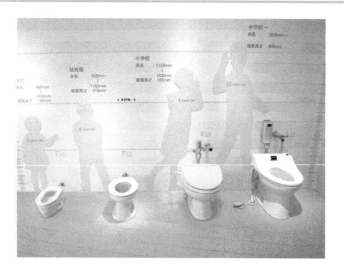

FIGURE 7.5 Japanese showroom display of a variety of toilet sizes

problems can occur between the user group and the architect. In this instance, the architect of record was appointed by the college administration after the program and preliminary design had been completed by the consulting design team. Although considerable effort was made by the design team and teaching staff to explain the rationale for the design decisions, the architect could not easily comprehend the nuances of the proposed design solution. Similarly, the architects' drawings were not understood by the teaching staff, since they were prepared for construction purposes. This created difficulty in the working relations with the client because the architect often urged quick approval to expedite the process.

The language of the program should reflect the concepts developed by the teaching staff and conveyed in terms of education goals and children's activities. The language of the architect—the floor plans and elevations—is the interpretation of verbal concepts and is often unintelligible to the user group, especially if it is not developed simultaneously with the program. The implications of these experiences are that ownership in the design process, achieved through active involvement in design decisions, permits the user/nonpaying client to exercise free and informed choice. The separation of the programming and design stages not only limits participation of a wide range of experts but also jeopardizes the ability of the product to fulfill the expectations of the program.

CHILD CARE ENVIRONMENTS

Facilities for children at various growth stages are available. In addition to the ubiquitous small tables and chairs, consider the design of bathroom fixtures, toilets, and sinks, found in schools, playgrounds, and hospitals (see Figure 7.5).

CHAMP HOUSE, A UNIQUE CHILD CARE FACILITY IN JAPAN

Noriko Yamamoto of Global Link, Inc. has been working with Japanese home builders for years to help them incorporate universal design principles in their projects. She has also been a long-time advocate of incorporating universal design into the creation of supportive environments for Japanese elders.

Yamamoto founded CHAMP, an organization for women that was patterned after AARP in the United States. In a showcase for the CHAMP community, she developed a unique housing environment for CHAMP members. CHAMP House started with a commercial building in Tokyo, in which the first eight floors were devoted to offices and other businesses. The 9th and 10th floors were for apartments rented to young families with children. The 11th floor (see

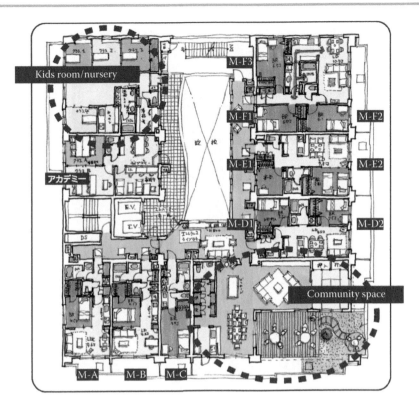

FIGURE 7.6   CHAMP House floor plan showing the kid's room, a community room, and individual apartments. (Courtesy of Noriko Yamamoto.)

floor plan in Figure 7.6) included six apartments for CHAMP members, a community space for residents, and a child care facility (kids' room nursery). The plan was that the CHAMP residents would care for the children of the residents from the apartments on the 9th and 10th floors while their parents worked. The conceptual image of the kids' room (Figure 7.7) was used in marketing literature for the building. Photos of the facility (Figures 7.8 through 7.10) show how folding wall dividers add flexibility (universal design) to the spaces.

FIGURE 7.7   Rendering of play in the kid's childcare nursery. (Courtesy of Noriko Yamamoto.)

FIGURE 7.8   Playroom in the CHAMP House in Japan. (Courtesy of Noriko Yamamoto.)

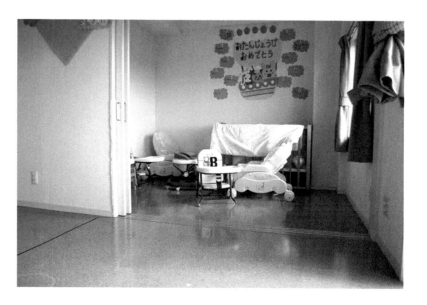

FIGURE 7.9   Nursery with movable divider walls. (Courtesy of Noriko Yamamoto.)

FIGURE 7.10 Detail of divider wall with windows. Child care facility for CHAMP House. (Courtesy of Noriko Yamamoto.)

Case Study: Airport Interior Design: An Integrated Approach

An environment as complicated as an airport terminal requires special emphasis on communication and wayfinding as well as on proper use of signage. The case study that follows shows how preliminary and ongoing research, development of evaluation tools, evaluation of use patterns, and the actual design of furniture and signage must all be integrated to create a successful environment for use by large groups of individuals.

Airport passenger terminals have grown considerably from the simple shelters constructed over 100 years ago. Their future growth will depend upon their success in maintaining an equilibrium among diverse passenger, airline, and operator interests as these interests expand and evolve within a truly globalized air transport system. Since the "accommodation of diversity" lies at the very heart of universal design, the fortunes of any given airport terminal may well be determined by the success with which universal design principles are applied to ongoing program development. This case study (1) describes an integrated design approach that facilitated the application of universal design principles to every interior design element in an airport terminal, and (2) presents a new seating concept as a way of illustrating the type of design innovation fostered by an integrated approach to airport interior design.

The ideas and methods presented in this study are the result of intensive design research originally undertaken to assist the Hillsborough County (Florida) Aviation Authority in designing Tampa International Airport's award-winning Landside Terminal. The concepts developed were then applied to the Airside A Terminal then under development. The coincident progress of Tampa's ADA Compliance Program expanded the scope of the Airside A work and also prompted Landside designers and developers to review that terminal's ability to meet the human, environmental, and economic challenges facing all major airports today.

THE SOCIAL CHALLENGE: MEETING HUMAN NEEDS

Although no one would question that airports serve people, this self-evident assertion has often been ignored in the final design of airport interiors. Confusing or lengthy traffic routes, cramped lining-up or seating space, harsh reflective light and sound, poor orientation cues, and related problems ignore basic physical and psychological needs, thereby making air terminals stressful, irritating places. Many of these needs are obvious: Airport visitors don't want to walk long distances or travel any distance in a direction they are not certain they should go; crowding is stressful, as is seating that forces people to look directly at one another (or else turn deliberately away to avoid doing so). Time is critical to most visitors, so no one wants to hunt for restrooms, concessions, or flight information displays. Nor does anyone want to leave his or her intended route, or stop, to scrutinize signage that is unintelligible at a glance. No one wants to move through an area where deteriorating finishes or obtrusive maintenance and repair pose safety hazards, and everyone wants clean restrooms. To respond to these needs, the interior design must be more than efficient; it must be hospitable, as well.

SPECIAL NEEDS AND THE AMERICANS WITH DISABILITIES ACT

The needs mentioned above are shared by all people. There are a great many more human needs that, although not universal, are critical to an ever-increasing number of people traveling through, or working in, air terminals. For instance, fixture heights and walking distances are formidable obstacles for children even when they are quite convenient for able-bodied adults. Signage, warning devices, and information displays that are readily intelligible to young visitors may be undetectable or illegible to older ones. Similarly, the elderly cannot bend as low or reach as high as they might once have done to select food or other merchandise. The very young and the very old frequently require assistance when using the restroom.

Furthermore, interior design elements that prove to be merely fatiguing or annoying, but accessible, to the young and the elderly may be entirely inaccessible to people of all ages with disabilities. The implicit exclusion of people with disabilities from public facilities such as air terminals has finally elicited a comprehensive legislative response in the form of the Americans with Disabilities Act (ADA). As provided under Title II of the ADA, "Nondiscrimination on the Basis of Disability in State and Local Government Services" (28 CFR Part 35), persons with disabilities can no longer be denied "full and equal enjoyment of goods, services, facilities, privileges, advantages, and accommodations."

The challenges that the ADA requirements pose to airport interior design were both immediate and profound. For example, a given design element (drinking fountain, telephone, merchandise, or storage shelf) cannot facilitate wheelchair access, yet prohibit access or use by anyone other than a wheelchair user. Nor could accessible elements be provided for visitors, yet denied to people with disabilities who worked in the air terminal. "Full and equal enjoyment" also means that accessible pathways and seating must be integrated with public-use pathways and seating, and not pushed to remote corners of the interior space. Maintainability and structural integrity became even more critical design values under ADA guidelines, since an accessible design element that is broken or inoperable no longer complies with the provision for "full use and access by people with disabilities." The ADA's mandate for a "barrier-free environment" extended beyond architectural design to cover wayfinding and communications systems to ensure that a building was sensorially, as well as physically, accessible.

ENVIRONMENTAL AND ECONOMIC DEMANDS

The basic structure of an airline terminal should be designed for easy expansion and modification to ensure that it has a very long and useful life. Finishes used in interior spaces should have the longest possible life cycles and cause the least possible harm to the air, water, and soil.

MEETING THE DEMANDS: THE PRINCIPLES OF DURABILITY, ADAPTABILITY, AND CONVENIENCE

An increasing sensitivity to human, environmental, and economic demands is changing the face of commercial aviation in this country. Airport interior design must move with this change in the direction of universal design, toward design solutions directed at a far greater percentage of people than heretofore attempted. Moreover, each universal design solution should solve more than one problem; it should address, for instance, cost-effectiveness and human needs at the same time. In this sense, "universal" applies to the comprehensiveness of the design approach as well as to the wide accessibility of the product designed. (There will, of course, always be someone with a combination of severe physical, sensory, and cognitive impairments who will not be able to access or use the design.) But a piecemeal, reactive approach that attempts special designs for each major design unit to accommodate each one of the immense varieties of disabilities or combinations of disabilities invariably ends up satisfying no one's demands. The same holds true with respect to environmental and economic demands: Universal design evolves from a careful balancing of diverse factors and constraints.

Universal design requires an integrated approach to the demands facing the nation's airports. From the preceding discussions of these demands, three principles emerge that demonstrate the interrelatedness of human, environmental, and economic imperatives and point to a design approach capable of integrating them all. Careful and continual attention to these three principles—*durability*, *adaptability*, and *convenience*—ensures that design development moves toward the desired goals of universal accessibility and use.

DURABILITY

Durability includes not only structural integrity but also aesthetic longevity. A given design unit must maintain a high appearance level for a very long time; its styling should never look dated, and its colors and finishes must resist soiling and wear as much as possible as well as mask whatever soiling and wear is inevitable. Moreover, these colors and finishes should be available for years to come. In short, *durability* encompasses any aspect of design that affects the useful life cycle of a design element.

ADAPTABILITY

Changes in airline operations, business activity, aesthetic concepts, the seasons—all of these combine to create an ironic "constant state of transition" at air terminals. Interior design units must be easily modified, expanded, adjusted, or extended as necessary to accommodate this continual state of flux. *Adaptability* also means that a modified or relocated design element should still be compatible with the surrounding space. From the perspective of design as problem-solving, an adaptable design is a solution that solves more than one problem. From a human factors perspective, an adaptable design accommodates the different manners in which a wide variety of users are accustomed to performing a given task. This includes the multitude of substitutions or coping skills developed by persons to compensate for deficits in one or more areas of physical, sensory, or cognitive functioning.

CONVENIENCE

As used in the Tampa International Airport design research, *convenience* embraced both functionality and aesthetics. Included under this principle were design unit safety and ease of installation, cleaning, and repair, in addition to the ease of use and wide accessibility typically associated with the term. Design features generally described as providing comfort or amenability also served the principle of convenience, since they made easier whatever activity the user was engaged in—whether it was a long wait in hold room seating or a short pause beside the concessions area.

APPLYING THE PRINCIPLES: THE AIRSIDE A DESIGN EVALUATION GUIDE

The principles of durability, adaptability, and convenience formed the basis for an integrated approach to interior design. Principles alone do not, however, produce an effective working design. Rather, an 18-section guide was developed to assist the design team in applying the previously mentioned principles. In the guide, each interior design element was evaluated with respect to the principles of universal design.

NEW DIRECTIONS IN SEATING DESIGN

The development of an integrated approach to airport interior design confirmed what had been suspected among certain design circles for many years previously: namely, that the conventional approach to interior design as "fixtures, finishes, and furnishings" was unequal to the social, environmental, and economic challenges now confronting airports. In very few design elements was this more apparent than in the case of hold room seating layout. Upon application of the three universal design principles of durability, adaptability, and convenience, the Airside A design team developed the following list of seating layout criteria.

Durability
1. Design must provide adequate seating for years to come.
2. Tabletops must resist damage and be easy to replace when damage occurs.

Adaptability
1. Design should pose no line or other traffic problems when flight delays or other conditions change hold room occupancy.
2. Seating should be easily added, removed, and reconfigured to conform to changes in hold room area.

Convenience
1. Seating design must provide ready access to all varieties of terminal visitors (including people with disabilities) and their luggage.
2. It must offer "natural" seat selection; it must not force strangers into proximity with one another.
3. Trash disposal receptacles should be incorporated into seating areas.
4. Seating must not block cleaning and maintenance of carpet or other flooring materials.
5. Seating must facilitate conversation and social cohesion among families and larger groups.
6. Every seat should have direct access to a circulation area.
7. No person should be seated where someone is looking at the back of his or her neck, or directly across from someone else.
8. Conversational arrangements must allow access and placement for wheelchairs.

No more than a cursory review of the typical tandem seating layouts was needed to see that what had become the status quo in airport hold room seating did not meet the wide range of human and facility needs facing the commercial aviation industry. To fulfill the seating layout criteria listed previously, industrial designer Joseph A. Maxwell developed a patented cluster layout that established a relaxed, conversational setting while facilitating the cleaning and maintenance of adjacent areas (see Figures 7.11 and 7.12).

Compared with the standard tandem seating featured in the nation's airports, the cluster seating arrangement offered the following advantages:

1. The arrangements were conducive to conversation.
2. Every seat was an excellent seat, with two arms and a table.
3. Any seat could be reached without climbing over people and bags.

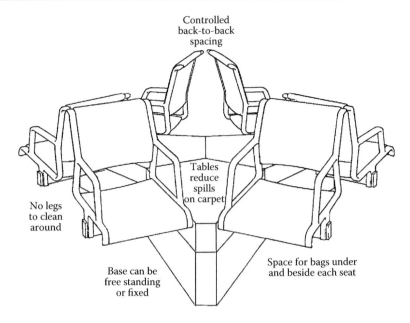

FIGURE 7.11   Airport cluster seating element (general representation) (Maxwell).

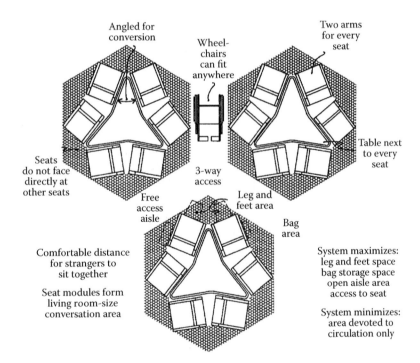

FIGURE 7.12   Cluster seating layout in an airport waiting area (Maxwell).

4. There was room for wheelchairs in every seating group.
5. There was full access for people with disabilities.
6. A table for every seat reduced floor-cleaning costs.
7. The table base and cantilevered seats simplified cleaning.
8. There was ample room to store carry-on items without taking up seating or circulation space.
9. The seating arrangement did not create tripping hazards.

10. There was room for free movement and stretching.
11. No one looked directly at another person; seats were not placed tightly back to back.
12. Seating was less stressful and more comfortable.

One simple fact confirmed in these studies of US airports was that people do not want to crowd or be crowded by others they do not know. The most acceptable way to seat the greatest number of people in a given space is to use a seating layout that works with, not against, people's natural seat-selection habits. The cluster arrangement provided just such a layout, as evidenced in a special trial installation conducted at Tampa International Airport after the seating utilization study there was completed.

In this test, a seating mock-up was installed in Tampa's Airside F terminal. For several days, the tandem seats in the mock-up area were videotaped for 2.5 hours each during the two peak periods of each day. The tandem seats were then replaced by cluster seating units featuring the same seat elements as those used on the tandem bases. This mock-up was videotaped during the same periods, for the same number of days as for the tandem seating mock-up.

Utilization of the tandem seats was the same as observed in the formal study. The cluster seating, however, filled to 100% as the area became crowded. In fact, the cluster seating mock-up was so well received that it is still in place. The videotape recorded wheelchairs being moved through these areas during active periods, and people were also observed putting their baggage in the space provided, rather than on a seat.

Conversational groups of cluster seating have been used in the public areas at Tampa International Airport for 30 years, and their success has been well proven. People tend to be friendlier, and they speak to strangers. The tensions of travel are reduced, and long waits seem shorter. In addition, access and use are increased for a variety of individuals, including the young, the elderly, and those with disabilities.

URBAN PUBLIC TRANSPORTATION

Public transportation is a major consideration for universal design. This is especially true in urban areas where a large percentage of the public depends on public transportation. Bus manufacturers have taken leadership in the design of accessible vehicles. See Figure 7.13 through 7.16 for pictures of several models from the Kumamoto prefecture in Japan.

A speciality bus that was designed to accommodate 10 wheelchair users was introduced at the UD 2012 Universal Design Meeting in Oslo, Norway, in June 2012 (see Figure 7.17).

FIGURE 7.13   Outside appearance of a low-floor bus with no step barrier (Kumamoto, Japan).

FIGURE 7.14   The movable sliding ramp is placed sloped at the bus's central entrance (Kumamoto, Japan).

FIGURE 7.15   The front exit with grab bars, special step surface, and visible signage (Kumamoto, Japan).

FIGURE 7.16   Three pedal-operated jump seats for passengers are sprung upward on the right side, to make space available for two wheelchair passengers. Three hooks on the floor fix a wheelchair to the floor (Kumamoto, Japan).

FIGURE 7.17   (a) Exterior, special bus that will provide access for up to 13 wheelchair users. The driver, who also designed the bus, is shown ready to operate the lift. (b) Several wheelchairs can be secured in an area with rails like these in Oslo, Norway. The special bus was introduced at the UD12 Oslo Universal Design meeting in June 2012. (c) Fastening wheelchair to secure clamps in the new bus.

A SUMMER CAMP WHERE DISABILITIES ARE THE NORM: IMPLICATIONS AND GUIDELINES FOR DESIGN

Children with chronic illness or physical and developmental disabilities are often shunned or ignored by others in our society. It is common for such children to feel worried, isolated, and unduly dependent. Fun, a sense of freedom, or shared adventures are rare experiences for many. The goal of summer camp at Whispering Hope Ranch (WHR) in the White Mountains near Payson, Arizona, was to provide the place and the opportunity for such experiences. This paper, prepared by Janetta McCoy is a report on the design program and process by the architectural team and the "user advocate."

The architectural design and research team attempted to understand the illnesses and disabilities of the children, as well as other needs and requirements of all the people who will spend time at the camp. Many design implications and guidelines emerged from the research. These guidelines were specifically relevant to the campers and caregivers of the 10 groups represented in Whispering Hope Foundation, but could also be useful in the planning and design of other, similar camps beyond WHR.

For the children who attend summer camp at WHR, celebrating childhood and having fun would be paramount. In an atmosphere of respect and dignity, in a place of safety and security, children would meet and learn from others who have similar illnesses and disabilities. Dealing with limitations and medications at WHR would be an accepted norm—not the isolating, differentiating secrets or problems often experienced in other social situations.

RESEARCH METHODS

For several months, the architectural team was immersed in the issues of children who will attend camp at WHR. Representatives from 10 different organizations sponsoring Whispering Hope Foundation were interviewed at length. Each interview required from 2 to 3 hours and followed a structured format with open-ended questions.

With representatives of Whispering Hope Foundation, two members of the architectural design/research team visited the proposed site and similar established camps in Texas, notably *Camp John Marc Meyers* near Dallas and *Camp for All* near Houston. Recent research literature on specific illnesses and disabilities was critically reviewed. Research relevant to children's issues of environment and behavior and health care settings was reviewed in the development of the interview protocol and for reference in the development of design guidelines (Connell and Sanford 1999; Evans and McCoy 1998; Harden 2000; Hart 1979; Kaplan and Kaplan 1989; Korpela, Kytta, and Hartig 2000; Moore and Young 1978; Passini 1984; Sanford, Story, and Jones 1997; Wohlwill and Heft 1987).

In addition to clarifying issues relevant to specific illnesses and disabilities, the experienced interviewers also asked practical, programmatic questions from which important, revealing discussions also evolved:

- Vital statistics, that is, ages of campers, numbers of campers and required staff
- Program philosophy
- Program of activities
- Current and ideal cabin attributes
- Current and ideal food service style, that is, cafeteria or family style
- Current and ideal medical facilities specific to the illness or disability
- Transportation to and within the camp

From these efforts, criteria critical to the success of WHR were made clear. The following design guidelines and implications emerged from this research.

WHR IS ABOUT PEOPLE

Understanding the people who attend the camp was paramount to understanding the camp. Summer camp at WHR provided services for children ranging in age from 5 to 18 years. Each of the 10 groups of campers represented a different illness or disability and therefore each group has a somewhat unique set of requirements. All requirements were to be met in a single facility. While the children of the camp are the major focus and the reason for the camp, it was also important to make appropriate provisions for caregivers and other operational staff. For some groups, the campers would come without parents. Other camps, those for autistic and HIV/AIDS children, would include parents and siblings. One organization encouraged siblings to attend with the designated camper.

The staff included volunteer counselors, medical volunteers, and permanent WHR employees. Some counselors were adults who have the same or similar disabilities as the children. Counselors' duties might include cabin leaders (who were positive role models for the campers). Other counselors might facilitate activities such as arts and crafts, theatrical performance, swimming, or stargazing. Some counselors were volunteers who might have encountered the children on a professional basis. For instance, counselors for the burns and trauma camp often included firefighters and paramedics.

Trained professional medical staff was required for all camps. Medical responsibilities included distribution of all medicines and treatment of minor emergencies including seizures as well as the more typical cuts, scrapes, and sprains. (There was a hospital within 20 miles in case of more dire emergencies.) Most children at WHR would require some medication or medical treatment on a daily basis.

Although many campers at WHR would arrive independent of their families, two camping groups encouraged parents to attend camp. Parents and siblings of autistic children attended camp at WHR and would require family-style cabins and sleeping arrangements. Parents and siblings of HIV/AIDS (infected and affected) children would be encouraged to attend the camp; parents would be individually assigned to same-sex adult cabins. Siblings of craniofacial campers will be encouraged to attend WHR; they would be assigned to cabins with children of same sex and age group.

WHR IS ABOUT CELEBRATING

In all interviews, every organization indicated that WHR should be a place where children felt a sense of normalcy by *not being different.* The intention of WHR is to provide opportunities for these special children to feel and be independent, to feel a sense of accomplishment and power, to make friends, to feel a sense of adventure, to expand their social and physical boundaries, and to expand their community of supportive adults and role models.

WHR was intended to be a place for having fun, "a place with no worries." Having fun would include full, busy days of varied physical activities pushing the boundaries of each individual's abilities. With safety considered at all times, many campers would return to their homes at the end of the week with "bragging rights" over accomplishments not previously known.

WHR IS A *HAPPEN'N'* PLACE

WHR would be a place for children to participate fully in fun activities and events, while learning about themselves and growing in self-respect and self-esteem. The design of WHR would include places for campers' eating, sleeping, health care, outside and indoor activities, and special events. It also included places for spiritual or private contemplation and places that provided opportunity for serendipitous socialization. One 16-year-old with advanced kidney disease said the one thing that could make him most happy at camp would be a dance floor, "I love to dance."

To maximize the opportunities fundamental to the philosophy of Whispering Hope Foundation, all facilities and all areas must meet or exceed recommendations of the Americans with Disabilities Act, with specific focus on the "kiddie code." For instance, although the mountainous terrain was rough in its current stage, the camp was to be designed with no grade greater than 5% to accommodate children in wheelchairs and on crutches, or those with breathing difficulties. Likewise, all facilities and all areas would be designed and constructed with the utmost concern for the safety and security of all campers of all ages and all levels of ability. There would be no places from which harm or danger could emerge, no places where danger could hide, or harm could be perpetrated. All areas of the camp should be visible from multiple lines of sight.

All facilities and all areas of WHR should follow guidelines for coherence and legibility, making the camp not only accessible and negotiable but also logical for campers to find their way. However, in keeping with the spirit of the camp's philosophy, this would also include elements of the unexpected or adventurous. Coherent and legible should not preclude the joyful elements of surprise and mystery.

Specific programmatic requirements included the following:

Main Lodge—a central building that sets the stage and tone for camp activities. This is probably the place where all meals (except for the occasional barbeque) would be taken and major announcements were made. For the campers at WHR, consuming food would be important. Likewise, the context of the eating ritual would take on significance and meaning. Campers would eat as a group with the others who share their cabin. Whether they would be served cafeteria style or family style, during meals they would practice social etiquette and responsibility, having meaningful conversation, taking responsibility for serving themselves, and bussing their own tables, as much as possible. A coherent, legible system of food service, table identification, and wayfinding within the eating area would be one key to its success.

The main lodge might be a multiple-use building and a crossroads for many levels of communication for both adults and children. This facility would provide opportunity for small intimate conversations between campers, small group educational or therapy sessions, and the three major dining events of each day seating all campers at the same time. It might also need to serve as an activity center on days with inclement weather.

The acoustical attributes and qualities of the main lodge would be very important. Autistic children react adversely to noise; the same noise could be disorienting to people who are blind; many camp leaders would find mealtime appropriate for making announcements and noise could compete with the effectiveness of such communication.

Because many of the campers at WHR are young, small in stature even for their ages, or are bound to the sitting position of a wheelchair, the scale of the Main Lodge should respect the scale of its campers. Likewise, as many camps would seat all campers at about the same time for each meal, the children and counselors are likely to assemble just outside the lodge, or nearby, for several minutes prior to each meal. Provision must be made for this assembly and to shade sun-sensitive campers while they wait.

Cabins—a place for sleeping, bathing, and socializing, and for some campers, a place to tend to medical requirements (i.e., peritoneal dialysis, catheterization). The cabins need to be appropriately flexible for both independent campers and families.

The campers sharing sleeping cabins at WHR would also represent the basic social unit of the camp. Eight to ten people would share a cabin and would typically include two counselors with six to eight children of the same sex, age group, and similar abilities. They would take their meals together, do most activities together, and learn to trust and depend on each other. More than just sharing a sleeping cabin, within this unit the development of close friendships and a unique, dynamic social life would be encouraged.

Each person with a personal cot or bed should also have storage or shelf space enough to accommodate the needs of a peritoneal dialysis machine or the organizational needs of blind campers.

The bathing and toilet requirements of many campers suggested the need for showers and toilets large enough to accommodate a wheelchair and an assistant, but private enough that the modesty of the child be protected.

Because toileting accidents do happen, soiled linens and clothing should be easily and unobtrusively removed with laundry facilities located nearby.

As the social unit of the camp, the sleeping cabin should be located in proximity to other cabins with additional opportunity for socializing between cabins.

Outdoor Activity Areas—a variety of places for formal, structured games as well as spontaneous activities. Outdoor activities planned by specific camps vary widely. All camps have stated goals for campers to stretch their own physical boundaries by doing and achieving more than expected—and they all want to spend at much time as possible in the outdoors challenging their large motor skills.

Outdoor activity areas might include a central plaza with a ceremonial place for arrival, departure, and awards.

Provision would be made for playing fields with nonabrasive surfaces. Multiple and ample sources of seating for players and observers, shade, and drinking water should be provided.

All children from the camps express an affinity for water play, but some should not or cannot swim or boat. Therefore, some opportunity should be provided for alternative forms of water play in which fear of drowning or infection would be minimized.

Organization and layout of outdoor areas should follow guidelines of coherence and legibility with easily understood directions to water, toilets, and primary landmarks on the campus.

In addition to sports and activities, the outdoor areas might also accommodate places for ceremonies, such as flag raising; places for reflection, such as an overlook with views of the mountains or stream; places for performance, such as an amphitheater; small areas appropriate for small group breakout sessions; places for bonfires; and places for dances.

Arts and Crafts Building—an indoor place for multiple, varied structured group activities involving artistic expression and techniques, which might require special materials and equipment. Activities might include ceramics, weaving, painting, cooking, photography, music, reading, wood shop, and theater. This facility might also be used for other activities on days with inclement weather when the outdoor areas were inappropriate.

In addition to producing arts and crafts, there might be a need to display and exhibit the items created. Substantial storage is required for the variety of activities organized by each camp. Attention should also be given to task lighting as well as lighting for the potential of exhibitions, and to acoustics for the sound sensitive as well as the musically inclined.

Infirmary—a place for distributing all camp medications and for treatment of minor emergencies; also housing for medical staff. The infirmary should be easily accessible and identifiable, but not the focus of the camp.

Some campers, such as those on hemodialysis, might spend many hours during their camping week in the infirmary. While this facility must maintain its sterile environment in order to function, it should also reflect the spirit of the camp: there is normalcy in taking these medications. The infirmary should allow the campers to continue to have fun and not be a reminder of their limitations.

Because the infirmary would be available 24 hours each day, and it would be housing for the medical staff, provision must be made to separate their personal and private space from the public areas of the infirmary.

Water activities represent some safety risk for all campers, so consideration may be given to a satellite first aid station at the pool or pond area. Because some campers, such as the children with diabetes who need frequent snacks, the first aid station might also be used to distribute selected snacks.

WHR was intended to be a place where children of all ages and abilities could feel free and happy. This was to be a place where the philosophy and programs of the camp were supported by the physical designed environment. This will be a place to have fun. This would be a place that will respect children and their families and their caregivers. This would be a place that nurtures self-esteem, fosters independence, and included many medical conditions in the definition of normalcy. WHR would be an important and memorable place in the lives of important and memorable people.

One of the important factors mentioned in the McCoy case study (a great example of universal design) was the need to help campers with wayfinding. Sharon Toji (Access Communications) has shared her expertise on signage and wayfinding in the following essay.

UNIVERSAL SIGN AND WAYFINDING DESIGN

In an ideal world, we could find our way around without many signs. Landscaping and architectural features would lead us to public entrances. Alternatives would be easily identified and easily accessed. For example, a curving ramp and stairs create an attractive entrance to the Cooper-Hewitt Art Museum (Figures 7.18 and 7.19). Once inside a building, stairways, escalators, and elevators would be visible as would restrooms, drinking fountains, and public telephones. Lighting, color, and floor coverings would all be used, along with "landmark" features, to lead us around the building. In facilities designed to be fully accessible for wheelchair users, no alternate path of travel signs would be necessary.

In the real world, however, a well-designed sign system to help us find our way can make up for many deficiencies. To do that well, the sign system itself must incorporate universal principles.

Wayfinding sign systems can be divided into three main categories: (1) Informational, such as general information about the layout of the site and its buildings, which tell us the entities and services we will find there. (2) Directional, which tells us what route to take to get to our destination (for instance, the restroom, Figure 7.20). They point us toward specific locations and services, and often use arrows to do so. (3) Identification, specifying destinations such as individual offices, rooms, or suites, or common use elements such as restrooms, stairways, and exits.

These three categories should work seamlessly together from the moment the site is entered to get people from the public right of way to their final destination. In order to do that, careful planning must go into the identification system that will be used for the various individual locations within the site and building.

In naming various locations, consistency and simplicity are vital. What will be most clear to people who have never been to the facility before? When people call for directions, what terms will the staff use? There is no point in sending people to "X-ray" when all the signs read "Imaging Technology."

FIGURE 7.18   Cooper–Hewitt Museum main entrance, site of the first UD exhibit.

FIGURE 7.19   Cooper–Hewitt Gallery alternative entrance ramp.

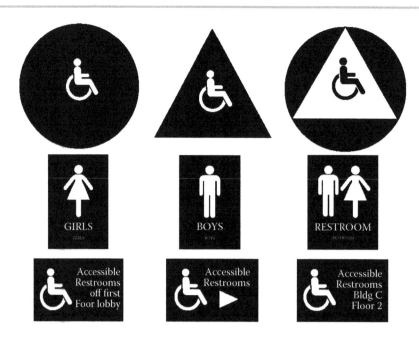

FIGURE 7.20   Recommended restroom signage. (Courtesy of Sharon Toji.)

Most individual suites or rooms within the building should be identified in a consistent and consecutive manner, usually by number. Room functions often change, and people, companies, and even entire departments come and go.

If the identification system is well designed, directional signs can be minimized and simplified. In a large facility, a site plan can tell us which building we want and where it is in relationship to our current location. Along the way, if there are junctions where a decision must be made, directional signs need only point to separate buildings, which should have been identified on the site map by a name, letter, or number. Indeed, if the buildings are all visible and there are easy-to-read identification signs on the buildings themselves, directional signs may not even be necessary. Once inside the specific building, a building directory should tell us what room number we need. In a large building, floor plans can point us to the correct part of the building and floor directories can jog our memories.

Even the best thought-out identification plan is of no use if people can't read the wayfinding signs. The great majority of people, even people who are legally blind, use their sight to read signs. Therefore, the most important attributes of signs are those that enhance visual legibility and readability.

There are four main considerations when designing a sign so that the majority of people can use it easily and efficiently. The first is contrast. Here, we are not speaking of color per se, although of course we are dealing with color. We are talking about light and dark. The best contrast is achieved with the use of very light text, along with essential symbols and pictograms, on a very dark background (see Figure 7.21). Of course, the ideal is white on a black background, but very dark shades of burgundy or maroon, charcoal, green or teal, brown or blue are almost as effective.

0% white on
100% background

0% white on
70% background

30% text contrast on
100% background

15% text contrast on
85% background

Minimum 70% contrast difference between characters/graphics and background

FIGURE 7.21   Contrast. (Courtesy of Sharon Toji.)

Besides bright white characters, we can use very subtle tints of off-white that harmonize with the background color—a warm white with dark brown or rust, for instance.

The next universal consideration has to do with glare. The sign designer cannot usually dictate the lighting that will be available in the vicinity of each sign. It may be artificial or natural, and will probably vary with the weather and the time of day. However, regardless of where it is placed or how it is illuminated, when a sign reflects light, it becomes unreadable to many people. Many "high-end" buildings pride themselves on the amount of polished brass, chrome, and stainless steel they display in their lobbies and corridors. A sign in which you can see your reflection may add to a feeling of luxury, or it may help you to check your makeup or see if your tie is straight, but it certainly will do a poor job of quickly and efficiently conveying information to many people. Most metals, unless they are painted, should not be used for either characters or background of essential wayfinding text. Even brushed metals usually have too much glare.

The third element is type. We need to be concerned with typestyles, sizes, spacing, and case. Although many serif and sans serif typestyles are very legible, it is important that when they are used for essential wayfinding information, they have very classic and non-distorted character forms. Condensed and extended styles, for instance, should be avoided, with condensed styles usually being the most unreadable because the negative spaces within characters tend to disappear. Decorative typefaces, such as scripts, should not be used, and even italics and oblique forms are very difficult for people with vision impairments to read. In short, any distortion of the character shapes should be avoided. Some of the new typefaces that have been designed for use on computer screens are excellent, and the Royal Institute for the Blind, in the United Kingdom, even supported research that resulted in a new typeface, Tiresius, which has a version especially for visual signs (see Figures 7.22 and 7.23).

Proper type size is a function of many factors. Are there any obstacles that prevent you from getting close to the sign? Is it high on the wall or suspended overhead? Do you need to see it from a great distance, such as in an airport or a convention center? Will you be under stress or in a hurry as you are trying to find your way? Is the sign in an exterior or an interior location? Will you be in a car, on foot, or perhaps in a wheelchair? Will there be a crowd of people around the sign (see Figure 7.24)?

One of the biggest mistakes people make when designing signs (and remember, most signs are not "designed" by professional designers at all, but by the person ordering the sign) is to try to cram the largest amount of text on the smallest possible sign. Text goes right out to the edges of the sign. The problem is compounded by the use of uppercase characters ("caps") for the entire

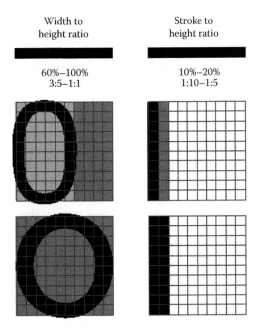

FIGURE 7.22   Font ratio. (Courtesy of Sharon Toji.)

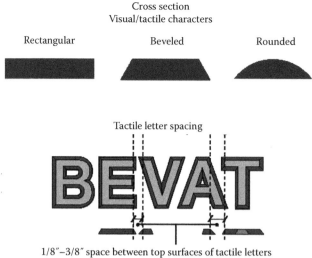

FIGURE 7.23   Character spacing. (Courtesy of Sharon Toji.)

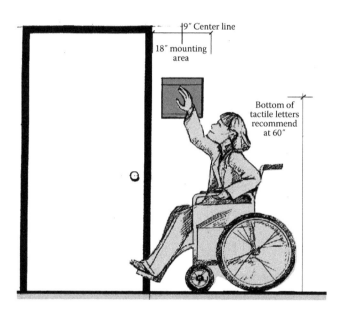

FIGURE 7.24   Mounting height. (Courtesy of Sharon Toji.)

text. Usually, if space to install a sign is a problem, it should be solved by simplifying the message or breaking it up onto more than one sign. Uppercase and lowercase characters should almost always be used on visual signs, with the possible exception of a one- or two-word heading. The word "shapes" assists people to read the sign and also create more negative space on the sign.

To this point, we have concerned ourselves with sign systems that can be used by people with usable vision and with those people who can read. Although that constitutes the majority of people who need wayfinding signage, it certainly does not provide a universal solution for wayfinding.

Signs may be one of those things where it is impossible to serve everyone with just one element. The original ADA Accessibility Guidelines (ADAAG) tried to do that, at least for signs that identify individual rooms and spaces in buildings, by writing standards for characters on signs that can be read both visually and by touch, and by adding Braille to these signs. However, such signs reduce the readability of the sign for the majority of the public because of the needs of tactile readers. For instance, the characters must all be uppercase, since lowercase characters are very difficult for tactile readers to make out. The characters cannot be larger than 2 inches, because large characters are also very difficult to read by touch (tactile characters are best at around 1 inch). The signs must be located where they can be reached and where the reader won't be hit by an opening door, even though an overhead sign or a sign directly on a door might be a better choice for most users.

Now, with the final adoption of the 2010 ADA Standards for Accessible Design, we have a solution that allows us to make a sign that is more readable for both those who read with their vision and those who can read only by touch. These standards allow us to separate the visual and tactile messages, which must correspond. They can be on one sign or on two separate sign blanks. The tactile portion can be "invisible," because it needs no contrast at all, and can even use one of those highly polished materials so favored by upscale establishments. The visual portion is also more readable, because larger and bolder typestyles can be used. The text can be uppercase and lowercase, and non-decorative serif fonts are fine, as well (see Figure 7.25).

Many people who are blind don't like tactile signs much, since reading them is usually quite laborious, and the problems are compounded by the poor tactile quality of most signs. They would prefer audible signs, both for directional use and to identify rooms and give additional information. There are some available audible systems now on the market. Certainly, combining one or more of those systems with both visual and tactile signs would provide the most accessibility to wayfinding information to the greatest number of people.

Some tactile signs are, however, required by law. Those are signs that provide permanent identification for rooms and spaces, including exits. Even though it represents a compromise, particularly for visual readers, the same set of characters can be used to provide both visual and tactile information. That means that the combined visual/raised characters have to be all uppercase and sans serif. They must be a minimum of 5/8 inch and a maximum of 2 inches high. It is important to apply standards for character and stroke proportion to these characters. Typefaces should be neither condensed nor extended, and stroke width should be fairly thin, rather than bold. The most readable tactile characters have a beveled or rounded profile, so that the top surface of the character is thinner than the base of the character (the part that rests against the sign surface). Of course, space between the characters is even more important than for visual characters. The new standards that require 1/8 inch space between the two closest points of adjacent characters is necessary for tactile readers. Characters less than 1 inch high may look as if they have too much space between them, especially to the trained graphic designer.

It's important to know that everything on a room sign does not have to be tactile. First, if rooms have been identified by number, most of the problems with combined visual and tactile characters disappear. Tactile numbers are very readable visually. One set is enough. The accompanying information, such as the room function, does not have to be tactile and can thus follow the rules for good visual sign design. If the sign has an audible component, this additional information is easily available for anyone who is blind or cannot read, as long as they understand some English. But, for rooms where the room function or name is essential to wayfinding, the "dual purpose" sign with separate tactile and visual sections is the best choice.

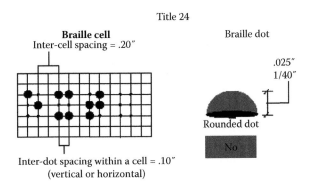

FIGURE 7.25   California Braille. (Courtesy of Sharon Toji.)

For those few signs that must use tactile text, such as restroom signs, or major destinations such as libraries, conference rooms, and cafeterias, two sets of text can be provided on the same sign. The visual text follows all the rules for contrast and is non-glare. An angled "shelf" or footer at the bottom of the sign holds the tactile text and Braille. Since no contrast is required, and glare is not a factor, this footer can even function as a decorative part of the sign. Besides being able to provide tactile characters that are designed solely for tactile reading, the hand position of the tactile reader is much more natural. The entire sign becomes easier for everyone to read.

The last consideration in sign design is the message itself. First, simple straightforward language is always best. Architectural signs are not there to impress the public with obscure terminology. The professional medical personnel at a hospital may understand long scientific terminology, but others who come there are in a hurry or under emotional stress and need signs they can understand.

Pictograms can be very helpful adjuncts to sign messages. Although there is a danger on relying too heavily upon them, when used in conjunction with text, they can aid those with cognitive disabilities or those who don't read English. The most effective pictograms are simple, with few "parts," and need to follow the rules for high contrast and no glare. If they are on overhead signs, they need to be quite large—usually much larger than the accompanying text.

One important set of pictograms is the set of international symbols of accessibility required by the ADAAG. These assist people with a wide range of disabilities, including those in wheelchairs, those with other mobility impairments, and people who are deaf or are hard of hearing in finding parking, accessible paths of travel, facilities, and accessible phones and listening systems. It is vital that these symbols be used correctly, consistently, and that they be presented in a manner that makes them easy to locate and see.

Our ideal environment, then, will provide architectural cues—visual, tactile, and audible—to help us get around, and they will be supplemented by an intelligently designed wayfinding signage system. The signs will present information in a simple and consistent manner, using characters and backgrounds with high dark/light contrast, no glare, and easy-to-read typefaces. Visual information will be supplemented with tactile information for those who are blind and may also be accompanied by audible information. Individual rooms and spaces will be identified with a logical, consecutive system, and directional signs will begin with general information and become more specific as the destination is approached. A wayfinding system such as that will be accessible to a great many people of various ages and abilities, truly a universal wayfinding design.

In Chapter 6, we identified some aspects of obesity that affected the work environment and described the design of special bariatric chairs. In this case study, a white paper titled "Obesity: Implications for Health Care Facility Design," we have addressed the impact of obesity on hospital and long-term care environments.

BARIATRIC DESIGN 101—AN INTRODUCTION TO DESIGN CONSIDERATIONS

Obesity has become an alarming epidemic with enormous implications for our health care system.

Obese people require more health care than average people, and there are increased physical problems for staff and attendants in administering that care across the spectrum of health care services.

Additionally, a boom in gastric bypass, stomach reduction, and banding operations has been bringing even larger numbers of bariatric patients to health care providers. Respecting patient dignity and delivering optimum clinical care are primary issues, as are establishing procedures for safeguarding the health and well-being of these patients and their caregivers.

Design is a critical tool in the care of and the improved long-term clinical outcomes for bariatric patients (Zasler and Calkins 2009). Design concerns include appropriate facilities and space, proper equipment, and furnishings.

We begin with a brief discussion of the obesity epidemic in our country, defining what qualifies as requiring bariatric care.

THE OBESITY EPIDEMIC

Obesity is defined by body mass index (BMI): Weight (lb.)/Height (in.^{2})
BMI ranges are as follows:

- Normal weight = 18.5–24.9
- Overweight = 25–29.9
- Obese 1 = 30–34.9
- Obese 2 = 35–39.9
- Obese 3 (also referred to as extremely or morbidly obese) = 40

5′2″, 190 pounds and 6′2″, 450 pounds could both be considered bariatric patients (Figure 7.26). Research indicates that overweight children are more likely to become obese as adults. Between 1976 and 2007:

- For preschoolers aged 2–5, obesity increased from 5.0% to 10.4%.
- For children aged 6–11, obesity increased from 6.5% to 19.6%.
- For adolescents aged 12–19, obesity increased from 5.0% to 18.1% (CDC 2010).

Obesity is becoming an epidemic with no end in sight and with significant implications on health care as it sets the stage for the occurrence of numerous medical problems including asthma, joint degeneration, sleep apnea, and lower back pain.

Severely obese people are more than twice as likely as people of normal weight to be in poor health and have about twice as many chronic conditions:

- *Type II diabetes*: 80% of patients with Type II diabetes are obese.
- *Gallbladder disease*: The incidence of gallstones soars as a person's BMI goes beyond 29.
- *Coronary heart disease*: 70% of diagnosed cases are related to obesity.

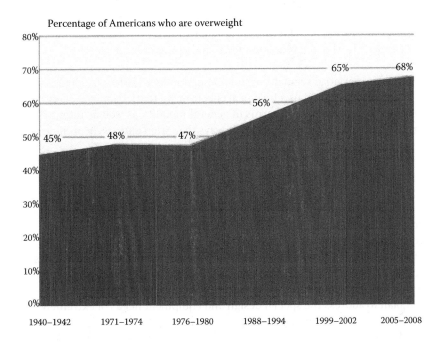

FIGURE 7.26   Percentage of Americans who are overweight, which includes both those who are overweight and those who are obese. As you can see here, only 32% of Americans are considered to be at or below normal weight. At the current rate of increase all Americans will be obese by 2059 (CDC 2010).

- *High blood pressure*: Obesity more than doubles one's chance of developing high blood pressure.
- *Breast and colon cancer*: Almost 1/2 of breast cancer cases are diagnosed among obese women.

Obese patients have longer, more costly hospital stays. In 2009, the Centers for Disease Control and Prevention estimated that overweight patient care costs averaged $1400 more per obese patient compared to a non-obese patient. The health care costs of caregivers are also affected. Health care workers sustain nearly five times more overexertion injuries than the average worker. And these costs and impacts are spread over the entire health care services spectrum, not just in bariatric units.

COMMON TERMS, DEFINITIONS, AND STANDARDS IN BARIATRIC DESIGN

Bariatric furnishings, fixtures, and equipment are not just "bigger" or "hold more weight." Bariatric equipment must combine load limit, appropriate dimensions, and a design aesthetic that blends with the environment by which a patient's and caregiver's comfort and safety are ensured. Common considerations are as follows:

Dimension and shape: Not all people have the same physique, so a person's shape must be considered.
Safe working load or working load limit: Rating for bariatric beds, lifts, and other equipment. It is the largest load (in pounds) that equipment can safely lift.
Static load: The maximum amount of non-moving weight a piece of equipment can bear. It would be applied to furniture, handrails, grab bars, and toilets, for example.
Dynamic load: Designing for safety, dynamic load accommodates the weight of a patient in falling motion. Dynamic load must always exceed the static load. This load rating is critical as unstable patients often will reach out to grab or lean upon items like grab bars, furniture, or railings to stop a fall. As a rule of thumb, a falling human is double his or her weight. If we are designing for a bariatric population of up to 900 pounds, anything we specify has to withstand an impact weight—or dynamic load—of 1800 pounds.
Functional load: A level of loading intended to be typical of hard use.

People within the bariatric segment of the population face numerous challenges in daily life. In addition to the emotional and psychological effects of stigmatization in society, there are many hardships that must be overcome in terms of general mobility as well as interacting with architecture, furnishings, and products designed in ways that often effectively exclude them.

THE TREND TOWARD BARIATRIC DESIGN

Design should emphasize the similarities between people rather than their differences. Keep patients from feeling alienated and restricted as they move through the facility. Design spaces and products that serve everyone.

ECRI—the Emergency Care Research Institute—recommends several strategies to effectively address the special needs of bariatric patients in terms of facility design and equipment.

Considering obesity trends and the skyrocketing increase in bariatric surgical programs, hospitals should plan to address the special design and equipment needs of extremely obese patients in both their short- and long-range planning.

These needs should be addressed not just in particular care areas, but throughout the health care continuum. Primarily, the ECRI says the design team should be charged with the overarching goal of providing safe, respectful, high-quality care for extremely obese patients.

And second, the ECRI states that a multidisciplinary team should be designated to assess bariatric-related facility design and equipment needs throughout the continuum of in-hospital care.

Most facilities currently do not have dedicated units for bariatric patients. In the past, the occasional severely obese patient was handled on an ad hoc basis with existing hospital equipment, reinforced or lashed together as needed. In existing facilities:

- The patient room is often not large enough to house bariatric equipment, beds, and required caregivers.
- Bathrooms are too small; doors are too small.
- Toilet fixtures are incapable of supporting bariatric patients.
- Handrails are pulled from the wall when used by bariatric patients.
- Floors deform or peel up from beds being moved.

As the average weight of Americans climbs ever upward, hospitals are rethinking facility design to accommodate not only larger patients but also the medical equipment needed to care for them. Safety, not only for the patient but also for the care team, must be factored into the design.

The ultimate design goal when designing for the bariatric patient: Provide opportunities and encouragement for increased mobility, independence, and strength.

Proactive solutions addressed as early as possible in the design process are important. It is essential to thoroughly discuss and plan for the following critical issues:

- A description of the patient population—physical and mental capacities, medical conditions
- The types of procedures being performed and what equipment is required
- How and where patients are transported (patient flow)
- Dimensions and storage of equipment

TAKING DIRECTION FROM EVIDENCE-BASED DESIGN

While there is no evidence-based research specifically pointing to the field of bariatrics, we can look to more general evidence-based design guidelines for:

- *Reducing pain*—Scientific studies have shown that exposing patients to nature can produce significant alleviation of pain. Other research also suggests that patients experience less pain when exposed to high levels of daylight in their rooms. Finally, some research also supports displaying visual art with nature as the subject matter as it helps reduce pain.
- *Reducing slip/fall risk*—In one Washington University School of Medicine study alone (2007), there were 8974 inpatient falls at just 9 hospitals tracked from 2001 to 2003. The slip/fall risk in health care continues to be a major problem, leading to injury, higher costs, liability, and protracted recovery.
- *Improving patients' sleep*—Sleep disruption and deprivation are common problems in health care buildings; increasing acoustic performance with reduced reverberation time increased sleep quality.
- *Reducing patient stress*—Patient stress is a significant negative outcome that bears many other health care negative consequences. A physical environment that contains stressful features makes a psychological patient state worse. Several experimental studies have shown that real or simulated views of nature can produce restoration from psychological stress in minutes.
- *Reducing depression*—Many studies show that exposure to bright artificial light and daylight is effective in improving mood and reducing depression, even for people affected by deep depression.
- *Reducing spatial disorientation*—Wayfinding problems in hospitals have an impact both on patients and visitors, who can be stressed and disoriented. Improved signage can greatly reduce the stress associated with moving through the health care facility.

■ *Improving patient privacy and confidentiality*—It is based on great evidence that the provision of single-bed rooms increases patient privacy. Furthermore, providing private discussion rooms near waiting, admission, and reception areas may help avoid breaches of speech privacy.

■ *Fostering social support*—Some studies recommend the provision of stays and waiting rooms with comfortable furniture arranged in a cluster, in order to encourage social interactions.

■ *Reducing staff injuries*—The risk of caregiver injury is multiplied in bariatric settings. While proper procedures play a role, staff injuries can also be reduced through adequate space for movement, proper equipment placement, and appropriate fixtures and furniture.

■ *Decreasing staff stress*—Stress is the most common cause of staff retirement. Environmental stressors include noise, light, and multi-bed patient rooms. In fact, survey research shows that single-bed patient rooms are perceived to be less stressful for both family and staff than ones containing multi-beds.

■ *Increasing staff effectiveness*—While most research is aimed at patients, there is a growing body of evidence suggesting to improve hospital efficiency through making the jobs of staff easier. This can be achieved by spatial solution, environmental factors, and technological devices.

■ *Increasing staff recruiting and retention*—Well-designed facilities that reduce stress, increase efficiency, and lower the risk of injury can be effective tools in recruiting, and lead to higher staff satisfaction.

DESIGN FACTORS FOR BARIATRIC SPACES

This next section addresses the essential design factors for the various spaces in a bariatric or health care facility, including:

■ Entrances and routes
■ Patient rooms and bathrooms
■ Diagnostic and treatment spaces
■ Lobbies and waiting areas

We're basing much of this on the 2010 FGI Guidelines for Healthcare Facilities for the design of bariatric care units and the 2010 Patient Handling and Movement Assessments.

ENTRANCES AND ROUTES

Providing adequate space and a safe environment begins at the outside of the clinic or hospital. Easy access with ramps and handrails, wide enough to accommodate bariatric wheelchairs, walkers, and other specialized conveyances, communicates to the patient that the facility is equipped to address and understand their medical needs.

Establish an accessible path from the hospital entrance to all major departments by accommodating for a 39″ by 49″ wide wheelchair (700 pounds capacity) with a 6′ turning radius, per the Facilities Guidelines Institute 2010 revised guidelines.

One of the most significant design issues is associated with the patient accessing various areas of the health care facility. Considerations: registration, physical rehabilitation, food service, common/family areas, emergency services, diagnostics, even the gift shop or vending areas.

Factor door width and elevator capacity to make common areas more easily accessible for patients, equipment, and caregivers. Most hospital elevators have an average weight capacity of 2000 to 3000 pounds, a capacity that may be exceeded when the weight of the bariatric patient, bed, transport staff, and specialized equipment are added together.

BARIATRIC PATIENT ROOM

Most examples of bariatric rooms currently in place are modifications to standard private or semiprivate bedrooms with a width (headwall-to-footwall dimension) of around 12 feet. The Bariatric Room Design Advisory Board (BRDAB) concluded that the optimal width should be 14 feet (13 feet was felt to be the minimum) and depth should be 15 feet (corridor to exterior wall).

Focus on the primary space drivers in these rooms, which include space for such specialized equipment and furnishings as bariatric beds, resident lifts, bariatric wheelchairs, or oversized chairs.

High-quality bariatric beds address the challenges inherent in bariatric care—patient comfort and mobility and mitigating the risk of injury to caregivers and patients. Such beds have built-in scales and can be converted to chair position, a configuration to facilitate patient transfer.

Other patient room space drivers include maneuverability needs for both the bariatric patient and for the care team who are trying to safely and ergonomically assist the patient.

One additional design consideration is the placement of cubicle track in relation to ceiling-mounted lift tracks. This is especially important for privacy curtain placements as the lift track runs from bed to bathroom.

Bariatric patients are likely to require assistance in transferring from a bed to a chair or from a chair to a toilet. Depending on the patient's ability to bear weight, level of cooperation, and upper body strength, two to three caregivers may be needed to assist the patient. Even with the most observant and diligent care, patients will fall. Wide spacing between the bed and other obstacles will facilitate the care team's effort in uprighting the patient and can mitigate the circumstance of the patient striking objects during a fall.

According to a 2008 white paper by the Houston office of Page Southerland Page (PSP) architects, the average hospital patient room requires a 16- to 19-foot unassisted trip from the bed to the toilet. Where possible in new construction, patient rooms should be designed to minimize the distance between the bed and the bathroom. In one project—St. Luke's Sugar Land Hospital in Texas—PSP was able to cut the trip distance in half.

However, in all cases of new construction, renovation, or retrofit, evidence-based design advocates are calling for handrails to be installed from the bed to the bathroom to reduce slip/fall risk. At St. Luke's Sugar Land, all rooms had a continuous handrail installed in the room, thus making all rooms handicap accessible.

THE BARIATRIC PATIENT BATHROOM

Sliding intensive care unit–style doors are employed in some bariatric room settings. A sliding door that will yield a 60-inch opening requires an overall opening of between 9 and 12 feet. The BRDAB preferred a pair of unequal-leaf swinging doors—one leaf 42 inches wide, the other 18 inches—as the optimal solution. This configuration will yield the desired clear opening with the least overall width, thereby giving more wall space to supporting functions.

Oversized toilet seats are another preferred option. Toilet fixtures and sinks should be mounted to the floor versus the wall, although care should be taken that floor-mounted sinks do not interfere with wheelchairs.

Bathrooms should be sized to allow for staff assistance on two sides of the patient at the toilet and shower. Dispensers should be flush mounted to aid in clearance and safety. The shower stall should have a sufficient opening and space for unrestricted movement by the bariatric patient and, if necessary, staff performing assists.

Consider combining shower and toilet into a central bathing room. The BRDAB concluded that a room of at least 45 square feet, with waterproof walls and floor, would be the ideal toilet/shower room. With strategic placement of the fixtures and sloping of the floor to a drain, the entire room becomes the shower. By not having enclosing walls around the shower and using a shower curtain instead, caregivers can offer maximum assistance.

The receptor should be tested for load bearing per ANSI Z124.1.2-2005. The receptor should also have a front transitional edge for easy access. Multiple grab bars should be available and

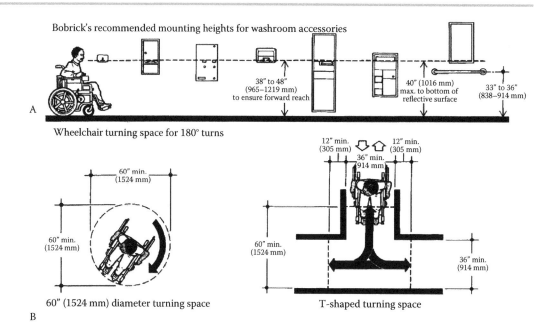

FIGURE 7.27   Wheelchair perspective, which also aides people using a walker, cane, or crutches. Bobrick Washroom Equipment, Inc. provides the mounting heights shown for washroom accessories and for wheelchair turning space. (Courtesy of R. Null.)

rated for 1000 pounds each. Also consider contrasting colors of grab bars for those who may have visual impairments.

In addition, space for adaptive equipment such as wheelchairs and lifts is essential to plan for. Consider tracking for ceiling lifts, accommodating ready access to shower, toilet, and sink (Figure 7.27).

Space design provides maximum safety and comfort—for both the patient and caregiver—and encourages some degree of participation by the patient.

LOBBIES AND WAITING AREAS

Basic room furnishings, such as chairs, may lack the size and weight capacity needed for bariatric patients. Therefore, it is important to provide a variety of seating options in waiting rooms.

As mentioned earlier, there are two types of severely obese people—pear shaped (weight concentrated below the waist in the hips and thighs) and apple shaped (weight concentrated above the hips in the stomach and chest). This is important when considering seating, as pear-shaped people cannot abide chairs with arms, whereas apple-shaped bodies will do well in seating with or without arms. Offering both types of seating would service the general population as well as the severely obese.

Also, a seat that is too low may result in the patient having trouble standing without assistance.

CONCLUSIONS

- Obesity has become an alarming epidemic with enormous implications for our health care system.
- Respecting patient dignity and delivering optimal clinical care are primary issues, as are establishing procedures for safeguarding the health and well-being of these patients and their caregivers.
- Design is a critical tool in improving long-term clinical outcomes for bariatric patients.

- Designers must focus on appropriate facilities and space, proper equipment, and furnishings in bariatric care facilities.
- The probability of greater wear and tear calls for stepped-up interior protection for walls, corners, door, and door frames.

Incorporating Universal Design in a Bariatric Care facility will provide the flexibility for that facility to meet the needs of the entire patient population.

The following case studies and chapter section are derived from evidence-based research done by Alzheimer design experts John Zeisel, Martha M. Tyson, Margaret Calkins, and Clare Cooper-Marcus.

The major resources include "Alzheimer's Treatment Gardens, Design Guidelines and Case Studies" by John Zeisel and "Healthcare Alzheimer Care" by Martha M. Tyson, Douglas Hills Associates, Inc., in Chapter 9 of the book *Healing Gardens* with editors Cooper-Marcus and Barnes (1999).

Margaret P. Calkins wrote an article titled "Evidence-Based Long-Term Care Design" that was published in *NeuroRehabilitation* in 2009. She summarized research on dementia and design conducted and published since 2000. These are integrated into the material from her article that is shown in this chapter. Because of the importance of this research, we have also listed part of the pre-2000 research that Calkins identified.

EVIDENCE-BASED LONG-TERM CARE DESIGN

Abstract: Research on the impact of the built environment in long-term care settings continues to grow. This article focuses on work conducted and published since 2000, when an earlier review on research on dementia and design was published. The vast majority of research that addressed neurological conditions in residents in long-term care settings (assisted living and nursing homes) relates to Alzheimer's disease and related dementias.

INTRODUCTION

The earliest work on the role of the built environment on the behaviors and emotions of individuals with neurological conditions—specifically Alzheimer's disease and related dementia—was conducted in the 1970s by M. Powell Lawton. The Weiss Pavilion at the Philadelphia Geriatric Center was a radical departure from traditional nursing home design, designed to address the needs and behaviors of residents as opposed to focusing solely on maximizing staff efficiency. While results of studies of this building were positive, Lawton noted "the independent variable itself (the total environment) was distressingly gross, in that the change in the treatment locale subsumed an immense variety of components whose effects are unquestionably related to one another in very complex ways." After this auspicious beginning, in the late 1980s, research again began to explore the role that built environment plays in the expression of anxiety, agitation, wandering, or other "challenging behaviors" in individuals with dementia, both at home and in shared residential long-term care settings. Importantly, providers and designers of settings for individuals with dementia are increasingly recognizing the importance of physical as well as social environment and are actively working to create supportive settings. The most significant evolution in research in the 90s was a shift from an almost exclusive focus on challenging or disruptive behaviors to a more inclusive approach that also explored positive emotional states and well-being. There have been new measures and methodologies that have allowed researchers to examine the expression of positive emotions.

Paralleling the increasing recognition of the importance of the physical environment as a component of care is a significant increase in the number of studies of the influence that the built environment has on the cognitive, emotional, and physical functioning of individuals with dementia in long-term care settings. Despite some weakness in this body of research, there is growing evidence that the design of the built environment, by itself and in combination with organizational policies and procedures, has a direct and measurable impact on the physical and psychosocial functioning of residents with dementia, which may translate into higher quality of life.

There are background (personal) and proximal (extrinsic) factors affecting behavior. Background includes characteristics of the individual such as demographics, cognitive status, and health status. Proximal factors are dynamic interpersonal and environmental factors, which includes the built environment.

While the focus here is on the physical environment, the world we inhabit is complex and reflects the interwoven effects of myriad discrete and interconnected elements that ultimately cannot be disentangled.

Tilly and Reed published two literature reviews, one on falls, wandering, and physical restraints and the other on general non-pharmacological interventions to improve quality care. Rather than replicate what has already been published, this article will only briefly summarize literature included in these previous reviews and will focus primarily on research conducted since these reviews were published.

RECENT RESEARCH RELATED TO DEMENTIA AND ALZHEIMER'S DISEASE

The trend over the past two decades has been to reduce the size of the nursing unit from 60 beds to generally less than 20–24, though some are as small as 9 residents…. Earlier literature reviews reported several studies indicating that larger units are associated with poorer resident outcomes, including higher agitation and aggressiveness, greater intellectual deterioration, and greater psychotropic and antibiotic medication use.

Smaller units appear to have a number of positive benefits, such as higher motor functioning, greater friendship formation, less anxiety and depression, and greater mobility. Smaller unit size was associated with greater positive and less negative activity, less sadness, and less restraint use.

Small group home residents experienced improvements in hygiene, dressing, and eating over the course of the 12 months of the study conducted by Suzuki and colleagues in Japan. Reimer et al. also found similar results, with residents who moved to small, one-bed households maintaining or improving on quality of life and ADL (activities of daily living) measures compared to residents who remained in traditional nursing home settings.

BUILDING CONFIGURATION

Traditional nursing home design typically comprised shared bedrooms (usually 2–4 residents per room) arrayed along a long, straight corridor. Letter plans (e.g., t, H, L) were common, meaning that shared social space location was often not directly visible to residents from their bedrooms. In the first specialty dementia care unit, which opened in the early 1970s, the plan was radically altered so that the bedrooms all opened onto one large, open shared social area that contained the dining room and activity spaces.

Residents in open-plan buildings showed less anxiety and greater interest than with other plans.

NON-INSTITUTIONAL VERSUS RESIDENTIAL DESIGN

It is interesting to note the prevalence of the term "non-institutional" in describing interior décor of nursing homes. This term is used frequently in the literature and in literature reviews. Yet, to design the "non" of something doesn't indicate what the design should be; only what it shouldn't be. The assumption is that people will know and agree on the "non-institutional" means, which is far from the case. Some designers and researchers who study these settings suggest more "homelike" furniture, which presumably means elements such as wood or upholstered covering versus metal and plastic, and a style that people recognize as being one that someone might put in one's own home. Others argue that it is color and pattern—that institutional is monotonous and therefore non-residential is colorful and with patterns. Still others suggest it means the inclusion of residents' own belonging and furniture—though this is generally limited to the residents' bedrooms, not shared social spaces. Earlier research reviewed by Day and Calkins suggested that positive outcomes, such as improved intellectual and emotional well-being, reduced agitation, and improved functionality, were associated with "non-institutional" environments.

WAYFINDING

Passini and colleagues reported that for newly admitted residents with dementia, learning new routes was a slow process. Residents who could not identify paths to desired locations exhibited anxiety, confusion, mutism, and even panic. [*The author, while visiting a new retirement facility years ago (one with many similar doorways along corridors), observed that one of the new residents had hung a necktie on the decorative knocker on the door to his room, making it easier to identify when returning from the dining room.*] They also noted that some residents perceived patterns on the floor as a barrier. They concluded that "capacity of decision-making is reduced to decisions based on immediate and visually accessible information whether that information was signs, landmarks, or direct visibility of the desired location." They also noted that the typical location of signs (at heights specified by the Americans with Disabilities Act Guidelines) is often not seen by residents whose visual field is low to the ground…. Many similar doorways along corridors, lack of windows to the outside, and ad hoc signage resulted in poorer orientation.

SAFETY

Kincaid and Peacock found that a wall mural over the exit door significantly reduced overall door testing behavior. It is worth noting that a mural that disguises the door is not allowed by fire regulations in many states. [*Zeisel makes exit doors less visible (see Figure 7.28) by placing them on the side walls and painting them to match wall colors.*]

There have been anecdotal reports for years that patterns or dark areas of flooring can cause ambulation problems and falls in nursing homes' residents in general and residents with dementia in particular. One study specifically examined different characteristics of carpet design on 107 individuals with dementia. Results support the anecdotal evidence and indicate that larger patterns and higher contract within the pattern were associated with greater problems with ambulation (side-stepping, reaching for handrail, or veering while walking).

Falls are an increasing concern, owing to both their prevalence and the high costs of care associated with falls. It is estimated that 89% of residents with dementia in long-term care have some degree of mobility impairment…. Falls are most likely to occur at night, when residents are trying to get out of bed to use the bathroom. There is clear and convincing evidence that use of bedrails, which are used to prevent injury in residents, is associated with greater fall and fall-injury risk…. Environmental interventions that have been explored and found to be effective in combination with other interventions include repositioning furniture to be used for steadying, using floor mats to cushion falls, and using anti-slip mats to improve traction and improved lighting, especially at night.

DINING ROOMS

With the move toward de-institutionalization and the creation of households, described above, comes the inclusion of kitchens and residential scaled dining rooms in areas adjacent to resident bedrooms. Reed et al. found that "non-institutional" dining rooms were associated with increased food intake in residents with dementia. Brush and colleagues found that increasing light in the dining room and contrast at the place setting increased caloric intake by approximately 1000 calories for a 3-day calorie count in residents with dementia. Resident–staff communication also increased.

BEDROOMS

There is a growing trend in the design of nursing homes and assisted living settings for the provision of a greater number of private rooms. Early research found that when residents with dementia

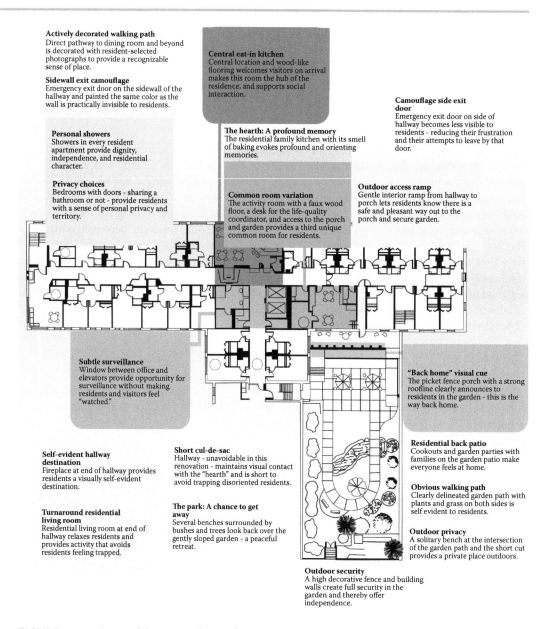

Actively decorated walking path
Direct pathway to dining room and beyond is decorated with resident-selected photographs to provide a recognizable sense of place.

Sidewall exit camouflage
Emergency exit door on the sidewall of the hallway and painted the same color as the wall is practically invisible to residents.

Central eat-in kitchen
Central location and wood-like flooring welcomes visitors on arrival makes this room the hub of the residence, and supports social interaction.

Camouflage side exit door
Emergency exit door on side of hallway becomes less visible to residents - reducing their frustration and their attempts to leave by that door.

Personal showers
Showers in every resident apartment provide dignity, independence, and residential character.

The hearth: A profound memory
The residential family kitchen with its smell of baking evokes profound and orienting memories.

Privacy choices
Bedrooms with doors - sharing a bathroom or not - provide residents with a sense of personal privacy and territory.

Common room variation
The activity room with a faux wood floor, a desk for the life-quality coordinator, and access to the porch and garden provides a third unique common room for residents.

Outdoor access ramp
Gentle interior ramp from hallway to porch lets residents know there is a safe and pleasant way out to the porch and secure garden.

Subtle surveillance
Window between office and elevators provide opportunity for surveillance without making residents and visitors feel "watched."

"Back home" visual cue
The picket fence porch with a strong roofline clearly announces to residents in the garden - this is the way back home.

Self-evident hallway destination
Fireplace at end of hallway provides residents a visually self-evident destination.

Short cul-de-sac
Hallway - unavoidable in this renovation - maintains visual contact with the "hearth" and is short to avoid trapping disoriented residents.

Residential back patio
Cookouts and garden parties with families on the garden patio make everyone feels at home.

Obvious walking path
Clearly delineated garden path with plants and grass on both sides is self evident to residents.

Turnaround residential living room
Residential living room at end of hallway relaxes residents and provides activity that avoids residents feeling trapped.

The park: A chance to get away
Several benches surrounded by bushes and trees look back over the gently sloped garden - a peaceful retreat.

Outdoor privacy
A solitary bench at the intersection of the garden path and the short cut provides a private place outdoors.

Outdoor security
A high decorative fence and building walls create full security in the garden and thereby offer independence.

FIGURE 7.28 Assisted living residence for people with Alzheimer's in Woburn, Massachusetts. (Courtesy of John Zeisel.)

moved from shared to private rooms, sleep improved, resident-to-resident conflicts were reduced, there was less rummaging behavior, and use of psychotropic medications was reduced.

CONCLUSION

The past five years has seen significant growth in studies that examine the impact of the environment on residents with dementia in long-term care settings. Some of the evidence is quite strong, reflecting similar results from multiple studies, while other evidence is not as robust, reflecting either a smaller number of studies or more contradictory results. The strongest evidence supports the positive benefits of private bedrooms on outcomes such as satisfaction of residents, families, and staff; quality of life; preference; and reduced nosocomial (hospital or group-living acquired) infections. There is similar strong evidence from multiple studies that smaller grouping of residents is associated with higher assessments on satisfaction, measures of quality of life, and ADL and MDS scores.... After years of anecdotal evidence about the negative impact of strong contrasts and patterning in flooring causing ambulation and safety (falls) risks for residents with dementia, there is finally one (but only one) well-designed study that supports this claim.

There is solid evidence that the built environment impacts the psychosocial–emotional well-being and probably physical health of individuals with dementia, although the complexity of these settings makes it difficult to come up with definitive or unquestionable results. That said, virtually no studies suggest that the traditional institutional setting, with multi-bed rooms, long corridors, and multi-purpose dining-activity rooms, results in overall better outcomes. It is time to stop allowing building of this style to be constructed.

This area of environmental research still suffers from some major limitations in terms of study design, replicability, and generalizability.... Designers and researchers need to work together to develop a common language and definitions to accurately describe the intent of the design of a nursing home for people with dementia.

It is also the case that design activities are moving forward, with or without empirical evidence to support it. Many professionals are increasingly purporting to engage in evidence-based design. However, the lack of consensus on what constitutes "research" and "evidence" leaves use of this term open to significant disparities.... Design is always about hypotheses—evidence based from rigorous research, assumptions from personal experience, or sometimes outright guesses—that people will see and use a building in the way the designer or provider intended it to be used. The value of conducting research, and of compiling the results of numerous studies into a single resource, is to help the field learn from what others have done. The impact of this effort would be significantly enhanced if designers and care communities made their design assumptions explicit. Knowing how a space is intended to function or a feature is expected to be used allows for more specific evaluation of the actual impact.... Hopefully, this will become increasingly popular for long-term care settings as well. In an ideal world, the design process would be guided by the best evidence available, design assumptions would be made explicit throughout the design process, and then evaluations would be conducted after the building is occupied to test those assumptions.

The comprehensive reference list for this article is located in Appendix A.

ALZHEIMER'S AND ENVIRONMENT-BEHAVIOR RESEARCH

What is Alzheimer's disease? Alzheimer's disease is a progressive, degenerative disease that attacks the brain and results in impaired memory, thinking, and behavior. It is the most common form of dementing illness. A person with Alzheimer's disease may experience confusion, personality and behavior changes, impaired judgment, and difficulty finding words, finishing thoughts, or following directions. A person with Alzheimer's disease has impaired abilities to comprehend present time, understand place, remember recent past experiences, and complete normal daily tasks such as dressing and bathing. From the first signs of impairment, the duration of the disease can last anywhere from 3 to 20 years (Alzheimer's Association, 1990, p. 1), with an average closer to 12 years.

The number of older Americans suffering from Alzheimer's disease or a related disorder is an estimated four million, and continues to grow (Alzheimer's Association, 1990). By the year 2050, the United States will have 67.5 million people over the age of 65, over two and a half times the 25.5 million there were in 1990 (Alzheimer's Association, 1990). According to the statistics supported by the Alzheimer's Association, if a cure or significant means of prevention are not found soon, an estimated 12 to 14 million Americans will be affected by the year 2040. Today, approximately 10% of the population over 65 have Alzheimer's disease with the percentage nearly five times that (47.2%) among the group over 85.

Normal aging. Old age is not a disease.

Historically, senility and old age were seen as synonymous. But dementia is a medical condition, while occasional forgetfulness and confusion are normal occurrences. Dementia has many causes, the major one being dementia of the Alzheimer's type. While increased age is associated with increased incidence of Alzheimer's disease, normal aging and the development of Alzheimer's disease and related dementias are different processes.

Presently, there is no cure for Alzheimer's disease. Nevertheless, well-designed environments, good planning, and medical and social management reduce symptoms and ease the burdens on family members.

Importance of environment. While the environment generally affects our state of mind and well-being, this is especially true for the elderly because of changes in our bodies and mind as we age. Decreases in visual and hearing ability that affect balance and orientation make older people particularly sensitive to the physical environment. (Please see Chapter 6 discussion of lighting by Eunice Noelle-Waggoner.)

Walking as well may not always be as easy when someone is over 80. Each step takes special care; gait and movement require more level surfaces, wider pathways, and defined edges in gardens to prevent falling, or to ease the fear of falling.

DEMENTIA-FOCUSED DESIGN RESEARCH

Environmental design research is playing an increasingly central role in creating quality living environments for older people with dementia. In the article "Evidence-Based Long-Term Care Design" earlier in this chapter, Margaret Calkins (2011) summarized research studies on Alzheimer-related environments before 2000 and then described research on the impact of the built environment with emphasis on work conducted and published since 2000. (See the Appendix.) Designs planned in response to such research are now often modeled after small villages or country homes, emphasizing elements like the old front porch, individual gardens, and family living. A familiar, homelike atmosphere, in addition to being therapeutic for older residents, may also reduce anxiety for their families and the staff who care for them.

Some researchers link behavior and design research with practical general principles for designing environments for people with dementia. They propose design and programming recommendations to meet specific therapeutic goals. Outdoor spaces are discussed as an important part of any complete therapeutic environment. In her research, Margaret Calkins clearly organized empirical experience and research into a set of behavior-based design guidelines for older people with dementia. Her later research focused on specific design solutions, showing how environmental adaptations like eliminating full views of exit doors can decrease the number of exit seekers and change the patterns of wanderers.

Wayfinding is helped by familiar destinations, layout of pathways, and surfaces that support residents' independent movement throughout the garden. Simple and understandable sequences of places along the path and significant landmarks help guide residents and make gardens more conducive to free movement and spontaneous use. One of the most difficult decisions a resident has to make outdoors is how to get back indoors, especially if there is more than one way to go inside. Reducing these choices by visually creating a single entryway to go back into the residence from the outdoor area aids wayfinding. This can be accomplished by the use of color and significant structural cues such as a built-in entry foyer that communicates clearly "this is the way home." For example, small goals next to the path draw people along the way. Birdfeeders, birdbaths, benches, and other familiar events cue residents to find their way through the garden.

Importance of gardens. One successful environmental element in treating Alzheimer's disease is a garden. People with Alzheimer's disease face problems of wayfinding, as well as object and place identification, whether inside or outside, so that the design of gardens for this group takes special effort and design research knowledge. Researchers have found that outdoor space use depends on resident motivation, comfortable seating, protection from the elements (Zeisel 2007), and staff-initiated use. Residents did not use spaces that were isolated, preferring areas of activity.

When people with dementia live in a group setting all day, every day of the year except when they visit or take trips with their families or caregivers, it is not surprising that they want to walk outside on their own sometimes. A safe and secure garden attached to a residence enables residents to go outdoors easily, especially if it is available to residents most of the time and if residents can use it freely.

Such free use for people with Alzheimer's can best be achieved by an outdoor space that is surrounded by a secure enclosure preventing residents from walking away—a fence, building edges, walls—visible to caregivers located inside the residence, with clear pathways for residents, with plants and planting areas, and with clear choices of where to go and what to do in the outdoor area.

It is necessary to understand the needs and desires of older people with Alzheimer's disease and their caregivers in order to design an appropriate and therapeutic outdoor environment. Research and clinical experience to date seem to indicate that appropriate and therapeutic outdoor spaces for people with dementia need to be both accessible and supportive. Gardens need to be accessible to residents physically, visually, and socially. They need to be designed to support physical needs as well as interaction with the environment and participation in social life. Gardens, combining the historic presence of horticulture and working the land, bring together two universal human activities that remain in deep memory and therefore are easily reintroduced into the lives of older people with dementia. Personally involving residents in a garden's stewardship adds a dimension of home that helps transfer the gardens' ownership away from the facility and designer, to the people who live there and their families.

Wandering and walking. Among the major activities that take place in gardens designed for people with dementia are walking and wandering. A great deal of environmental design and social behavior research addresses these issues. While there are numerous explanations for wandering behavior, the direct causes are easily identified and can be directly applicable to design. These include the following: reactions to crowds or noise, searching for someone, boredom, disorientation, searching for something to handle, and need for exercise. Research in wayfinding for the normal elderly (Weisman 1987), as well as for older people with dementia (Calkins 1989), demonstrates that the physical layout of buildings, circulation patterns, and the location of landmarks clearly affect behavior and independence of residents. The need for wandering is more than a physical need to move from place to place; therefore, pathways without destinations or landmarks along the way do not respond to that need. Safe walking paths that provide opportunities for interesting experiences appear more responsive and help reduce wandering behavior. Fenced-in grounds made safe for wandering is one effective design response. When the weather is appropriate and the patient's clothing is adequate, there is no reason why a wandering individual should not have the freedom to walk outdoors in all weather. Wandering, which is so disruptive in conventional nursing-home settings, can be ameliorated by supervised walks, which in turn lead to better general health, increased appetite, and better sleep patterns.

The use of the outdoor environment also benefits caregivers. Because a comfortable outdoor area provides a place that is different from the indoor rooms and spaces, when working with residents who become agitated or tend to wander in the unit, staff or family can use outdoor spaces to walk or involve residents in activities such as potting plants, sweeping the patio, hanging laundry, or going for a walk.

When working with elderly people suffering from dementia, simple gardening provides opportunities to include familiar activities that can be presented in easy stages. Environments that provide more than usual information and cues about time, place, and purpose are even more therapeutic. A richly planted garden area can provide material for reality orientation, providing the therapist with opportunities for verbal cues "such as 'Look at the beautiful yellow daffodils,' 'What a lovely spring day' and for the use of actual objects in activities such as picking the fruit and then being able to eat it." Little things can make a big difference. A garden that accommodates familiar daily activities enables even demented residents to water a window box or re-pot a geranium.

Daily life activities like setting a picnic table, feeding birds, watering plants, sitting in the sun, listening to music, preparing a meal, working on projects, or simply watching out a window are all ordinary everyday events for older people with and without dementia. Simple activities that can make the day worthwhile for many older people need to be seen as an integral part of design for their lives and health. After moving to a nursing home or to an assisted care or other health care facility, older people often feel a loss of control over their surroundings. Taking part in ordinary daily life activities with which they are familiar inside as well as in gardens helps ease the transition and improve residents' well-being. With Alzheimer's disease, although short-term memory is affected, daily life activities and quality living environments help draw on deep memories of their earlier life. The most successful Alzheimer's programs systematically introduce tasks that reflect common activities from everyday life, and gardening can be among these. Researchers also suggest that caring for dogs, cats, and birds can be therapeutic for some residents. They suggest that exercise can improve the physical and emotional health of older people, especially those with Alzheimer's disease. Walking in a garden provides this as well.

EXPLORING HEALING ENVIRONMENTS FOR PEOPLE WITH ALZHEIMER'S DISEASE

Designing residences for people with Alzheimer's disease and related dementias provides designers with special challenges and opportunities to explore the question of healing environments. These diseases are not mental illnesses, they are diseases of the brain—brain matter is lost over time. It is difficult for persons with these diseases to lay down new memories like remembering a message just taken over the phone or the name of a person just met, but it is easy for them to draw on deep memories of the past, of their long life. It is difficult for persons with these diseases to carry out complex tasks, like organizing a multicourse meal or balancing a checkbook, but it is easy for them to understand environments that are presented clearly and legibly.

Environments that heal—healing environments—for people with Alzheimer's disease clearly have to represent and reflect deep memories. What are these deep memories? They may be memories of workplaces, of traditional houses, of streets they have lived on, or they may be even more profound environmental memories.

What are profound memories? Fireplaces represent warmth, safety, and food. In traditional house settings, the Inglenook (the covered hearth with built-in benches on both sides) evokes such profound feelings. The kitchen is one place that strongly evokes food, family, and friendly communication. One's own personal objects, such as a familiar photography or an old housecoat, bring back memories of friendship and caring to demented residents. Music evokes great joy and profound sadness among Alzheimer's residents, letting them know that they are still alive. And just outside the front door—like in the clearing in front of a cave—people know they can feel the elements but are still close enough to home to be safe.

DEEP MEMORIES TRANSLATED INTO HEALING ENVIRONMENTS

How do such deep and profound memories translate into healing environments—homes and gardens—for people with dementia? Certain key areas and issues help translate these principles into physical form:

- *Exit control* provides residents with enough actual security to enable them to feel and be safe in the entire planned environment and thus to be able to be free and independent in their actions.
- *Social places* planned to appear different and unique, and not too numerous to be confusing, help cue residents to appropriate behaviors—helping them to draw on deep memories and be competent in their actions.
- *Private away places* provide residents with the chance to collect themselves and get away from the pressures of spending all day and night with a community of the same people.
- *Walking paths* with clear destinations and visual diversions to stop and look at along the way create environments where residents' need to wander becomes purposeful activity—taking a walk.
- *Healing gardens* accessible to residents all day in all seasons, sometimes to plant and sometimes to shovel snow, give residents contact with another deep memory—the outdoors, the seasons, the weather, flowers, and trees.
- *Residentiality*—scale, furniture, and décor that convey the feeling of home and homeyness—relaxes residents by tapping into their deep understanding that at home everything they do is okay. They can relax because it's safe and friendly.
- *Independence* is something that residential living provides residents by focusing environmental design on supporting what residents can do instead of what they cannot.
- *Comprehensibility* of the ambient and spatial environment—colors, sounds, sights, smells, space—is important for people who are confused by disordered and strange complexity. In understandable settings, demented residents can handle themselves well and are much less upset (Zeisel, Hyde, and Levkoff 1994).

DETAILED CRITERIA FOR HEALING ENVIRONMENTS

Exits. The brains of Alzheimer's residents cannot hold cognitive maps and they frequently forget how to return home. Therefore, people with dementia should only leave their homes when accompanied by someone else. Doorways from a residence that open to the larger public community, therefore, need to be controlled. Residents who spend so much time indoors become agitated by doors with mixed messages: On the one hand, windows and hardware on the door attract residents and seem to invite them to go out; on the other hand, locks and keypads prevent their use. Exit doors that are less visible—more unobtrusive—with no attracting hardware reduce agitation. Increasing the visibility and making more inviting any safe door to a secure healing garden further diverts attention from doors that exit to dangerous areas.

In this century, other signals can be chosen so that they do not disturb the ambiance of the residential setting—such as chimes rather than alarms. The less obvious the door, the signal, the hardware, and the other side of the door, the greater independence will the resident have in his or her safe environment.

Walking Paths. One of the symptoms of Alzheimer's disease for certain people is the desire to walk, perhaps looking for something without knowing precisely what. While aimless wandering can be a problem for staff in a facility that has no place for this activity, a well-designed pathway can transform wandering into walking. A pathway can achieve this goal if it is interesting and does not dead-end. Such a pathway need not be a specially designed circular track, but rather can be the thoughtful connection of corridors that pass through common areas and connect up again to corridors going in another direction. Interest along the path is important so that those walking always have some goal in sight—the next interesting picture, view, or plant. And interest at the end of the path, a social space or a fireplace, provides a place to walk to, a destination that gives purpose to each trip.

Common Space. Residents in Alzheimer's assisted living facilities spend almost all their time in the facility and together. To satisfy their need for diversity and to reduce boredom and agitation, it is essential to have at least two if not three different common spaces—dining room, kitchen, living room, foyer. The more the settings of these rooms are different and interesting, the easier it is for staff to manage smaller family-like activities there and for residents to feel stimulated by the differences in ambiance they can sense.

Private Areas. Because residents spend so much time together, they also need places to be alone, to avoid the pressures of social interaction. Just because someone is demented does not mean that they can stand being together with others 24 hours a day. Individual spaces that residents can use to get away by themselves can include private bedrooms or small, out-of-the-way corner sitting areas in a living room or garden. Residents with visiting family members who just need to sit together quietly can also use places like this. Private areas are also places residents and their families use to decorate and furnish personally, thus creating a soothing mood that triggers positive memories.

Healing Gardens. Not every residence is able to provide its residents a safe and secure outdoor area immediately adjacent to the residential area. Yet, this ideal gives residents a sense of nature, weather, and plants. If nothing else, Alzheimer's residents enjoy being outdoors and are relaxed by being able to get out of the confinement they feel inside. Yet, a healing garden is even more than a place to get out; it is a sanctuary where a basic drive to have contact with normal forces can be met.

Later in this chapter in the case study, "No Ordinary Garden," Alzheimer's and other patients find refuge in a Michigan Dementia Care Facility. Environmental psychologist Clare Cooper-Marcus describes a garden that has provided a successful healing environment for a specific population of Alzheimer's patients.

Residentiality. Home, fireplace, front porch, and garden are residential environmental design elements that create a positive mood in residents by touching deep-seated memories. The familiarity of residential furniture, spaces, decorations, and lighting fixtures relaxes everyone in Alzheimer's facilities—residents, their families, and staff members. Managing the size of features to be residential—a scale people can relate to and grasp easily—can be soothing itself. Residential furnishings create settings with manageable elements for demented residents and everyone with whom they interact.

Independence. Details in the environment such as handrails and floors that prevent slips and falls contribute to the independence and autonomy of residents, because they support each person's ability to do things on their own. It may seem obvious that a toilet so low that it prevents an older person from standing up alone limits independence, but it does. Any non-prosthetic or unsafe design element has this effect. The safer the environment, the more likely staff are to permit residents to move about by themselves and make independent choices.

Comprehensibility. Alzheimer's residents are not confused by everything around them. When the sounds, sights, and smells they experience are familiar, they can cope with them and enjoy them. A common myth and mistake in design for dementia care is that if everything is sedate and bland, residents will be soothed. This is not the case. Soothing can be anxiety producing if taken to an extreme. What is needed is to create enough activity to keep residents interested and to make sure that the activity provided is understandable to them. Colors are fine, and traditional patterns for wallpaper are better than abstract patterns. A television is fine, and recorded films that have fewer rapid changes for advertisements are more satisfying than random violence and loud noises from the television. Comprehensibility comes from sensible, common-sense management.

ALZHEIMER'S FACILITY AND ATTACHED GARDEN

One of the consciously designed projects is located in a renovated hospital building in Woburn, Massachusetts, about 20 minutes north of Boston (see Figures 7.28 and 7.29). The facility is the work of John Zeisel and the attached garden was designed by Martha Tyson. The assisted living residence for people with Alzheimer's disease houses 26 people cared for by a 24-hour staff, with several people who come to participate in the program just for the day. The treatment residence is located entirely on one floor and is secured for residents with magnetic door locks controlled by a coded push pad limiting egress and access to those who know—and can remember—the code. This code can be changed if a higher-functioning resident learns the code. The doors have no windows to the outside, and if they do, these are painted over. A tall, decorative fence surrounds the garden. The fence provides safety, and also potentially reduces agitation. It prevents views out to activities that would, by their interest, present attractive nuisances that might encourage residents to leave.

The main central pathway in the residence is essentially a straight line from one end to the other. But this straight corridor provides an interesting walking path with its many wall decorations—photographs chosen by residents and thus understandable, reminiscence shadow boxes

FIGURE 7.29 Assisted living residence for people with Alzheimer's in Woburn, Massachusetts. (Courtesy of John Zeisel.)

with mementos of residents' lives, decorated boards announcing events, staff members' names, and residents' faces. The walking path ends at one end with a fireplace/hearth and living room where resident meetings are held, small group activities are organized, and there is a television set. Almost at the other end is a room with a tile floor in which painting and other messier activities can be run, and in the middle, one passes a large dining room and residential kitchen.

The common spaces are each different from one another and, thus, can stimulate different moods in residents' minds. The living room is carpeted, has a unique decorative border near the ceiling, and has white flowing curtains. The kitchen/dining room has windows along one side, dining chairs and tables, a faux wood tile floor, and a residential kitchen with wooden cabinets and a breakfast counter at one end. Another common room is adjacent to the porch and garden and has less light and more active furniture. While residents *may not remember* the precise attributes of each room, they are likely to remember the feel of each.

The bedrooms generally provide the opportunity for residents to have *privacy* and be surrounded by their personal furniture and mementos. Every bedroom has a door and is therefore private. Most have a dedicated bathroom for each bedroom, while three sets of bedrooms each share a common bathroom—behind the entry door to the apartment. Residents all have their own furniture, wall hangings, and other decorations—all memory cues that are intended to reduce agitation and improve memory.

There is a lovely therapeutic healing garden accessible from a wide porch adjacent to the common room (Figure 7.28). The outdoor porch is covered and wide enough to sit on—although it is cold in winter in New England—and provides a view over the garden. A gentle ramp leads down to the completely enclosed and safe garden—a half level below—designed with both landscaped care and residential and wayfinding principles. There is a clear walking path, planting boxes, benches, and landmarks to help orient residents.

Each of the rooms in the assisted living residence is scaled like a residential space—perhaps with the exception of the dining room that can seat all 26 residents. The ceilings are low, the furniture residential in style, and unique decorative borders reflecting the use of the room grace the walls near the ceiling. And the residence—with 26 people—provides the opportunity for everyone who lives, works, and visits there to get to know each other. While the number is larger than a nuclear family, it is about the size of an extended family unit or a small residential community.

Because the entire residence is safe, both inside by virtue of the finishes and fixtures, and outside by virtue of secure doors and fences, residents are generally free to be as independent as their physical capacity enables them to be. Staff members, secure in their knowledge that residents will not wander away, do not feel they have to constantly follow and hold residents up. The "lean rail" along the walls in each hallway even enables residents who might otherwise be unsteady on their feet to make their way to where they are going by themselves.

There are no strange sounds, views, or other sensory stimuli in the residence. Furniture is familiar, the arbor in the garden is the same as in many residential yards, and the photos on the wall present comforting and interesting sights. There is no overhead public announcement system, and no strange and shiny floors waxed to meet regulations for cleanliness, as might be found in long-term care institutions. And the radio and television are not left on all day—programs are chosen and videotapes and audiotapes that present familiar tunes and shows are used.

In sum, the entire design and layout of this assisted living residence for people with Alzheimer's—its architecture, landscape architecture, and interiors—are planned to augment residents' memories and ability to function on their own. By taxing the parts of residents' brains that are still working well, and relieving the parts that are damaged, the whole person is supported. He or she feels at home, in control of himself or herself—as much as his or her age allows—and competent.

GENERAL LESSONS LEARNED FOR UNIVERSAL HEALING DESIGN

When taking care of people with Alzheimer's and other dementias, one learns to listen carefully to what they say so that one can understand what they mean even when the words are not quite clear. One learns not to say "No," but to gently divert a confused person from difficult situations.

When words do not seem to make sense to the other person, one learns to replace words with hugs and touch. These practical management lessons can be usefully carried over to daily life with colleagues, family, and friends. Using them in "normal" society helps people get along with others better, and generally be better people.

Design that enables people to achieve a profound understanding of their surroundings is equally powerful for those without dementia as for those with such a disease. Healing design principles for people with Alzheimer's disease can be translated into archetypal deep healing design principles:

- Feeling safe, secure, and free
- Understanding what is expected in one's community
- Being able to get away by oneself and unwind
- Knowing where one is going and having fun getting there
- Enjoying the outdoors and the changes of seasons and weather
- Knowing that anything one does is okay because the place is safe and familiar
- Celebrating what one can do, not what one fails at
- Not having to struggle to understand one's surroundings

NO ORDINARY GARDEN: ALZHEIMER'S AND OTHER PATIENTS FIND REFUGE IN A MICHIGAN DEMENTIA-CARE FACILITY

In the Sophia Louise Dubridge-Wege Living Garden at the Family Life Center (see Figure 7.30) on a sunny morning in early October, one is immediately entranced by the colors of autumn flowers, the sounds of falling water, and the feel of a secure, restful refuge. This place is exactly that—a refuge for the people who live at home with their families but who come to the garden almost every day. Some were professionals, even heads of corporations; others never worked anywhere outside the home. Now, approximately half have Alzheimer's disease; the remainder have other forms of dementia, schizophrenia, multiple sclerosis, Parkinson's, or Huntington's disease. The oldest patient is 90; the youngest is 36.

This garden, designed by Martha Tyson of Douglas Hills Associates in Evanston, Illinois, was the first of its type in Michigan. There are two main components to the half-acre site: the main strolling and viewing garden and the working garden. The working garden is a rectangular area east of the building with raised beds and trellises for horticultural therapy, a potting area with shade and a sink, a garden shed, a small orchard, a butterfly garden, and an umbrella-shaded area for seating near the atrium entry door. The larger component, the main garden, is entered via an arbor from the working garden and consists of lawns, paths, perennial beds, gazebos, a waterfall and pond, and various places to sit. The garden opened in 1999 and is maintained by a staff gardener, who works three days a week, and a landscaping service that comes once a week.

Founded in 1991, the Family Life Center is a health care facility housed in what was once a convent. A dining/activity room and a large, glass-roofed atrium with a fountain, a tile floor, and many plants have views and access to the garden. Overlooking the entire garden is a large conservatory, heavily used in colder weather for indoor horticultural activities—sorting and sowing seeds, pressing flowers, and turning gourds into bird feeders.

As with any health care facility garden, it is very important that potential users know it is there and have easy access into it. This is especially so where users are individuals with dementia: For them, graphic signage or a circuitous route would be daunting. The Family Life Center has spacious views to the garden and entries into it from the atrium and the dining room. The doors remain unlocked during the day.

Daily life at the center includes two visits each day (weather permitting) to the garden. Some patients who do not need to be watched constantly go freely back and forth from the building to the garden. But many do need constant care. In late stages of Alzheimer's, for example, people regress to a state not unlike infancy. They have to take liquid food and will put anything they can into their mouths, including flowers and plants—hence the necessity for nontoxic plants.

The design of the garden works extraordinarily well for programmed activities: chairs in a semicircle on the lawn for conversation, a garden house for listening to music and singing, a flat

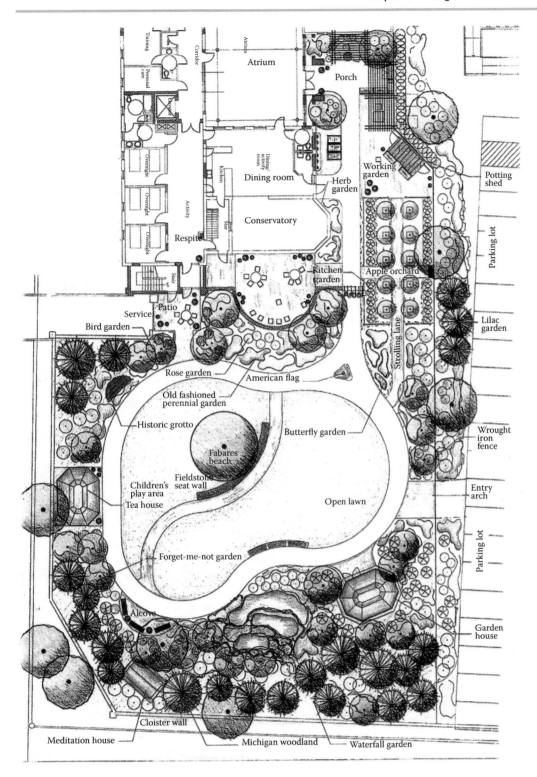

FIGURE 7.30   The plan of the Family Life Center garden shows entries from the atrium, dining room, and residential wing; the working garden; the main garden with looped pathway; and teahouse, garden house and plantings shielding the view of the perimeter fencing and walls. (Courtesy of Martha Tyson Garden for Clare Cooper-Marcus.)

lawn for croquet, a concrete path for wheelchair races, a number of faucets for watering the flower beds, and raised beds, tables, and a potting shed for gardening. "The garden also works very, very well for physical therapy staff working with people who have had a stroke or need help walking," says Sherry Gaines, the program and activity manager. "It has a wonderfully calming effect on people who are agitated. We have two customers who are totally blind and two who are legally blind. They like to hear the waterfall and birds."

In a facility with a high staff-to-patient ratio, the designer must bring staff into the design team. "We have a ratio of staff to customers of 1 to 5; when you factor in volunteers and interns, it is 2 to 1," says Gaines. A garden such as this is an important component of daily therapy; for some, it may be just as important as a dose of medication. The design of the Living Garden is successful because so many significant actors were involved in the process: representatives of patients' families, the center's staff, and horticultural therapists from the Meijer Botanical Garden in Grand Rapids.

Consulting advice has also been sought from Gene Rothert, manager of horticultural services at the Chicago Botanic Garden; landscape horticulture students from Michigan State; garden designers from a Grand Rapids nursery; and a local garden contractor. Input from such a range of experts has ensured that the garden serves the needs of patients, families, and staff; that the selection of plants is appropriate; and that built and movable features in the garden meet patients' needs.

Martha Tyson was uniquely qualified to design this type of garden. During her BLA studies at the University of Minnesota, she visited elderly relatives at nursing homes. Their yearning to be outdoors aroused her sensibilities. Not only did these homes provide meager outdoor space, but the doors to the outside were often kept locked. Her senior project was a nursing home garden.

For her MLA thesis at the University of Illinois Urbana-Champaign, she studied an Alzheimer's garden in Urbana that was functioning poorly. Using behavior-mapping techniques, she pinpointed the roadblocks, redesigned the space, and led a team of students who installed new elements in the garden.

When she went into practice, Tyson says, "sometimes I had a hard time finding employers who wanted to follow this approach." She adds, "In the mid-1980s, one of them asked me, 'Why bother to design for Alzheimer's people when they don't even know what's going on?' But I also found good people willing to let me search out clients amenable to a behavioral and participatory emphasis."

Tyson has had her own business since 1989 and has designed approximately 30 gardens for nursing homes and Alzheimer's facilities in Minnesota, Michigan, Illinois, and Massachusetts. Her current projects include gardens for a hospice and a home for developmentally disabled men as well as a line of outdoor furniture for people with disabilities.

At the Living Garden, the path offers a walking route and a fundamental orienting element; specific places such as the working garden and the lawn provide gathering places encouraging physical and social interactions; the entry arbor, the flagpole, the grotto with its statue of the Virgin Mary (50% of the center's users are Roman Catholic), and bird feeders hanging from trees provide landmarks to aid orientation; the teahouse and garden house create nodes or hubs of activity; and the building, walls, fence, and peripheral plantings provide edges defining the space.

The main garden, entered via a wooden arbor, has a clear perimeter path of tinted concrete, bisected by a curving brick pathway, thus allowing patients (who tend to be restless) a number of alternatives for moving around the garden. Because too many choices, however, can lead to confusion, agitation, and even aggression, one trip around the 6-foot-wide loop path provides a 300-foot route with changing details but no anxiety-provoking choices in wayfinding. The garden has plenty of seating in two wooden gazebos and elsewhere. There are three comfortable gliders alongside the perimeter path and one near the potting shed, patio seating with tables and umbrellas outside the conservatory, a curved stone seating wall, and numerous movable chairs scattered throughout the garden.

One of the characteristics of Alzheimer's patients is that they frequently try to "find their way home." Thus, an outdoor space or garden needs to be visually enclosed so that people are not exposed to tempting or frustrating views of "the outside world." Tyson's design fulfills this need in a functional and aesthetically pleasing manner. The garden is bounded on the north side by the building and conservatory; on the west and south sides by high walls of mellow, buff-colored brick; and on the east side (facing the parking area) by a steel fence. The walls and fences are virtually invisible, screened by a variety of trees—mostly evergreen—so that even in winter, the boundaries of the garden are blurred. One exception is a steel gate allowing entry from the parking lot for service personnel and people arriving for an event in the garden.

A garden, in contrast to an unchanging building's interior, is also a cogent metaphor for observing and talking about growth, blossoming, maturity, decay, and renewal. While the boundaries of the garden are marked by moderately tall trees stepping down in height to shrubs (lilac, roses, rhododendrons, and dogwood) and perennial borders, the center of the garden is open lawn that is in turn bounded by the circular concrete path and bisected by the curving brick path. The slightly mounded western lawn is partially bounded by a stone seating wall (permitting transfer from a wheelchair onto the grass) and punctuated by a beech tree and a playhouse with steps and a slide. Though this might seem a surprising feature in a garden for (mostly) elderly clients, those who live in the residential wing of the facility have family visitors. For lively grandchildren, a visit to the center may seem boring, or troubling, even frightening. A play feature in the garden provides a welcome diversion.

Not only does the Living Garden work extraordinarily well and provide a beautiful, colorful refuge for those who use it daily, it also provides a soothing milieu for occasional users and important ceremonies. Hope Network, which owns the Family Life Center, also owns other facilities in the area. In one such facility, for example, the seriously mentally ill live locked up, with their only outdoor space a parking lot. Sometimes, residents of these facilities come to the Living Garden for a picnic.

A recent study shows just how important a garden can be in the care of dementia patients. Julie Galbraith and Joanne Westphal, ASLA, studied the Martin Luther Alzheimer Garden at an Alzheimer's Dementia Unit in Holt, Michigan (2004). They examined nursing records for eight variables (aggressive and nonaggressive behavior, physician-ordered and as-needed medications, pulse rate, diastolic and systolic blood pressure, and weight change) during spring and summer 2000. When they compared the variables with the amount of time residents spent outside, residents showed significant improvements on virtually every parameter (one stayed the same; none deteriorated) with as little as 10 to 15 minutes of unprogrammed activity in the garden each day during the summer months.

A significant result like this can translate into substantial savings in staff stress, the cost of medications, and so on. Landscape architects need to have this kind of information ready to quote to potential clients to convince them that investment in a garden for a dementia unit not only will create a restorative setting for patients and staff but also is likely to lead to long-term financial savings.

With the aging of the US population and larger numbers afflicted with dementia, restorative settings such as the garden at the Family Life Center are going to be more and more in demand. Today, approximately 10% of those over 65 are affected by Alzheimer's disease, with the percentage nearly five times that (47.2%) among the group over 85. In 2050, the United States will have 67.5 million people over the age of 65, a great increase over the 25.5 million there were in 1990. If a cure or significant means of prevention is not found soon, the Alzheimer's Association estimates that 12 to 14 million Americans will be affected by the year 2040.

Landscape architects need to become familiar with and advocate for the particular needs of dementia patients and their caregivers so that appropriate, supportive environments become an important component in the whole spectrum of care. Gardens for people with Alzheimer's disease present clients and designers with a very special set of opportunities and challenges.

DESIGN ELEMENTS FOR PEOPLE WITH DEMENTIA

- Visibility and accessibility. Patients need to know that a garden or courtyard is available for use by seeing it from a well-used interior space (e.g., an activity room or dining room). There must also be easy access from that space into the garden. A less clear or more circuitous access route may frustrate the patients. And there should be only one access door. Patients can easily become confused if the door they exited is not the one they return to. The whole garden should also be visible to staff who bring patients outside and, ideally, from an interior nurses' station. In some situations where this is not the case, staff keep the exterior door locked, thus precluding patients from the benefits of spending time outdoors.

- A covered patio or terrace. A feature that is often overlooked is a shaded patio or terrace right outside the access door. Many patients like to sit and view the garden while remaining close to the building's amenities. Many programmed activities need a hard surface area with tables, provisions for sitting in the sun or shade, and enough space for group activities and people in wheelchairs.

- A simple path system (see Figures 7.31 through 7.33). Alzheimer's patients often engage in "wandering" behavior, so places to walk are essential. Because many have cognitive problems with spatial orientation, the circulation should consist of a simple returning path system (looped, circular, figure eight, bisected loop) with no dead ends and a minimum of choices that may cause confusion. Where space is tight, a continuous loop can be created with an outdoor path and a parallel interior corridor.

- Path design. As we age, our eyes are less able to cope with the glare caused by light bouncing off light-colored materials. Hence, outdoor spaces for older adults should have paths of tinted concrete with strong edge delineation. Ideally, paths should be a minimum of six feet wide and made of brushed concrete for good traction.

- Destination points. To encourage people to walk, a garden for dementia patients needs visible and memorable destination points such as a seating arbor or gazebo at one or more places along the looped path system.

- Ample seating. Dementia patients are often restless: They might pace, rest briefly, and then pace again. It is important to provide plenty of seating of different types (e.g., fixed seats, movable seats, and a glider) in different locations, and with a choice of sun or shade. All seating should have sturdy arms projecting just beyond the seat edge so that older people can push themselves up from a seated position.

- A homelike space appropriate to local culture. Research indicates that residents exhibit fewer dementia symptoms in more familiar, homelike environments than in larger institutional settings. This finding should also apply to outdoor space, as the Living Garden illustrates. Since many Alzheimer's patients recall details from their childhood, the design should incorporate plants and other elements that might trigger long-ago memories.

- A lawn. A lawn will provide an area for programmed events such as a barbecue or a croquet game, as well as being a reflection of the home settings of many patients.

- Activities reminiscent of home. Features that will remind patients of their daily routines at home are often provided in Alzheimer's gardens. These include lines for hanging clothes, cans for watering plants, and brooms for sweeping paths. In one facility in Victoria, British Columbia, an old red Buick is bolted down on a concrete slab; residents can wash the car and "go for a drive."

- Space for gardening. Simple gardening tasks allow people suffering from dementia to engage in familiar activities. A small working garden, a garden shed, raised beds, a potting table accessible to a person in a wheelchair, and outdoor faucets can support gardening, whether in a formal horticultural therapy program or as an informal activity with family members.

- Ample shade. Since one characteristic of Alzheimer's patients is that they have difficulty recognizing when they are too hot and would not think to put on a hat or sunblock, adequate shade is essential. This can be provided by umbrellas, trees, canopies, solid-roofed arbors, or gazebos. Slatted roofs should be avoided since the shadows of slats on paving can be perceived as depressions and thus be confusing, as can the juxtaposition of dark and light paving.

- Shade provided by a canopy or vine-covered arbor is essential at all exit doors since the eyes of Alzheimer's patients—as is true for elderly people in general—have difficulty adjusting from indoor light levels to bright sunlight.

- A wide variety of perennials. Funds for garden maintenance are often limited; the planting and replanting of annuals can be costly. Perennials should be selected to reflect flowers familiar to the residents from their youths, to create color throughout the seasons,

FIGURE 7.31   Alzheimer's garden benches and paths. (Courtesy of Martha Tyson Garden for Clare Cooper-Marcus.)

FIGURE 7.32   Alzheimer's garden gazebo. (Courtesy of Martha Tyson Garden for Clare Cooper-Marcus.)

FIGURE 7.33   Alzheimer's garden. (Courtesy of Martha Tyson Garden for Clare Cooper-Marcus.)

and to help with crafts. In areas with snowy winters, select trees and shrubs with interesting and colorful branching systems.

- Nontoxic plants. In late-stage Alzheimer's disease, people tend to put everything into their mouths, whether suitable or not. To relieve staff of the responsibility of watching every patient at every moment, plants that have any toxic component should be avoided. Commonly used poisonous plants include azaleas, black-eyed Susans, bleeding hearts, chrysanthemums, daphnes, daffodils (bulbs), English holly, English ivy, foxgloves, hydrangeas, lilies of the valley, lupine, mistletoe, monkshood, oleander, poinsettias, St. John's wort, spider plants, wisteria, and yew. Facilities with separate units and gardens for those in the early, moderate, and late stages of Alzheimer's disease should note that plant toxicity is not an issue for those in the early stage.
- A sense of enclosure. An outdoor space for dementia patients needs to be securely enclosed with a fence, walk, or railings screened with vegetation so that patients are not tempted to leave. A gate to enable maintenance staff to enter, or residents to leave in an emergency, needs to be subtly located or designed to look like part of an opaque fence.
- Inclusive design. Involve management and staff in the design of the garden. Research indicates that management policies and staff attitudes and training affect the success of outdoor space in a facility for dementia patients as much as the actual garden design.

BUILDING AN ALZHEIMER'S GARDEN IN A PUBLIC PARK

The American Society of Landscape Architects (ASLA) in 1999 initiated a program to build 100 parks through volunteer efforts to commemorate their 100th anniversary. The "100 Years, 100 Parks" program included nine therapeutic garden projects in different states in collaboration with the Alzheimer's Association. Two of the nine Alzheimer's gardens were planned for public spaces. In this paper landscape architect Mark Epstein describes the unique public process undertaken for one of those public gardens, the Portland Memory Garden, in Portland, Oregon. The Portland Memory Garden illustrates an innovative process of site selection and site design of an Alzheimer's garden within an existing City of Portland park (see Figure 7.34).

The multidisciplinary partnership for the Memory Garden had the following objectives:

- Create a restorative environment meeting the needs of those with dementia, their family, and their caregivers.
- Promote the project as a model or demonstration garden to teach how to design restorative gardens.
- Use the garden as a tool for public education about Alzheimer's disease.
- Provide increased recreational opportunities to the citizens of the Portland metropolitan area.

FIGURE 7.34   Alzheimer City Garden, Portland, Oregon.

Approximately 40,000 people were living with Alzheimer's disease in the Portland metropolitan area, and about 75% of those were living or being cared for at home. Caring for a loved one with Alzheimer's is a demanding, stressful task, and this garden was envisioned as a safe place to supervise a loved one in a relaxing outdoor environment. The garden would also provide an opportunity to educate family members about the disease, through signage or exercise suggestions, in a passive, nonthreatening manner.

Construction projects typically proceed in a linear sequence of funding, site selection, design, and then construction. The Memory Garden diverged from the typical path, resulting in a unique public education process that engaged and informed a wider body of design professionals and lay public than the usual construction project.

Early in the planning process, it became evident that the Portland Parks Department would require a lengthy public involvement process to ensure community acceptance of the project. They were wary from stiff public resistance to two recent proposals for memorials in existing parkland and did not wish to repeat the negative process. The steering committee therefore decided to pursue three tracks independently but simultaneously: site selection, garden design, and fundraising.

Open dialogue between the decision-makers and the potential users of the garden, and the active participation of the public in the site selection process, eliminated any "not in my backyard" sentiment experienced by the other park proposals.

The winning park site was selected in October 1999. The Parks Department Director and the Park Board quickly approved the project without dissent.

DESIGN PROCESS

The site selection process took 10 months. To keep the project moving ahead, the steering committee decided to start the design process before the site selection was completed. If the site selection process was used as a *public education* tool, the design process was used as a *designer education* tool.

The archetype design was created through ASLA workshops and community meetings that involved about a dozen local landscape architects. These sessions informed the participating designers of the specific needs of Alzheimer's patients and generated creative ideas and thoughtful discussions on the merits of proposed elements. The archetype design included a circular main path with secondary paths leading to a central lawn area and small intimate seating areas off the main path.

A trellis marks the garden entrance and allows a vista over the entire garden (see Figure 7.34). Plantings are lush, yet ordered, including some in raised planters of different heights. A "rain-catcher" water feature, a touching garden, and a symbolic orchard were proposed as memorable features within the enclosed space of the garden.

The archetype plan became useful for subsequent publicity of the project, for early review by the Parks Department, and was especially helpful, opening a design dialogue with the selected neighborhood association. Landscape architects and community members refined the archetype design a few times through open community design meetings after the site was selected, culminating in a final site design.

PUBLICITY AND FUNDRAISING

The project generated supportive enthusiasm by all that were involved. The open process and clear objectives for the garden contributed to the broad acceptance and continued donation of time and resources from the all-volunteer effort. The various skills and abilities of the steering committee members allowed the project to proceed smoother and with greater success than originally anticipated.

In addition to the professional in-kind donations, more than 50 local contractors and suppliers donated labor and materials for the construction of the garden, along with three youth programs. Portland State University applied for neighborhood grants to support the project, and university students were preparing to perform post-construction evaluations of the garden.

Fundraising and volunteerism still continues through a nonprofit group, the Friends of the Portland Memory Garden, formed in 2001 and incorporated in 2008. They hold garden work parties twice per month to care for the plants. Park maintenance crews only cut the grass and sweep the paving. The garden has become popular among area nursing homes for day activities as well as individuals bringing their friends or relatives for passive enjoyment. The garden has also held many events, including parties, seminars, and weddings (Figure 7.34).

CONCLUSION

The Portland Memory Garden, dedicated in May 2002, was an extraordinarily successful volunteer effort because of the clear direction and goals of the project, the multidisciplinary talents and commitment of the steering committee, and the unusual process of site selection and project design.

The community effort put into this project yielded educational benefits before any construction began and continues to contribute a valuable new resource to the city and the region.

DESIGN FOR ASSISTED LIVING: GUIDELINES FOR HOUSING THE PHYSICALLY AND MENTALLY FRAIL

Humanitas Bergweg has been classified as a "leveensloopbestendige" or apartment-for-life project. In a book he wrote about designs for senior housing, SC architect, Victor Regnier described "Humanitas.Berweg." The basic idea is to create a housing and service system that supports older frail people in a normal apartment. The building is located in the central city of Rotterdam. In the surrounding neighborhood, 27% of the population is over the age of 65. The 195-unit housing development has a dramatic form that is recognizable from the freeway several miles away (see Figure 7.35). It consists of two housing blocks separated by a [climate] conditioned atrium. One housing block is a tower that rises from 4 to 12 stories in its 450-foot length. The other is a four-story, single-loaded corridor housing block. The atrium is three stories in height and is open on both sides to single-loaded corridors. A third of the population receive nursing care, a

FIGURE 7.35   Bergweg Apartments, Netherlands, exterior of the atrium. (Courtesy of Victor Regnier.)

third receive assisted living, and a third receive no helping services. The rooms are designed to be modified as residents age in place and need more support.

The bathroom is large enough to accommodate not only a wheelchair but a stretcher bather as well. The philosophy of care encourages residents to be highly independent. Residents are asked to take as much responsibility as possible for their self-care needs. Given the emphasis on independence, management believes that the third of the population that receives nursing care services lives far more independently than they would if they were in a nursing home. The three types of residents are scattered throughout the complex rather than segregated on separate floors or in one particular area of the building. The atrium is a public space open to everyone in the neighborhood. A restaurant, lounge, and bar are available for residents and community members, along with an array of professionals such as doctors, physical therapists, and occupational therapists.

PROJECT FEATURES

THE "APARTMENTS FOR LIFE" PHILOSOPHY IS ADVOCATED

The apartment-for-life idea is gaining strength in the Netherlands, which has had a history since World War II of placing older residents in relatively small service flats. This new approach allows residents to move to an apartment and stay there as long as they live. It is particularly appealing to couples who want to stay together, especially if one of them is chronically ill. Residents adhere to a "use it or lose it" philosophy and are admonished when they ask for help in tasks they can carry out themselves. The "apartment for life" philosophy extends the idea of "op maat" of the tailoring of services to the needs of the individual. Care managers are employed to coordinate care from a variety of sources, including spouses, friends, volunteers, and family members. Home and health care agencies that operate in the building provide formal services.

THE BUILDING WAS CREATED AS A NEIGHBORHOOD CENTER WITH HOUSING ATTACHED

The site was formerly a hospital. The neighborhood has many services located within walking distance of the building. The building concept centers on the creation of a large public space where everyone, including people from the neighborhood, is welcome. This atrium is located on the second floor, above a grocery store. The grocery store has a strong retail presence on the street level. Escalators on the northern corner of the site transport participants up to the atrium floor. The restaurant, lounge, and bar located here are open to everyone. Residents can take meals in their own rooms or they can come down to the restaurant. The units are mostly one-bedroom suites with a full kitchen. Although these are rental apartments, Humanitas is currently developing other projects where the units will be sold. Behind the grocery store on the ground floor and accessible by elevator from the dwelling units are a clinic, therapy office, and day care center. These services are shared by residents in the building and older people from the surrounding neighborhood.

THE COVERED ATRIUM IS VISIBLE FROM THE UNITS

The atrium is the main meeting space for everyone and resembles a shopping mall. Reverse steel bowstring trusses support glass roof panels. Elevators are centralized, and bridges connect the two sides. The atrium floor is dotted with tables and chairs. The atrium also contains a fountain, a 52-foot koi pond, and mailboxes for residents. It has an open plan, with exotic plant materials designed to be seen from the single-loaded corridors that surround and overlook it (see Figure 7.35). Opening the atrium to the public caused problems at the beginning, when vagrants and drug users frequented it. However, management was committed to keeping the atrium an open public space rather than a private one. Eventually, these problems were resolved, and the atrium is highly successful today (Christophersen 1997).

FIGURE 7.36   Senior housing reminiscent of a regular Norwegian house design. (Courtesy of Jon Christophersen.)

CARE AND SERVICES ARE GEARED TO SUPPORT AN AGING-IN-PLACE STRATEGY

The approach to this care philosophy relies on separating housing and services. This is done to deinstitutionalize the setting and to open it to a range of residents; rich and poor, healthy and sick, and young and old. Management believes in helping residents do things for themselves—often referred to as "helping with our hands behind our backs." They also believe that too much care is worse than too little. This system works well because it takes advantage of assistance from a spouse, family members, friends, or volunteers. It generally keeps care costs low and keeps residents more motivated to live an independent life. Residents and their families work with a care manager who coordinates needed services. If residents need nursing home care, it is delivered by a home care provider. Some services are planned, while others are ordered on an as-needed basis through the emergency call system. Hospice and day care are available, as well as night care for neighborhood residents. Demented older residents have been a challenge. Approximately five mentally impaired dementia residents were moved because their spouses couldn't manage their care. Twenty-five younger individuals with developmental disabilities also live here. The apartment-for-life program is very popular. Humanitas claims to have a waiting list that includes thousands of people. Although the units are much larger than those in a nursing home, the costs are about 35% less. This is due in part to the peripatetic approach to caregiving that characterizes this system. A total of 90 FTEs (full-time equivalent employees) operate in the building.

DWELLING UNIT FEATURES STRESS ADAPTABILITY

One of the main attractions of the building is the fact that residents need to move only once. Dwelling units are designed to adapt as residents age. If a resident experiences a severe chronic illness, services are provided. Dwelling units are arranged along single-loaded corridors throughout the building. Windows in the kitchen and a glass panel in the entry door allow borrowed light to enter the unit from the corridor side. The entry door is a 36-inch residential door, but adjacent to it is a 10-inch hinged movable panel that, when opened with the door, provides a space wide enough to accommodate a hospital bed. The units are relatively uniform in size. Most of them are 750 square feet with a bathroom in the middle. A few of the corner units are as large as 850 square feet. This location allows the bathroom to be accessible from the bedroom and the entry hall. An enclosed balcony or serre is provided adjacent to the bedroom but is also accessible from the living room.

DWELLING UNITS ARE DESIGNED TO ENCOURAGE AGING IN PLACE

A number of features have been added to the units to make them adapt easily to disability. The kitchen and the bathroom were given the greatest scrutiny. The bathroom is very large and has a roll-in shower. It is big enough to accommodate a stretcher bath, which is typically given to bedridden patients who cannot take a shower. The sliding door to the kitchen and living room is quite large and can be left open. The toilet has grab bars that can be moved, and there are places where additional grab bars can be added. A drop-down shower seat is also provided. Throughout the unit, there are wide doors and an absence of thresholds. The kitchen is adaptable and the counter can be moved up or down, depending upon the needs of the resident. The kitchen is complete with rollout shelves. The unit also has lever fixture taps and door handles, smoke detectors, easy-to-use windows and doors, and computerized door keys.

HOUSING THE COMING GENERATION OF ELDERLY PEOPLE

Researcher Jon Christophersen wrote on an issue looming before us, the care of an aging demographic.

In 1994, the Norwegian government initiated a housing program designed to provide for the projected rise in proportion of the elderly in the population. Aase Ribe outlined the workings and the results of the program including some of the built results.

DESIGN OBJECTIVES

Current ideals depart from the traditional. The care ideals and consequently also the design models take the notions of "home" as basis, and the design ideal is thus domestic architecture.

Designs were modeled on ordinary dwellings (see Figure 7.36) but were also adapted both to the needs of the elderly and to local conditions. Local conditions were anything from a city center or a small town to a rural, forested, mountainous, or coastal region, and the architectural setting varied correspondingly. As regards the users, personal needs included the basic need for a home, but ranged from a low level of accessibility to space-consuming technical aids and personal assistance round the clock. The latter posed severe planning problems: The dwellings must function equally well for both the elderly inhabitants and the care providers. Designing for both embodies the conflict of creating a workplace without compromising a home environment.

As regards dwelling standards, a main objective was that nobody should have to share a room and that everybody should have access to a private bathroom without leaving the private unit. Earlier standards were considerably lower. In addition, the Housing Bank required that

■ The drawings document accessibility and usability for disabled inhabitants (regardless of the type of disability) as well as the care providers
■ Dwelling functions are supplemented by communal areas if the individual dwelling units are smaller than 55 square meters

DESIGNING FOR THE ELDERLY—RESOLVING CONFLICTS

Good design has to resolve a number of conflicts, some of which can be serious enough to severely hamper the design and planning processes. Planning for elderly people will often bring out particularly severe conflicts:

1. Conflicts between the notions of home and the efficiency expected at a place of work. Balancing and resolving these very different interests have obvious design implications.
2. Disagreements reflecting different traditions in the care professions often result in clashes between the traditional institutional mode of thought and the more modern approach of providing care without invading the private sphere or jeopardizing the elderly people's right to independence and social life. The need for efficient and economical running of the care services adds to the problem.
3. Many design considerations might well be seen as expressions of value judgments reflecting attitudes to care and concern for elderly people. The building becomes an

expression—in physical form—of these attitudes and concerns, which further complicate choices of architectural form and expression and may create aesthetic conflicts.

4. Conflicts between short- and long-term costs. Constructing new housing strains most local authorities' tight budgets. There is a need to provide new housing cheaply and at the same time to build for the future and to limit the need for future alterations. Building cheaply obviously means using space economically. However, economy of space conflicts both with the need to plan for flexibility and the free floor space needed to assist elderly inhabitants.

Economy and efficiency also have consequences for both the choice of site and building form. Constructing new housing adjacent to existing institutions or building concentrated but fairly large new structures may contribute to keep running costs down, but is not in tune with the objectives of a domestic-scale architecture; the dominating form of Norwegian housing is low- to medium-density, single- or two-story timber frame structures.

The combined effects of economic and practical considerations relating to land ownership and site availability as well as efficient, large-scale service production methods are often that new projects are planned as extensions to existing institutions or on adjacent sites. The resulting concentrations are in some cases dramatic, creating mini-ghettos for the elderly, often including some dwellings for the disabled or other special needs groups. Such concentrations are widely discussed and criticized, but avoiding them seems hardly possible in the present economic climate.

WAYFINDING

Most elderly have sensory impairments, and quite a large proportion have cognitive disabilities. Most architects do not, however, know much about the aspects of wayfinding in buildings, and many designs show this clearly. A number of projects exhibit complicated circulation systems.

COMMUNAL AREAS

Communal spaces are generally seen both as an asset and as a necessity in both new housing and institutions for the elderly. There is a general feeling that communal facilities are needed, partly to create a good environment for therapy and care for confused or very frail elderly, and partly as a means to create and maintain good relations between less handicapped residents.

There are basically four different types of communal rooms:

- Lounges
- Dining spaces
- Kitchens
- Smaller areas off circulation paths

DWELLING TYPES AND STANDARDS

For the purposes of allocating funds and control of quality, the possible varieties of dwelling types are split into four categories: (1) self-contained dwellings, (2) communal living (Norwegian: bofellesskap), (3) collectives (Norwegian: bokollektiv), and (4) nursing homes.

SELF-CONTAINED DWELLINGS

Independent living apartments must conform fully to the general Norwegian criteria for barrier-free housing ("life span" dwellings). The standard implies that the dwellings occupy a minimum of 55 square meters and include entry/hall, kitchen, bedroom, and bathroom. Communal sitting room and dining facilities are more often than not provided as an optional extra.

COMMUNAL LIVING PROJECTS

The individual units are between 40 and 55 square meters. As they are smaller than the 55 square meters required for a fully functioning, self-contained dwelling, and the policy is that a fully functioning home should be available for every elderly person, the unavoidable shortcomings of the private unit must be compensated in communal spaces.

Private units in these projects resemble self-contained dwellings; space reductions affect mainly the kitchen. Thus, only the food preparation and eating functions need to be located in communal spaces. However, most schemes include eating spaces, communal kitchen, a sitting room/lounge, and an accessible toilet.

REFERENCES

Calkins, M. P. (2011, January 1). Evidence-based design for dementia: Findings from the past five years. (DESIGNMATTERS). *Long-Term Living, 60*(1), 42–45.

Christophersen, J. (1997). *Varieties of barrier free design: Accessible housing in five European countries, A comparative study*. Oslo, Norway: Norwegian Building Research Institute.

Marcus, C. C., & Barnes, M. (1999). *Healing gardens: Therapeutic benefits and design recommendations*. New York: Wiley.

Sanoff, H. (1979). *Design games* (Experimental ed.). Los Altos, CA: W. Kaufmann.

Sanoff, H. (1994). *School design*. New York: Van Nostrand Reinhold.

Zasler, N., & Calkins, M. P. (2009, January 1). Evidence-based long-term care design. *Neurorehabilitation, 25*(3), 145–154.

Zeisel, J. (2007, January 1). Creating a therapeutic garden that works for people living with Alzheimer's. *Journal of Housing for the Elderly, 21*, 13–33.

Universal Design in the Home

8

CONTENTS

UNIVERSAL DESIGN IN THE HOME

Perhaps no other environment so readily reflects the underlying principle of universal design (UD) as the home. The environment should adapt to fit people, not vice versa. When you think about "home," it is that fit that most readily comes to mind; home is where you are surrounded by the possessions you most value and enjoy, organized in a spatial arrangement that best reflects your personality and desires. Housing, however, has never really been designed to support the wide range of human beings that exists.

As the population ages, there are more and more examples of a disconnect between creating supportive environments and the aging user group.

Extensive research tells us that people desire to age-in-place, and current transition health care and community service emphasize aging-in-place. However, the structural barriers in much of the existing housing can prevent older adults and people with disabilities from leading independent lives and participating fully in their communities.

Many houses have steps at all entrances, narrow doorways, long and narrow hallways and lack an accessible bathroom on the main floor. TV ads for electric scooters and other mobility aids ignore these problems. Suburban house styles with large lots and few or no sidewalks create additional architectural barriers that make it difficult for nondisabled people including older friends and relatives who need basic accessibility.

If at some point your body changes because of accident, illness, or simple aging, then normal houses become less like homes and more like obstacle courses.

Another reason for creating more universal housing is one that is seldom considered: Inaccessible housing isolates individuals with disabilities, subtly segregating them from the mainstream. Even if an individual's own home has been adapted for his or her use, if friends and relatives live in inaccessible environments, there is little room for the type of socialization most people take for granted. This point is argued eloquently by Eleanor Smith, who convinced the Atlanta, Georgia, branch of Habitat for Humanity to make all their new homes more universal. Her grassroots organization, Concrete Change, has lobbied for basic access to every new house, including 32-inch-wide doors and one no-step entrance (Smith 1993).

It's important to remember that most people are only temporarily able bodied. Nearly all people will experience limits to dexterity, sight, hearing, and their senses of touch, taste, and smell. Therefore, homes must be designed with universal features so that families will not be disrupted or uprooted because of sudden disability or the natural effects of aging.

A home should allow for each individual's uniqueness, and UD offers the platform from which adaptations can be made to support that uniqueness throughout the life span. The 21st century will find many Americans not only living in their homes but working there as well (see Chapter 6, The Universal Designed "Smart" Home Office by Charles Schwab); thus, design principles fundamental to the workplace will also be applicable to spaces within the home. Requirements for accessibility and adaptability will need to be transferred to home construction and design (Kira et al. 1973).

Some general suggestions for universal residential design in new construction or major renovations include the following:

- Doorways with at least 32-inch clear access
- Level or ramped living spaces
- Powder room, or full bath, with 32-inch clear doors and wheelchair maneuverability on every level
- Room on main living level that can be converted into a bedroom (multilevel homes)
- Staircases that can accommodate a lift, or space for a home elevator (multilevel homes)

However, it is not just builders, designers, architects, and product manufacturers who are responsible for enabling or blocking universal environments. People are psychologically most comfortable with the familiar; they want to continue to experience life in similar ways—especially within their homes. But there's a danger to this, and within that danger is a pressing reason for approaching UD in the home as a priority.

A striking observation from an early study was the degree of difficulty subjects displayed in performing even the most fundamental everyday tasks, including opening doors and windows, maneuvering through the house, and wayfinding, with a concomitant inability to recognize and report on their limitations. Many subjects reported that they were quite capable of acting independently, when in fact they were almost completely dependent upon a spouse or relative for support of these activities.

To a large degree, these activities could have been conducted independently given relatively minor and inexpensive changes to the subjects' environments. However, in almost every case, home adaptations and behavioral changes had already been suggested but not followed. The individuals stubbornly refused to change either their homes or their methods of operation.

The need for adaptability—and for making it the norm—is the focus of this chapter. Underlying the call for UD in the home is the belief that homes should be able to change gracefully to meet changing needs. To this end, the case studies within this chapter serve as guides to creating fully adaptable and supportive universal homes.

Architect Charles Schwab (2004) developed a home plan book titled, "Universal Designed Smart Home for the 21st Century." The 100-plus home designs are truly innovative in that, along with traditional UD home design features, they include energy-efficient construction, security, and sustainable "green" building products that provide clean indoor air. In the following UD strolling tour, architect Schwab has described the UD features he has incorporated in each home plan.

TECHNIQUE: STROLLING THROUGH THE UNIVERSAL DESIGNED "SMART" HOME

UD is not simply accessible design but is "Design for all." It is truly inclusive, and when properly designed, it functions just as well for people of tall and short stature, people with visual or auditory impairments, overweight people, or anybody who uses mobility devices. It is appreciated by people of all ages. A person pushing a baby stroller will agree that no-step entries make it easier for strolling and an elderly person will enjoy being able to live in his or her own home longer and "age in place." A child or seated person can help prepare meals in the kitchen with varying-height countertops. Wider doors and hallways will be appreciated on moving day. Paramedics and firemen will be able to professionally do their job (without turning a gurney sideways to get someone through the door).

Ease of use, safety, and convenience are the by-products of UD. It is not a series of codes and it does not use signs or other designations to identify it. The purpose of UD, like assistive technology, is essentially the same: to reduce the physical and attitudinal barriers between people with and without disabilities. Well-conceived UD should be stealthy and can be invisible; it is quite simply good and economical design. UD in housing provides the freedom to live or work at home safely and independently. It means living in freedom in your own home for as long as you would like.

LET'S BEGIN OUR STROLL (BOTH WALKING AND ROLLING) THROUGH THE UD *"SMART"* HOME

Achieving UD is more than including a checklist of features. I encourage your interaction here and begin your own list of UD home features and the rationale behind them as you read the rest of this text. I would like you to think of any and all additional benefits in each room or detail and add to your list anything else you can think of. In my opinion, UD in housing features should have at least two uses for various demographic populations and situations. As an example, think of closed-captioned TV that was intended for use by the hard of hearing and is also very helpful when waiting for your table at the bar or riding a bike at the gym. *Universal design is a virtuous way of thinking. It is the creative realization of how one detail or spatial arrangement can work for various populations, or how it can be redesigned to do so.*

So let's look at the big picture by exploring spatial planning and circulation in UD housing as one strolls (both walking and rolling) from the garage into the main living areas. We will explore how a thoughtful floor plan design is a must when successfully incorporating UD in any home.

LET'S START AT THE GARAGE

We exit the car in a garage that has a 5′–8′ center aisle, so one can use a wheelchair lift with adequate space for exiting the car. Each side passenger will have a minimum 2′-6″ space to exit. There will be at least a 1/20 slope on the garage floor to a 5′ × 5′ flat area in front of the entry door. Prior to entering, you will notice a 2′ × 2′ flat metal floor grate with a drain inside that can be used to clean your boots after deer hunting, washing the dog, or detailing the wheelchair.

The switch to the remote and battery backup garage door and lights is no higher than 42″ above the floor. An electrical panel may be located nearby with adequate clear space in front that is no higher than 48″ to the highest switch. The outlets in the garage and home are a minimum of 18″ above the floor for ease of use without bending.

It will be easy to open the door as there will be at least 1′-8″ on the opening side of the door. Hence, if you use a wheelchair, there will be adequate maneuvering space. If you are carrying groceries, you can even use your elbow to open the door with the lever handle. Lever handles are easier to open for everyone and are especially appreciated by those who have limited hand motion or arthritis. A package bench 30″ above the floor enables you to set down your bags and then easily open the door. This is a convenience and a safety feature as it promotes stability for everyone entering the home.

THE UD EVOLVING "DROP ZONE" AND "COMMAND CENTER"

You will enter the home from the garage into a foyer that is at least 4′-0″ wide. The drop zone is an area between the entry door, mudroom, and the kitchen. It has counters and storage where the entire family can drop keys, mail, cell phones, laptops, wallets, and purses. The UD transition space or drop zone should have varying-height cabinets and full-extension storage just like the rest of the house. It should also have front-of-counter electrical outlets to charge phones and computers. These everyday items will then be handy and easily accessible on the way out in the morning. This is a convenient space and is also helpful for those who are forgetful.

Another growing concept is the "command center." It may be located in the same area. This takes the place of the traditional desk with the planning calendar, shopping lists, laptop, phone, and so on. Stamps, envelopes, and bills, along with short-term emergency supplies such as flashlights, batteries, and Band-Aids, are also kept here. This is the place for general knickknacks. The drop zone and the command center may be combined but must work for both fast- and slow-moving people.

Adjacent to the drop zone and command center is the utility room or half bath. This room may also contain the front-loading washer and dryer. When possible, use a pocket door so you don't have to consider space for a door swing. It is convenient to have the water closet near the garage without having to move through the house just to get to it. Many clients have told me that it is often the first stop when arriving home after a nice meal with friends.

THE EASY-USE KITCHEN

As you move through the home, you will notice an open floor plan. The kitchen is centrally located and near the garage entry, making it easier to unload groceries. Counters there are placed at various heights, making it convenient for everyone, young and old.

From the kitchen, you can view most of the common areas of the house. The main functions of the sink, refrigerator/freezer, and the cooktop were typically placed in a work triangle in the typical post-World War II home. Each leg of the triangle connecting the three should not be longer than 9'-0" for a total of 26'-0" for maximum efficiency. This is still good basic kitchen design; however, when multiple people are cooking and people are also mingling in the kitchen, new UD strategies need to be considered (see Kitchen Design for a Family of Cooks later in this chapter).

The new UD kitchen is being designed to accommodate these concerns, but a large area can be at odds with our 26-foot work triangle guideline. The solution can be to design combinations of triangles to intertwine with each other for a complete multitasking, optimum functioning area. For example, the main kitchen triangle may be a lower "work zone" area for a child, grandparent, or wheelchair user. A lowered sink in an island or peninsula with an under-counter drawer microwave and a cutting board would serve this area well. It can be tied into the other triangle with a two-door, side-opening refrigerator/freezer combination unit or the wall oven.

This "zone" may even have its own drawer dishwasher and refrigerator for complete working independence. These drawer units can be expensive; an alternate option for the drawer dishwasher is to raise a standard dishwasher so that the middle of the two dish racks is even with a seated user's elbow height for easy use.

I like to include lower-level seating near this second "work zone" to enhance the social setting and provide seating for company near the seated kitchen helper. When possible, I include a higher 42" serving bar near the main sink. This provides high and low seating in the overall kitchen design and gives one a choice of breakfast bars. These varying-height serving areas intermingled with the preparation zones make for pleasant casual entertainment areas within the kitchen itself. This will surely prove to be a popular "hangout" spot.

MAKING THE CASE FOR THE UNIVERSAL DESIGNED OPEN HOME PLAN

An open plan makes it easier for wheelchair users and people who use sign language as a form of communication. It is advantageous to be able to see people within the common areas of the home. Communication is made easier without having to move around corners while constantly navigating doors to have a word with someone. For wheelchair users, it also has the practical advantage of not bumping into walls and scratching the door casings as one is not turning around as much. An open plan can also be beneficial for elderly people who wish not to move around a lot but want to be included in general activities and easily approached.

I want to be careful however not to imply that open plans are the only plan type that will work for a UD home. An existing home remodel can also have room separation walls "opened up" and doors widened with appropriate structural analysis. It is however much more expensive to provide home modifications within an existing structure than it is to include UD home design throughout in the first place. In most cases such as an existing home, a combination of home modifications and a full home addition may make more financial sense. A home remodel with UD features can cost as much as 75% more than if you built new with full UD. Open plans have advantages but they also pose significant design challenges.

Defining room functions, differing floor materials and surface transitions, furniture arrangements, cross-area acoustics, colors, and circulation patterns are all aspects of the interior that must be considered. These aspects are only but a few of the design parameters that would also include natural and artificial lighting, windows, and design elements (Raschko 1982).

SMOOTH FLOORING AND SURFACES ARE CRITICAL IN THE UD HOME

Differing floor surfaces combined with varying ceiling heights are a fun and effective way to define different functional areas in an open floor plan. Many people, especially older people, take cues from the floor. Differences in texture are important for locomotion and depth perception. Floor coverings for the main circulation patterns through the home should be hard-surfaced materials such as ceramic tile, impregnated wood, rubber, and solid vinyl floor coverings to name a few. These should all have a nonskid, matte finish for minimum glare. According to the Americans with Disabilities

Act (ADA), a nonskid surface must have a coefficient of friction no less than 0.6 (0.8 for ramps) whether wet or dry. A rubber floor often exceeds this specification and can be a good selection. The joints of a rubber floor can also be seamlessly welded. Rubber floors also cushion falls. Although I mentioned the ADA above as a good base reference, it does not legally apply to private housing.

The finishes noted are easily used with wheelchairs, scooters, walkers with wheels, and rolling carts. For people with allergies, adhesives and leveling compounds should be avoided if possible. At the very least, make sure all sub-flooring materials are volatile organic compound free. Deep joints wider than 3/4″ may cause wheelchair wheels to turn or hold them up so joints should be 1/4″ to 1/2″ wide and of minimum depth. I specify tile on the diagonal, making it easier to navigate any device with wheels. This also creates visual interest and actually adds an additional tolerance level for the tile setter, as few would notice a nonperfect diagonal as much as they would a square pattern. Floors with a high percentage of vinyl resin are most resistant to stains while rubber and solid vinyl floors are the most resilient and dent resistant.

The color of floor coverings should contrast with the wall color to highlight edges of a room. It is also a good idea to contrast the floor with the furniture finishes. This can prevent falls and collisions for those with low vision while sitting or moving through a room. Take care to avoid high-contrast patterns in all floor coverings that can make small objects on the surface difficult to locate, especially to those with limited vision.

Carpeting with strong patterns or too strong color contrasts within the pattern may be visually confusing for the elderly and lead to misjudgments of spatial distance. Light-colored carpeting increases light quality throughout the spaces without increasing glare. Carpet also reduces the incidents of falls and cushions them if they do occur. A pile height of 1/4″ to 1/2″ should be high density and commercial grade. Carpet of this height is also best for wheelchair use and offers less resistance for all rolling traffic. A carpet surface that is too soft is too easy to sink into and may cause loss of balance.

A sense of touch is sometimes substituted when other senses decline. Carpet generally feels warmer and textural cues like carpet and fabric are stimulating and can be used to signal spatial changes in an open floor plan from one area to the other. Antimicrobial carpeting can be selected to stop bacteria growth and is waterproof. These types of floor coverings are most often used in health care settings.

DON'T FORGET THOUGHTFUL FURNITURE PLACEMENT

Furniture should be placed so as to allow access to each item of furniture, storage, windows, and appliances. Ideally, there should be 3′-0″ around each piece of furniture and a clear circulation path of 3 to 4 feet through each space. Floor outlets can create problems, but if used in central living areas, they allow lamps (and other accessories) to be plugged in. To avoid tripping hazards, a sofa or chair may be strategically placed next to them. I specify in-floor outlets for central living areas. A person may be limited in hearing or sight; thus, the seating should be arranged at right angles to each other and about 3 feet apart.

In typical furniture layout, the coffee table height is 12 to 14 inches high, but a higher table, one that is roughly 24 inches high, reduces the possibility of tripping over the table while lessening the amount of bending required.

THE UNIVERSAL BATHROOM

The UD bathroom is a pleasant departure from the bathroom of the 20th century. Upon entering through the 3-foot door with lever handle, you will notice a minimum 5-foot turnaround space in front of the sink. The floor is nonslip as discussed above. The faucet is a single-handle, lever type and can be located on the wall side toward the front of the sink. It should not be located further than 21″ from the front of the counter. This makes it easier for seated people, wheelchair users, and even small children to reach.

Pedestal sinks do not work well for wheelchair users, so a sink base cabinet with open knee space and doors that slide back within the box is a good solution. Another UD feature of the bathroom in Figure 8.2 is the wall-hung sink. The sink should not be lower than 34″ above the floor and the mirror should sit right above the back splash. Cabinet door hardware with open D-shaped pulls and slide-away hardware is similar to pocket door hardware but it slides into the cabinet as opposed to the doorjamb. This feature allows the vanity to have an open knee space within and still look the same as the others when the doors are closed. The hot water pipes have scald protection covers that protect legs from hot pipe burns. The sink is a maximum 7 1/2″ deep and can have an elongated bowl that extends past the counter front for greater accessibility.

Vanity lighting is side mounted so as not to cast shadows on the face. Full-height cabinet drawers with full-extension hardware make it easy to view and retrieve the contents within the most accessible heights; 22 to 48 inches above the floor. Electric, fog-free mirrors have benefits for the person who may not have the use of their hands and thus cannot wipe the mist off the mirror in a steamy wet room. Bathrooms also have a full-length mirror for grooming and dressing. A side-swing mirror is also helpful for close grooming (see Figure 8.1).

I like to specify UD bathroom tubular soar skylights with built-in fans for natural daylight and ventilation. These tubular skylights, if not too long, are a green feature and provide natural light without the deep skylight shaft. Natural but translucent light is important in my UD bathroom designs. I have found that many of my clients spend longer amounts of time in the bathrooms owing to disabilities; thus, natural light can be positive and stimulating.

The toilet is centered 18 inches from a sidewall. Wood blocking or marine-grade plywood is installed beside and at the back of the toilet for future grab bar placement. The blocking is strong enough to hold a 250-pound vertical and horizontal force when a user pushes or pulls on a future installed grab bar. It needs to be of use for heavy people too. It is important to include a 3′ transfer space on one side of the toilet and in front of the toilet. This will also be of use if a caregiver is needed. If possible, discuss this with the end user if there is a specific disability need as many people have a dominant stronger side (of the body) and will require access from a specific side.

Toilet height is perhaps the most difficult detail to make universal. Older folks who do not use a chair will appreciate a higher seat as they don't have to push up as far while a wheelchair user may want a toilet that is at the approximate height of his/her wheelchair seat. The good news is that the toilet itself is a fixture and not a spatial design matter so it can be changed out accordingly. There are also self-washing toilets that clean, wash, and even warm your posterior without the use of your hands. So, as I said, the toilet is a very personal matter and not easily universal. The solution is often the use of varying-height toilets in different bathrooms. There are also accessible toilet lifts and raised seats for specific needs (see Figure 8.2). When you have the opportunity, always interview the end user on this detail.

Every home should have at least one curbless shower and many have a "wet room" area where the toilet and the shower are adjacent to each other. On one side is a curbless shower and on the opposite side is a toilet; in between is another floor drain (see Figure 8.2). Many wheelchair users have shower chairs that carry water with them as one rolls out of the shower, creating a wet and

FIGURE 8.1   Side-mounted Baci mirror.

FIGURE 8.2   This bath utilizes several recommended features: color contrast, smooth floor surfaces, curbless shower with flexible water dam, grab bars, a wall-hung sink, and a raised toilet seat.

possibly dangerous situation. Hence, the wet room serves as a transition space from the shower proper to the rest of the bathroom.

The floor is sloped to drain but a flexible water dam can be used in between spaces to keep any water out of the rest of the bathroom.

The handheld shower units and the controls are located on the front side of the wall adjacent to the wet room. By providing a 7-foot shower hose, one can also use it in the wet room area if needed. I specify a product that provides control of the water flow from the handheld device. The pressure-balanced shower control provides scald-free protection. Grab bar units are included in prefabricated units or marine-grade plywood backing is installed behind the tile. This allows grab bar placement in any and all locations. The shower has a vapor-proof light/fan above. For those who still prefer tubs, automatic tub lifts are available to lower you after you have made side chair transfer. Walk-in tubs can be provided but with much thoughtful consideration. Phones are included in bathrooms and may be connected to an emergency alert system.

A SAFE HALLWAY LEADS TO PRIVATE ROOMS

Don't forget that smooth circulation as we "stroll" through the house is critical for a successful UD Smart Home. Finished hallways cannot be narrower than 3'-8" in order to turn a wheelchair into a room from the hallway. A 3'-6" finish still won't cut it. I design them 4'-0" when possible. This allows a caregiver to walk alongside a person using a walker if needed. It is also large enough to locate a whole house fan/ventilator above.

A chair rail is a nice added touch at both sides of the hall with a grab-able revealed surface to guide a casual walker down the hall. I like to provide a 5-foot wheelchair turn diameter at the end of the hall with a tubular skylight above to brighten up the hall. This brightens up and adds to the safety of the hallway, making it a room type in and of itself. The floor type should have the same nontrip or nonslip qualities discussed previously. Consider a differing color/texture edge at the hall sides so as to visually differentiate the wall from the floor. The wider hallway will also assist emergency responders should they ever need to access a bedroom with a gurney.

A 12″ tile base along the wall is a prudent detail if a person is already a wheelchair user. Wheelchairs can destroy doors and walls throughout the house. For this reason, I like to angle the corners of the hallway walls (and all interior wall corners when possible) and provide a tile base with an extra indestructible concrete board behind it that will take wheelchair punishment. Angled wall corners are also good feng shui practice (Chinese design practice that harmonizes energy for balance in a home environment). Feng shui notes that sharp corners "cut through" a home, thus unsettling and creating havoc. A curved or angled corner also provides a more blunt or rounded surface as opposed to the sharp cutting edge if one should fall.

YOUR UD STAIRWAY TO HEAVEN

Homes with stairs are well lit and designed for maximum safety. Stairways have light switches at both the top and bottom of the stair run. Tubular skylights are nice for improved visibility. The treads are a minimum 11″ and the risers are no more than 7″. This goes beyond code and makes for a less steep staircase. Treads have a closed riser and a nonslip surface to help avoid tripping. The stairwell is a minimum 3′-6″ wide but again I prefer to design them 4′-0″ for future installation of a stair lift. This makes the stairs still usable by ambulatory people when the stair glide is installed at the top or bottom. The handrails extend 1 foot past the top riser, and 2 feet past the bottom riser. This is a safety feature that accommodates the inclination to "reach" for the handrail that can cause a fall. This detail needs to be considered from initial design or the extended rails could intrude into circulation paths.

A HEALING INCLUSIVE BEDROOM

As you enter the bedroom, you will notice that the sills are no higher than 36″ above the floor. This allows for clear viewing outside while seated and can aid in emergencies if egress is required. Windows are located near a closet with a 5-foot turning diameter located in front. Locating the closet near the window provides daytime lighting near the closet. Closets also have automatic light switches that go on when the door is open and off when closed. The minimum size of a bedroom should be 13′ × 15′ in order to work for a wheelchair user. It is preferable to have a 5-foot-diameter turning radius at each side of the bed.

Closets have adjustable rods or universally designed closet systems. Optional sensor switches that turn lights on and off in bedrooms are specified for those who are forgetful. They also conserve energy. Walk-in closets are designed large enough for wheelchair use.

Every home should be wired with category 6 wiring for new "Tele-wellness" technology. (This requirement should be re-evaluated in view of the advances in wireless technology.) It can be expected that people will be healing at home, as opposed to lengthy hospital stays, and long-term care with visiting nurses and doctors that make house calls will be more commonplace. In fact, in-home health care provider is one of the fastest growing occupations at this time.

There are more rooms and areas in the UD Smart Home including the home office, the exterior front approach, interior foyer, double stacked storage closet areas for future elevators, storm safe rooms, and even lower levels (basements) designed for everyone. There are also exterior gardening, outdoor entertaining, and patio areas that need to be inclusively designed. Electrical systems and controls also have their own requirements and the main power panel should always be in an easily accessed location.

I hope you have enjoyed our stroll through the Universal Designed *Smart* Home. When it comes right down to it, universal design in housing results in personal and family freedom and accountability. UD is better living home design with self-reliance, independence, and wellness at its core. It's about active living for all household members who will have peace of mind knowing their home will serve them in good times and bad, and indeed for the rest of their lives.

Remember, true sustainability in housing must not place a greater burden on future generations, whether financial, physical, or emotional. Universal home design provides the best possible opportunity for all household members to be self-reliant and independent (Schumacher and Cranz 1975). Universal design in housing also has the opportunity to save taxpayers billions of dollars in long-term health care costs.

I wish you many years of pleasure in your 21st Century Universal Designed *"Smart"* Home.

Charles Schwab's Plan book of over 100 universally designed homes is a valuable resource for designers and builders.

LIFESTAGES: REDEFINING THE KITCHEN—A COMPREHENSIVE STUDY ABOUT GENERATIONAL AND SOCIETAL INFLUENCES ON KITCHEN DESIGN

Evidence-based research studies the design requirements of different people. Those who espouse the universal design philosophy must be willing to prioritize the needs of different user groups and work hard to find the closest approximation to a universal design. An example of evidence-based design, a lifestage research study conducted by National Kitchen & Bath Association members follows. This study influences design choices for kitchens.

The phrase "new normal" has been used to describe how society (from government to financial institutions to individuals) deals with the prolonged uncertainties of today's economic climate.

Society itself is less homogeneous in the new millennium with the rise of "generational power" and conflicting agendas. Advocacy groups like the American Association of Retired People now represent this generational power and are largely institutionalized. At the younger end of the spectrum, an entire generation has come of age with computer technology at its fingertips. Technology is more affordable and available than ever, even in developing nations. This shift is leveling the playing field dramatically between generations. In the United States, the collision of external and internal forces is influencing change on an industry that has been traditionally slow to change—home building.

THE NEW NORMAL

For the nation's homeowners and homebuyers, the "new normal" is not only affecting the way we live but also influencing how the spaces in which we live evolve, including, but not limited to the following:

- Delaying the sale of a home because of decreased property values
- Cooking at home more often to stay within a reduced monthly budget
- Renovating an existing home to have more room to entertain or provide daycare for grandchildren
- Having an elderly parent move in with adult children (vs. living in a nursing home) in order to provide at-home care, or owing to the uncertain future of the health care system
- Spending more time at home with family and friends versus going out

These are just a few of the real-world examples that emphasize how the "new normal" has shifted attitudes and behaviors in and around the home. Knowing that the kitchen is the heart of the home, the question becomes

"How will the collision of these social, economic, and generational influences impact kitchen design?"

In order to better understand where kitchen design is headed, Masco Cabinetry—home of the KraftMaid, Merillat, QualityCabinets, and DeNova brands—has taken a closer look at today's societal and generational differences, and how they affect kitchen design. The result is GenShift 2011: Lifestages Redefining the Kitchen—a comprehensive industry study with two main objectives:

1. To determine how the "new normal" is affecting generational needs as they relate to kitchen layout, specifically those of Baby Boomers, Gen X, and Gen Y.
2. To provide architects and designers with the most current insights into clients' generational needs in terms of kitchen design. The GenShift 2011 study confirms some widely documented generational stereotypes and relates them to the kitchen. It also spotlights some surprising findings as well—like how much time each generation is spending every day in the kitchen beyond just cooking and eating and how pets are influencing kitchen design.

The GenShift 2011 study incorporates an online survey conducted on behalf of Masco Cabinetry, from February 28 to March 2, 2011, by Harris Interactive, among 1027 US adult homeowners ages 18–65, as well as Nielsen Spectra data compiled from various reports from 2005 to present.

The study identifies changes in lifestyle, typical kitchen layout, and the function of the space that have implications for the aesthetics, storage, and other factors of today's ideal kitchen, as it relates to different generational segments and subsegments.

THE GENSHIFT 2011 REPORT

The GenShift 2011 report defines Baby Boomers, Gen X, and Gen Y as indicated below.

BABY BOOMERS (BORN BETWEEN 1946 AND 1966)

Born after the end of World War II, the Baby Boomer generation tops 76 million people. Baby Boomers are at a lifestage where they are now often in charge of their parents' care, which has allowed them to better plan for their own senior years. In fact, many Boomers are finding aging in place to be more desirable than relocation.

GENERATION X (BORN BETWEEN 1967 AND 1976)

Of the three generations, Gen X represents the smallest group at 50 million. Gen X is currently busy taking care of children and advancing their careers and trying to find that ever-elusive work/life balance. This generation is always on the go, and specific design elements can help complement a fast-paced lifestyle. Gen X tends to look to friends for advice.

GENERATION Y (BORN BETWEEN 1977 AND 1993)

Also known as Millennials or Echo Boomers, Gen Y consists of as many as 84 million young consumers. Even though they're somewhat new to the workforce, they're in the market for big-ticket purchases like homes—for many of the Gen Y, as early as age 26.

Armed with the latest technologies, Gen Y looks to third parties on the Internet for advice, information, and the lowest price for getting what they want… now.

KEY FINDINGS

Kitchen design can be affected by everything from market differences (southwest vs. New England) to budget constraints and peer influence to personal preferences. Every client and project presents different challenges and opportunities. Specifically, we wanted to learn

- Whether homeowners expect their parent(s) to live with them in the future
- Whether homeowners have a multigenerational household
- Likelihood to change home size over the next 5 to 10 years
- Whether homeowners feel their kitchen is universally designed for living

- Amount of time homeowners spend in their kitchen doing various activities
- Homeowners' opinions on the top three "extras" for kitchen layout
- Floor plan of homeowners' ideal kitchen
- Suitability of current home for children

KEY FINDING 1: HOMEOWNERS MAY NOT HAVE THE HOME THEY REALLY NEED

Twenty-four percent of homeowners ages 18–65 indicated that they live in a multigenerational house, and 73% indicated that their current kitchen is not universally designed for living (e.g., children AND older adults, aging in place, mobility, accessibility, functionality).

What does this mean?

- Identifying clients' current lifestages is important for architects and designers in order to best suggest products/solutions to address current and future kitchen functionality needs. The following are examples:
  - Older or smaller hands may have trouble grasping or pulling certain types of kitchen hardware. Consider larger drawer and cabinet handles that are easier to grasp and more ergonomically friendly.
  - Bending down can become more difficult as we age. Pull-out shelves in kitchen cabinets that fully extend make for easier access and can reduce the chance of injury.
  - Lighting is an important consideration. Different kitchen tasks (and different generations) require various levels of lighting. A combination of recessed, pendant, and under-cabinet lighting provides both aesthetics and functionality. Adding dimmer switches is a way to add even more flexibility.
  - While monochromatic color schemes have been popular in recent years, older clients may need more contrast between countertops and cabinets in order to maximize visual acuity.

KEY FINDING 2: GEN X HOMEOWNERS ARE STAYING PUT

What does this mean?

- Real estate statistics indicate that people spend an average of five to seven years in their home. The fact that a majority of Boomers and half of Gen X homeowners surveyed think they will spend up to a decade in their current residence indicates a major shift, which will most likely result in additional home improvements to enhance functionality and quality of living.

> Nearly two-thirds (63%) of Baby Boomers and nearly one-half (49%) of Gen X homeowners said they planned to stay in their current home over the next 5 to 10 years.

- Because of a combination of factors, Boomers are aging in place instead of moving. This means they are more likely to spend money on kitchen renovations if they're going to remain in their current home for several years.
- Since people are living longer, many Boomers will face the reality of having their parents move in with them. Floor plan modifications, including wider walkways and more clearance around kitchen cabinets and islands, will help improve accessibility for multigenerational homes.
- If a family is unable to sell its home owing to decreased property values, they may feel "stuck." However, this should be viewed as a positive instead of a negative and as a great opportunity to improve their current living space to create a home that truly fits their

needs. Architects and designers can guide them through the process by understanding generational needs for the kitchen.

KEY FINDING 3: DOG FOOD MAY BE MORE IMPORTANT THAN WINE

What does this mean?

- Pets are often considered members of the family, which increases the focus on their needs versus other creature comforts. In fact, as families grow and the nest becomes empty, pets often become the kids in the household.
- Homes need spaces for pets, including a pet feeding station with pet food storage.
- Mudrooms or laundry rooms sometimes double as "bedrooms" for the pets. Integrated cabinetry or storage for pet food, supplies, and even pet washing stations could become more commonplace.
- An emerging trend suggests that homeowners are moving away from the traditional kitchen desk. This space could be used for a pet feeding area.

> Twenty-two percent of all homeowners feel that a place for feeding pets is a top three kitchen extra, compared with only 10% who felt they needed a place for wine storage.

- Designing an area in the kitchen for pets and their supplies doesn't necessarily require ordering custom cabinetry. Removing a single base cabinet, with a narrow base cabinet next to it, could allow room for a food/water bowl at floor level, with nearby storage for extra food and supplies.

KEY FINDING 4: MULTITASKING REQUIRES MULTIFUNCTIONAL SPACES

> On a typical weekday, Gen Y spends more time entertaining and watching TV/using a computer in the kitchen than Baby Boomers or Gen X.

What does this mean?

- The kitchen is truly the hub of the home, especially with Gen Y homeowners.
- The kitchen is more than a room to prepare food—it's essential for everyday functionality, efficiency, and time management.
- Gen Y (44%) spends more time in the kitchen watching TV and working on computers than Gen X (36%) and Baby Boomers (33%). This generation in particular might appreciate a computer station near the kitchen (either desktop or laptop), which allows these generations to continue cooking dinner, answer homework questions, read the latest Facebook updates, and check their bank account balance. An additional consideration could be a charging station for all the technology devices that keep these generations connected.
- For homeowners without children, the functionality of the space changes slightly. This group is more focused on friends, cooking and wine clubs, and intimate dinners for two and baking. An entertainment space adjacent to the kitchen might be an important addition.

KEY FINDING 5: MULTIGENERATIONAL HOMES ARE ON THE RISE

The "multigenerational home" is a home that houses more than one generation, other than the typical parents with children under 18. Multigenerational, in this case, might include parents living with their adult children, grandchildren moving in with their grandparents, or adult children moving back in with their parents. A multigenerational home may include three generations living under one roof and sharing one kitchen.

What does this mean?

■ This surprising statistic highlights a significant shift in attitudes. While it may be several years before their parents actually move in with them, Gen Y homeowners appear to be more aware and accepting of the concept. As the idea of multigenerational homes becomes more commonplace, so too will the desire for kitchens that meet those specific needs.

■ The Gen Y kitchen may be required to be the most versatile. It must meet young child needs—like a space for high chairs and storage options for small, kid-friendly dishes—as well as adult needs, such as an entertainment area for the occasional get-together with family or friends.

■ An overlapping subset of the Boomers and Gen X has been described as the Sandwich Generation. They are sandwiched between their aging parents and children at home. When three generations are living under one roof, a consequence of the "new normal" for some families, they have the most demanding challenge of all—making a kitchen work well for so many family members at so many ages.

■ The aesthetics for this group follow a "keep-it-simple" design philosophy. This includes cleaner lines, less detail, less countertop clutter, and easy-to-clean surfaces for the parents of young children and the adult children of live-in parents. An example might be pull-out drawers that provide easy access to snacks, pet food, and small quantities of prepared items.

■ Although focused on their own lives, Gen Y embraces the concept of multigenerational living. Forty percent of Gen Y homeowners said they expect their parents to live with them in the future.

KEY FINDING 6: DO SWEAT THE SMALL STUFF

From holiday gatherings to dinner parties to weeknight homework assignments, the kitchen continues to be the heart of the home. The GenShift 2011 study found a common theme when it comes to kitchen cabinetry accessories—more storage in a clean design style.

> The top three kitchen extras for homeowners in all generational groups: a place for hiding small appliances, a place for bulk food or cleaning supplies, and a place for waste/recycling.

What does this mean?

■ Creative storage options are a plus for food processors, coffee grinders, and other gadgets, helping to create uncluttered countertops. Today's homeowner wants the latest kitchen gadgets, and it's important to provide convenient, easily accessible storage places for these "must haves."

■ Coffee drinking is more than a habit, it's become a national obsession. Keeping all the coffee supplies in one place, and perhaps in proximity to the sink, is something today's homeowners might appreciate.

■ Hiding places can take many forms: appliance garages are being replaced by slide-out storage and countertop cabinets with doors that rise up and out of the way. In addition, intelligently placed outlets can help keep cords under control (see Chapter 5, Figure 5.15, cabinets from Dai Sogawa).

■ Innovative under-sink solutions make storage and cleanup more convenient than ever.

- In the 70s and 80s, trash compactors could be found in many custom kitchens. Today, slide-out trash and recycling bins are making kitchens even more functional. It's important to consider placement (under the main sink, but perhaps a second trash can under the bar/prep sink) and which way the cabinet doors open in relation to one another.
- While traditional pantries are still fairly common, more homeowners are choosing tall, slide-out cabinet storage. Tall, adjustable shelves can accommodate big cereal boxes, heavy canned goods, and even large appliances like slow cookers. Oftentimes, this slide-out storage is positioned near the entry to the garage, which helps make unloading groceries more convenient.

KEY FINDING 7: THE BEST FLOOR PLANS ARE OPEN TO INTERPRETATION

What does this mean?

- Features are important, but the floor plan is the foundation for everything. Household demographics are critical and can include multiple generations—adult children, aging parents, multigenerational marriages, and children and grandchildren.
- There is a significant gap between how today's kitchens are designed and what is needed for this aging population, especially those who want to age in place. While 63% of Baby Boomers plan to stay in their current home over the next 5 to 10 years, 75% don't feel their current kitchen is universally designed for their living needs.
- Since Baby Boomers frequently use their kitchens for entertaining friends, church or card groups and grandkids, a standard-height table near or connected to an island is an ideal extension of the kitchen space. Multiple countertop and table heights in the same area offer the most options for universal design, whether it's multiple generations living together or one generation aging in place.

> Eighty-seven percent of all homeowners would incorporate a semi-open or completely open floor plan if they were creating their ideal kitchen. The number is even higher for Gen Y: 92%.

- Eliminating walls or creating half-walls can help make small areas feel bigger, in addition to providing better flow.
- Even with open floor plans, peninsulas and islands can help enhance the functionality of the space. They need to provide adequate room for cooking and entertaining, often at the same time.
- The size of the island isn't as important as its proportion and relation to the rest of the kitchen. If space allows, there should be at least 42 inches of clear floor space around islands and peninsulas, and even more room (48 to 60 inches) between the counter and kitchen table.

ADDITIONAL CONSIDERATIONS

The GenShift 2011 study generalized Baby Boomer, Gen X, and Gen Y populations. Each demographic group has subsets, which could have implications for kitchen design. Design professionals who spend time understanding the specific lifestage of clients are more likely to have a higher degree of success in designing beautiful kitchens that work.

LOOKING TO THE FUTURE...

The "new normal" has shifted attitudes and behaviors in and around the home for Baby Boomers, Gen X, and Gen Y. In addition, the collision of social, economic, and generational factors will continue to influence kitchen design. Despite these changes, the kitchen is still the heart of the home and the most important room for homeowners. Design professionals who understand the needs of each generation, and all generations, will be more likely to create beautiful home spaces that work.

SOURCES

For further information about the GenShift 2011 study or to speak with one of the design pros, please contact Marketing@MascoCabinetry.com or (734) 205-4600.

1. GenShift 2011 Study, Harris Interactive, February–March 2011 for Masco Cabinetry
2. Compiled Reports, Spectra Marketing, 2005–present for Masco Cabinetry
3. Facebook is a registered trademark of Facebook Inc.

THE AMERICANS WITH DISABILITIES ACT AND UNIVERSAL HOUSING DESIGN

The Americans with Disabilities Act (ADA) was a bold step toward universal design. It was significant first, because it defined a set of design requirements for a previously neglected segment of the population and made them readily accessible to the design community, and second, it legislated the priority that must be given to these functional requirements in designing public accommodations and commercial facilities. Every designer in this country needs to develop and apply a "functionality mind-set"—an ever-present awareness that their designs must meet the functional requirements of certain disabilities. At minimum, this mind-set will motivate designers to design public accommodations and commercial facilities to meet the needs of people who use wheelchairs, but many may also take on the larger challenge of universal design for all environments and all physical capabilities (Department of Justice 1991).

As designers accept this challenge, design researchers must progress in developing design requirements for many specific physical limitations (e.g., hearing problems, use of a walker, use of crutches) and many functional spaces (e.g., bedrooms, living rooms, garages) that are not yet well understood. Further, they must publish such information in places that are familiar and accessible to designers. A review of housing design recommendations for ambulant older adults (the majority of the over-65 population) revealed that many publications that purport to present housing design guidelines for the physical limitations of older adults, in fact, focused on design guidelines for wheelchair users, leaving large gaps in the existing knowledge of how to design for ambulant older adults. Those who believe in and design for the universal design philosophy must realize that although the design requirements of wheelchair users are well represented in the ADA and in design research literature, there are many other physical capabilities that must be considered in creating truly universal design. Legislators wisely limited the ADA to public accommodations and commercial facilities, recognizing that in environments such as housing, the needs of people who use wheelchairs might not always take priority over the needs of those with different capabilities.

The chart done by Dr. John Gill at the Helen Hamlyn Research Centre in London that appeared in Chapter 2 shows approximate percentages of each impaired group. Wheelchair users are only a small percentage of that population.

Many of the case studies in this book reflect some of the available research. And now, the Internet has created a nearly endless set of resources. The challenge for designers comes from the need to evaluate and prioritize these resources.

APPLYING UNIVERSAL DESIGN PHILOSOPHY TO THE RESIDENTIAL KITCHEN

Every physical limitation, whether a normal part of the life course, normal variation in the size of the adult population, or the result of a disabling condition, has different design requirements.

As one learns more about the physical capabilities of different people, one will also identify some conflicts between design requirements that make the goal of universal design so challenging. Such issues may be more numerous in the households of young veterans returning from the wars such as those in Iraq and Afghanistan because the survival rates have enabled many amputees and others to resume family life. Their earnest desire to live independently fuels a need for successful modifications for accessibility in the home. The following are examples of such conflicts in the domain of residential kitchen design (Koontz and Dagwell 1994).

SAFETY CONFLICTS

Safety is a particularly important criterion for universal design because an unsafe environment can lead to accidents that can result in physical disability or even death. When deciding where to locate potentially dangerous kitchen design features, one encounters several conflicts in the design requirements of different potential residents:

- *Location of electrical outlets:* Electrical outlets located at the front of the kitchen counter are much easier for short adults and wheelchair users to reach. But this location is more accessible to children than the usual back-of-the-counter location. It is also more

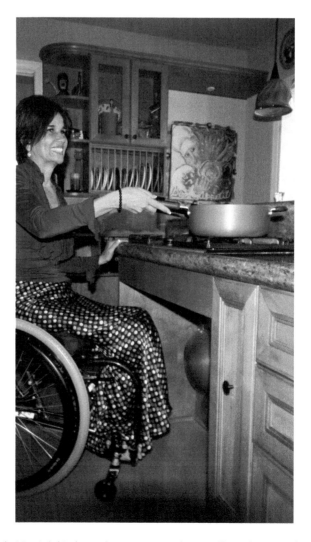

FIGURE 8.3   Valinda Martin's kitchen plates are stored vertically in the rear of the picture; her cooktop is over an open space, which not only provides storage for hanging pots, but also allows her to move into a more accessible position.

dangerous for ambulant adults who may become entangled with appliance cords hanging off the counter.

- *Location of stove controls:* Controls at the back of a stove reduce the chance of injuries to small children by making the controls inaccessible. But controls at the front of a stove reduce the chance of injuries to adults whose clothing may catch on fire when they reach over a hot burner. Front controls are also accessible to those using wheelchairs. Some designers compromise by providing stove controls at the front of the stove that require a degree of manual dexterity that children are presumed not to possess. However, this solution creates a problem for those adults with severe arthritis in their hands, who also lack manual dexterity. (For example, the "child-proof" lids on medicine bottles; children are not the only ones to lack the dexterity to open them.)

- *Location of microwave oven:* A microwave/range hood combination locates the microwave oven above the cooktop in a position that is safely out of the reach of children and that is in a convenient location for tall, healthy adults. However, it is inaccessible to those who need to work from a seated position, and it increases the chance that frail adults will spill a pan of hot food down over themselves (the National Kitchen & Bath Association specifies a maximum comfortable reach of 42.35 inches). The under-counter microwave oven (Figure 3.31a and b) is a solution similar to that addressed in providing a variety of counter heights in the kitchen.

- *Location of kitchen shelving:* Kitchen shelves designed for easy access by a wheelchair user also insure that everything will be easily accessible to children, including items such as cleaning products that must be stored out of their reach. Technology has created more accessible cabinet storage alternatives that are convenient. The vertical storage of plates adds to their accessibility (see Figure 8.3 or Figures 3.26 and 3.40).

SUPPORTIVENESS CONFLICTS

Universal design is also supportive of the physical abilities of all residents. Some examples of the conflicts that must be resolved in designing a universally supportive kitchen are as follows:

- *Kitchen counter height:* Ideal counter height is based on the user's elbow height and the requirements of the task. Thus, a compromise must be made in order to accommodate people of different heights (who may be working together) or the same person performing different cooking tasks. Current recommendations include:
  - Use of a variety of counter heights in the kitchen
  - Refrigerator and freezer access at all heights

  New configurations are designed for flexibility and good access (Figure 8.4).

- *Type of refrigerator*: Side-by-side refrigerator/freezers with new French door configurations are manufactured with a lower drawer-type freezer (Figure 8.5). These not only provide tempting access to icy treats for the younger generation—not necessarily desirable—but also enable wheelchair users ready access to frozen goods. Another benefit is the ease of storing large sized items in the refrigeration area because both refrigerator doors can be opened at the same time.

- *Clear door openings:* The ADA allows doors to have clear openings of 32 inches for wheelchair users, a problem in older homes. However, 36-inch-wide doors provide easier access to those who use crutches and walkers. Expanding hinges increase the size of door openings in older homes. Valinda Martin (Chapter 2, Figure 2.19) added expanding hinges when she renovated her home. She found that with popularity, the cost had come down in recent years.

FIGURE 8.4   Flexible and accessible kitchen countertops; St Louis 100% universally designed apartments.

FIGURE 8.5   French door refrigerator.

The housing designer needs to integrate design recommendations for wheelchair users with design recommendations for infants, young children, teenagers, and adults of all ages and abilities. Where a truly universal design solution cannot be found, priorities must be established among different user groups to guide design decision making. In describing the Seven Principles of Universal Design, Wally Duetcher said that "Equitable use" always was a priority for him because he is a wheelchair user. Universal design is a challenge even in the absence of budgetary, aesthetic, structural, and other important considerations. Nevertheless, the process of recognizing conflicts and creatively resolving them should lead to better housing design than would be achieved if such conflicts were overlooked. The teaching kitchens at the Center for Real Life Design at Virginia Tech illustrate good work center arrangements (Chapter 3, "The Universal Design Process").

THE FACTS OF FLOOR SPACE

The universal kitchen must incorporate enough space to allow use of the kitchen by an individual who may require some type of assistive device such as a walker, wheelchair, motorized scooter, or crutches. A universal kitchen should also provide enough space to accommodate multiple users, a lifestyle option or necessity for many of today's households. The two primary areas that must be considered in order to provide this space are the entrance or access to the kitchen itself and the clearances between opposing cabinetry and appliances. Criteria for doors and entrances into the kitchen are the same as for those in other accessible spaces (see ANSI 1992).

A clearance of 60 inches between opposing appliances or cabinets is most desirable as it provides for ease of maneuverability using assistive devices. Some standards for accessibility (e.g., ANSI 117.1, 1992) allow for a T-turn with arms at least 36 inches wide and a length of 60 inches. Under any conditions, a clear floor space of 30″ × 48″, either parallel or perpendicular to the appliance or counter, must be allowed in front of all appliances, counters, and storage. Provision of generous clearances is especially important in kitchens incorporating an island or peninsula, as the inappropriate placement of these features can effectively negate the entire basis of the universal kitchen concept. Also note that simply enlarging the kick space from 3 inches to 6 inches or 9 inches will allow one to cut overall dimensions.

Four Common Work Center Arrangements

Some work centers are easier for certain people to use than others. If the existing workstation is not appropriate for use by the family, the descriptions below suggest ways you can modify the design to make it more appropriate. However, unless the limitation is only temporary (such as a broken leg), the best solution may be to redesign the kitchen to use a more suitable work center.

U-SHAPED WORK CENTER (FIGURE 8.6)

Appropriate:
- Provides room to maneuver wheelchair
- Provides room for two people to work
- Reduces kitchen traffic flow problems
- Reduces risk of bumping into appliances

Inappropriate for:
- People who have difficulty maneuvering or seeing across wide open areas, such as people with walkers, crutches, or low vision.

To adapt a U-shaped work center for their use, place appliances closer together to shorten the work triangle.

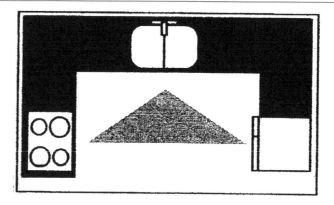

FIGURE 8.6   U-shaped work center: stove–sink–refrigerator.

L-SHAPED WORK CENTER (FIGURE 8.7)

Appropriate:
- Provides room for kitchen traffic to flow through the room without interfering with work triangle
- Provides ample room for storage next to each workstation
- Provides sufficient room for two people or a wheelchair

Inappropriate for:
- People who have difficulty maneuvering or seeing across wide open areas, such as people with walkers, crutches, or low vision

To adapt an L-shape work center for their use, place a cooking work station closer to the corner of the L.

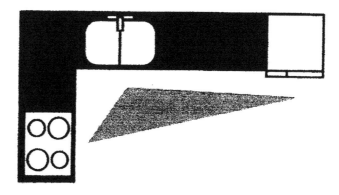

FIGURE 8.7   L-shaped work center.

ISLAND AND PENINSULAR WORK CENTERS

Appropriate:
- Shortens work triangle for easy use by people with low vision, walkers, or crutches

Inappropriate for:
- Open appliance doors may partially block aisle space needed for wheelchair.

To adapt a work center for use with a wheelchair, move the island/peninsula further away from the main wall to increase aisle space.

CORRIDOR AND PULLMAN WORK CENTERS (FIGURE 8.8)

Corridor work centers put appliances across an aisle away from each other. Pullman work centers place all appliances along the same wall (Figure 8.9).

Appropriate:
- Shortens work triangle for easy use by people with low vision, walkers, or crutches

Inappropriate for:
- Open appliance doors may partially block aisle space needed for wheelchair.

To adapt a work center for a wheelchair, widen aisle space.

*Note: The distance between the sink and appliances can make these work centers tiring to maneuver when using crutches, a walker, or a wheelchair. Shortening the distance will help eliminate inefficiencies.

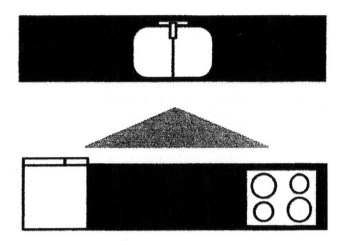

FIGURE 8.8   Corridor work center.

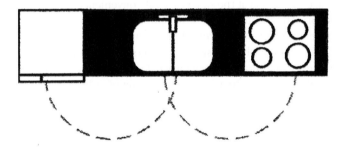

FIGURE 8.9   Pullman one-wall work center. The Pullman configuration with all appliances on the same wall can present problems because of the distance the user must travel or reach to perform his work tasks.

RECOMMENDED WORK TRIANGLE DIMENSIONS

	Standard	Wheelchair	Walker/Crutches
Total distance connecting refrigerator, range, and sink	12'–22'	14'–24'	10'–12'
Refrigerator to sink	4'–7'	6'–9'	2'–5'
Sink to range	4'–6'	6'–8'	2'–4'
Range to refrigerator	4'–9'	6'–11'	2'–7'

WORK CENTERS AND COUNTER HEIGHTS

The conventional approach to the design of kitchens focuses on five centers related to the food preparation process: refrigerator, mix and preparation area, sink, cook, and service area. A right-to-left sequence is generally preferred, although this may be reversed when the cook is known to be left handed or when architectural constraints prohibit a right-to-left order. The important point is for the centers to be in the order listed so that the food preparation process is facilitated. This is simply good universal design that helps create a safe work environment. (Specialized work centers are found in Figure 3.30, Virginia Tech's GE Center for Real Life Kitchen.)

An additional consideration in the design and layout of the universal kitchen is the height of the work surfaces at each of the centers. The customary standardization of all counter heights at 36 inches does not allow for comfortable use by seated users, shorter or taller people, people with neck or back problems, or children. In order to provide work surfaces at suitable heights for these users, some of the surfaces must be adjustable or the kitchen must include countertops at several heights.

Since side-by-side refrigerators provide both types of cold food storage at all heights, they are the best choice in the universal kitchen. The refrigerator should be located so both doors open to a full 180° with space for a lateral approach by seated users. Door ice and water dispensers make this type of refrigerator even more convenient.

The mix and preparation center should be either adjustable in height or permanently lowered to a height comfortable for seated users. Permanently lowered work surfaces can be achieved through provision of a mix center, desk area, or eating area (peninsula, island, and so on)—all areas usually built at lower heights. Pull-out cutting boards, drawers equipped with cutouts to hold various sizes of mixing bowls, and pull-out or fold-down tables are other methods of providing a variety of heights of work surfaces. (Specialized work centers are found in Chapter 3, the Virginia Tech reference.)

Convenient, comfortable access to the sink cleanup center is essential in the universal kitchen, just as it is in conventional kitchens, because of the many food preparation and cleanup activities that take place there. If the primary cook is seated, the sink area may be made adjustable in height or may be permanently lowered with open knee space below. In either case, the sink should be no more than 6 1/2 inches in depth, with a rear drain and pipes that are insulated or covered with a panel to prevent burns (a panel is preferred for both safety and aesthetic reasons). An instant hot water dispenser operated by a lever located at the sink is a great convenience for those with limited abilities.

Two appliances normally located at the sink center, the garbage disposer and the dishwasher, present problems when the sink and surrounding countertop are lowered and open knee space is provided below the sink. The garbage disposal restricts access to the sink when installed with a single-bowl sink and enclosed in a standard base cabinet. Knee space is provided under the second bowl and adjacent countertop to achieve an opening of at least 30 inches in width. Another alternative is to use a separate garbage disposal sink that sits at the back corner of the sink (available from Kohler and others).

The problem created by the dishwasher (which should be front loading in design) is that it is available only in heights that finish to 36 inches to conform to standardized kitchen counter heights. When the dishwasher is placed adjacent to a lowered sink area, care must be taken to provide enough lowered work surface so that the user's elbows do not bump the higher side of the dishwasher. Dishwasher drawer models add flexibility and options.

An alternative solution (and one that shows a good example of choice) to the problems created in the sink center by the garbage disposal and dishwasher, and one that also meets the lifestyle consideration of multiple cooks, is to install a secondary sink center. The primary sink center can then be at conventional height and contain both the dishwasher and the garbage disposal. A secondary sink, in a lowered counter (with knee space below and pipe protection), can then be available for seated users, children, or another cook and, with the addition of a microwave oven, can form the basis of a second cooking station.

The provision of a separate cooktop and oven in preference to a standard range not only provides greater flexibility and accessibility but also accommodates two or more cooks with greater ease. Cooktops with burners arranged in a straight line or staggered format enhance safety by reducing or eliminating the need to reach across the front burners to attend to food cooking on the rear burners. Controls for the cooktop unit should be along the side or front for better accessibility.

Design of the cooktop area with open knee space below will allow easier access to the cooking elements by a seated user (Figure 8.3). The safety of this arrangement is sometimes questioned—the potential for spills on the cooktop overflowing into the lap of the user is obvious. Consequently, some individuals may prefer a cooktop placed with a base cabinet and space provided for a lateral approach to the cooking elements. In this case, the straight line configuration of burners would provide the highest level of both accessibility and safety.

Ovens with side-hinged doors (currently manufactured by Gaggenau and Frigidaire) provide the greatest ease of access for both the standing and the seated user. For these ovens, experts suggest that a pull-out shelf be located immediately under the oven (a landing pad).

A microwave oven can be of great assistance to persons with reduced abilities. The location of the microwave should be determined on the basis of its most frequent type of use. It may be placed on the countertop or above the countertop on a microwave oven shelf, or it may be installed below counter height. Installation of the microwave with the controls at an accessible height (no greater than 48 inches above the floor) assumes special importance with the realization that the microwave is often used by children on an independent basis.

For the serving center to meet the needs of the universal kitchen, space for a lateral approach using a wheelchair or scooter should be provided along with storage features that enhance accessibility. Also, it is helpful to provide a continuous surface from the preparation/cooking-to-serving area to allow people with limited strength to slide dishes, pots, and so on from one area to the next.

STORAGE

Sufficient, well-designed, and properly located storage is essential to the success of any kitchen. The universal kitchen demands a consideration of ease of access to storage.

Traditional base cabinets with fixed shelves make access to storage difficult even for the ambulant, able-bodied user. Improved access to storage in base cabinets may be achieved by specification of units with drawers and pull-out shelves or trays (with a lip on the front to prevent items from sliding off). Corner base units become more effective and accessible when provided with swing-out shelves or revolving shelves. Visibility in wall units is enhanced by the use of either wire or clear plastic shelves.

In a conventional kitchen, some of the most easily accessed vertical space, the space between the base cabinet units and the wall cabinets, is usually left empty. Some of this space may be designed for use in the universal kitchen through the following means: a shallow shelf (8 inches in depth, wall mounted), a portable U-shaped shelf that rests on the countertop, or appliance garages.

Adequate pantry storage for food items is important in any kitchen design, but it is even more critical in the universal kitchen as persons with limited abilities often prefer to minimize grocery shopping trips by keeping larger quantities of food on hand. Walk-in pantries should be avoided as they generally do not provide adequate maneuvering space for a person using assistive devices. Shallow closets with bi-fold doors provide an alternative that is easily accessible for most persons. A second suitable alternative for pantry storage is the full-height or pull-out pantry units available from many manufacturers of kitchen cabinetry. These units capitalize on the underutilized but accessible space between the base and wall cabinetry. Some of the new drawer configurations provide safe access to heavy plates and so on and support the trend to fewer wall cabinets.

LIGHTING AND ELECTRICAL CONCERNS

In a universal kitchen design, lighting assumes greater importance because the aging eye needs more illumination to perform effectively. Installation of rheostats and flexible lighting controls allow individual users to adjust the lighting to their particular needs.

When elderly or low-vision users of the kitchen are present, it is desirable to provide both ambient and task lighting at the upper ends of the suggested ranges. Switching should be carefully planned so that controls are located near the lights they operate (Null 1987).

Switches and receptacles that are usually placed on the wall at the rear of the countertop may need to be relocated to a panel on the front of the base cabinet unit in order to accommodate seated users or children. This may be done by replacing a drawer unit with a panel designed for the necessary switch and receptacle plates. In households where young children are present, receptacle covers should be used at all times to ensure the children's safety. Another alternative for making receptacles accessible for a seated cook is the use of a power strip with an extension cord. All receptacles in a kitchen designed for elderly or disabled users should be equipped with ground fault circuit interrupters (mandatory by code in some areas).

CONTROLS AND HARDWARE

Faucets in the universal kitchen should be controlled by a single lever for greatest ease in use. In addition, a spray attachment on a long hose (separate or part of the faucet head) eliminates much lifting and carrying of heavy pans.

Controls for ovens and burner units are easier to operate if they are of the blade, extended-blade, or small-lever type. Knobs that audibly click into operating position are helpful for users with reduced visual abilities. Many microwave ovens are controlled by touch panels that are operated relatively easily by most users, including those with mobility limitations (Moore and Ostrander 1992). For low-vision users, several manufacturers of kitchen appliances make available, at no charge, Braille or raised letter panels to add to the control area of both these and other types of appliances. All appliance controls and switches should be located no more than 48 inches above the floor.

Hardware on cabinetry units should be selected so that operation is possible with minimum strength and gripping ability. This is best accomplished by selecting hardware in a D or loop shape. Knobs and recessed pulls and handles should be avoided.

ADDITIONAL CONCERNS FOR LOW-VISION USERS

A common hazard to the low-vision user in a standard kitchen is the use of swinging doors on wall cabinets. Safer choices are to use open shelves; wall cabinets with sliding doors; or doors that are counterbalanced so that they open up, out of the way, and then close down very easily (see Figure 5.15).

Utilization of the concept of value contrast can also facilitate use of kitchens by those with limited vision. A light-value countertop contrasted with a dark-value backsplash and countertop front edge facilitates perception of the beginning and end of the countertop space as seen in the case study of the San Diego Center for the Blind in Chapter 3.

ADDITIONAL RESOURCES

As the market for products that enhance accessibility continues to expand, new products and additional resources for product information are constantly becoming available to assist the designer. Web sites of the National Kitchen & Bath Association and individual appliance and fixture websites are very helpful. Virtual planning tools can also be found online.

TECHNIQUE: UNIVERSAL BATH DESIGN

Until recently, the bathroom in the United States has been truly the "Necessary Room" and nothing more except for the very wealthy users. Indoor plumbing with a tub, water closet, and lavatory in one room was not an option for even the wealthy until the early 1900s; the modern bathroom did not emerge into general housing until the 1920s. At that time, well-designed spaces were not high priorities for architects and builders, so often Americans still have the 5′ × 7′ bathroom that currently challenges so many when they are remodeling. As the population has become more sophisticated, bathroom requirements have changed. Today, most new buildings have at minimum a family bath, a master bath, and, in many cases, a powder room.

Function and universal design as it relates to the bathroom can be divided into two categories: *safety* and *accessibility/comfort*. In order to help designers provide safe and accessible bathrooms, the National Kitchen & Bath Association (NKBA) has established 27 rules of bathroom design. These rules are merely the starting place for exceptional universal design, but they require the creative input of involved designers. (The illustrated NKBA guidelines reference can be found in the appendix of this book.)

- A clear walkway of at least 32 inches must be provided at all entrances to the bathroom. (Ideally, doors would be 36″ wide.)
- No doors may interfere with fixtures.
- A mechanical ventilation system must be included in the plan.
- Ground fault circuit interrupters must be specified on all receptacles. No switches may be within 60 inches of any water source. All light fixtures above tub/shower units must be moisture-proof, special-purpose fixtures.
- If floor space exists between two fixtures, at least 6 inches of the space should be provided for cleaning.
- At least 21 inches of clear walkway space should exist in front of the lavatory.
- The minimum clearance from the lavatory centerline to any side wall is 15 inches.
- The minimum clearance between two bowls in the lavatory center is 30 inches, centerline to centerline.
- The minimum clearance from the center of the toilet to any obstruction, fixture, or equipment on either side of the toilet is 15 inches.
- At least 21 inches of clear walkway space must exist in front of the toilet. (Note, however, that the American National Standards Institute (1992) requires 48 inches and a wheelchair width of 26 inches to 28 inches.)
- The toilet paper holder should be installed within reach of the person seated on the toilet. The ideal location is slightly in front of the edge of the toilet bowl, the center of which is 26 inches above the finished floor.
- The minimum clearance from the center of the bidet to any obstruction, fixture, or equipment on either side of the bidet is 15 inches.
- At least 21 inches of clear walkway space should exist in front of the bidet.
- Storage for soap and towels should be installed within reach of the person seated on the bidet.
- No more than one step should lead to the tub. The step must be at least 10 inches deep and must not exceed 7 1/4 inches in height. (Note that it is better to avoid steps if at all possible, and if they must be used, grab bars should be in place to facilitate safe entry and exit.)
- Bathtub faucets should be accessible from outside the tub.
- Whirlpool motor access, if necessary, is included in the plan.
- At least one grab bar is installed to facilitate bathtub or shower entry.
- The minimum usable shower interior dimension is 32″ × 32″.
- A bench or footrest should be installed within the shower enclosure.
- A minimum clear walkway of 21 inches should exist in front of the tub/shower.
- The shower door swings into the bathroom.
- All shower heads are protected by pressure balance/temperature regulators or temperature-limiting devices.
- All flooring is of slip-resistant material.
- Storage must be provided in the plan, including counter/shelf space around the lavatory, grooming equipment storage, convenient shampoo/soap storage in shower/tub area, and hanging space for bathroom linens.
- Heating system must be provided.
- General and task lighting must be provided.
- A no-threshold shower should be considered.

SAFETY

When designing and specifying products for bathrooms, it is the designer's responsibility to address safety before beauty. It is known that 25% of all accidents and 3% of all fatalities that occur in the home take place in the bathroom. Some of the hazards include falls, electric shock, scalding, broken glass, poisoning, and door swings.

Slippery surfaces—on the floor and in tubs and showers—are a major cause of accidents. Shiny tile and polished marble should be avoided. Throw rugs—even those with slip-resistant backing—can cause a fall if tripped over. Bathroom carpet, although generally believed to be the least sanitary flooring, can provide a warm, slip-resistant flooring if installed properly. Most tub and formed shower manufacturers offer nonslip surfaces in at least some, if not all, of their units. If unavailable, the next best thing is for the client to use a rubber mat or "place and press" rubberized flowers. For site-built showers, be sure to specify a small (no more than 2″ × 2″) nonslip tile for the base. To further prevent a fall, grab bars have been proven to enhance everyone's safety. Note that grab bars must be able to support a 300-pound static load.

The National Electric Code requires that, in any room where there is a combination of high humidity, water, and electricity, the electric circuits must be protected with ground fault circuit interrupters. These help prevent electric shock. Many local governments have passed plumbing codes that require pressure and temperature balanced tub and shower controls to prevent burns from scalding water and falls caused by "shower shock." These are all single-handle configurations. Specifications should include only tempered glass or shatterproof plastic glazing on all glass areas to protect against broken glass.

Finally, in terms of safety, consider the way the door swings. Most bathroom doors swing in so as not to block hall traffic. However, if someone falls in a bathroom (especially one that is small), they may block the door, preventing entry. Also, in small spaces, open drawers from vanities can block the door from opening. If swinging the door out is not an option, then try to have it swing against a blank wall or in front of the tub. Definitely avoid having it open into the water closet or lavatory. Sliding barn doors or pocket doors are options, but pocket doors cannot be put in a plumbing wall.

Doors within the bathroom itself must also be considered for accessibility and safety. Shower doors that open out into the room or sliding doors are an essential safety feature. When using sliding doors, triple-panel sliders or accordion doors allow maximum access to the space. In a compartmentalized bathroom, door swings for the various compartments will have major impact on the room design.

ACCESSIBILITY/COMFORT

It wasn't until 1976, when Alexander Kira wrote *The Bathroom*, that people started thinking seriously about this space. Although this wasn't a best seller when it was first published, it is now recognized as the first scientific bathroom reference. In looking for comfort in the bathroom, the most obvious example is the height of the standard vanity. How did this uncomfortable standard come into being? Washstands were approximately 30 inches to 32 inches high, with a wash basin sitting on top, creating an upper measurement of 34 inches to 36 inches, which is almost ideal for most adults. Somewhere in the translation to indoor plumbing, instead of raising the washstand around the basin, the basin was dropped into the stand. Nearly everyone can be made more comfortable by raising the vanity height. Other areas where height could affect the comfort of the user are the water closet, shower head, towel bars, soap dishes, and toilet tissue holders.

Adequate lighting is critical to a well-planned bathroom. A variety of types of lighting on separate switches and dimmers for incandescent or halogen lights should be designed into all bathrooms. General room lighting is the first area of concern and can easily be achieved in many aesthetically pleasing ways. Task lighting should include lighting for the tub and shower (see Figure 8.1) as well as for the vanity and water closet areas. An adjustable, lit, magnifying mirror should be included so that anyone with visual impairment can comfortably use the vanity without having to be a contortionist.

CONCLUSION

Safe, accessible, functional, comfortable, and aesthetically pleasing spaces to live in; homes that will change with people as they change so that individuals will not have to leave if injured or disabled—these are the goals of residential universal design.

UNIVERSAL DESIGN FEATURES FOR AGING IN PLACE—THE BATH

Lack of information and awareness, not lack of need, contributes to low demand. Most people still do not know about home modifications and assistive technology. They do not know where to buy quality products, where to get financial assistance, and where to find skilled practitioners. As a result, they do without, move away, or put up with ugly, inappropriate, inadequate, or unsafe alterations. An attractive bathroom remodel created by interior designer Carol Lamkins features nonslip tile flooring, a drive-in shower with a double drain, Corian walls, and a skylight. Corian grab bars and accessories are also included. A "comfort-height" toilet adds to the universal design features (see Figures 8.10 through 8.13). Lamkins cites the following checklist of UD features.

Universal Design Features for Aging in Place—the Bath

- General
  - Slip-resistant porcelain tile flooring with a thermostat-controlled floor heating system installed under the tile
  - 32″ clear doorway with easy-glide pocket door
  - Electrical switches and outlets at convenient heights and locations
- Vanity area
  - 36″ height for adult comfort
  - Doors in lavatory base cabinet are convertible to be pocketed if needed in the future
  - Cameo White, easy to care for Corian coved countertop and dual lavatories
  - Single-lever faucets with pull-out heads for versatility
  - Swing-out mirror with X and 3X magnification opposite the vanity for close viewing (makeup) and to work in tandem with the full wall mirrors over the vanity for grooming hair
- Shower area
  - Multiple slip-resistant, Americans with Disabilities Act–approved, Corian grab bars
  - Slip-resistant porcelain tile flooring
  - Dual drains for more capacity in case of plumbing backup in house
  - 18″ deep by 60″ wide bayed shower seat
  - Accessible storage with multi-height inset storage shelves in shower
  - Two single-lever valves servicing slide bars with adjustable height, removable shower heads
  - Pressure balanced valves to avoid scalding
  - Dual towel bars at accessible height
  - Operable roof window for ventilation and natural light
- Toilet area
  - Handicapped 17″ toilet height
  - Ventilation fan with easy-access switch
  - Horizontal and diagonal Corian grab bars service each side of the toilet
  - Toilet tissue with extra shelf for additional roll and magazine inset storage unit at accessible height
  - Double sash windows with four surface mount pulls for easy operation

---

- Lighting for safety
- Energy-efficient fluorescent general lighting
  - Two dimmable moisture/vapor-proof incandescent fixtures for safety in the shower
  - An alabaster wall sconce with soft night lighting
  - Heat lamp over toilet can be substituted with a standard incandescent lamp during warm weather

FIGURE 8.10 Color contrast on bathroom counter tile (Carol Lamkins home).

RESIDENTIAL REDESIGN FOR ACCESSIBILITY

The following case study describes two design projects that architect Lee Meyer undertook to provide accessibility for wheelchair users. In each example, the changes made were relatively inexpensive and added to the overall value of the residence.

Many people are beginning to ask for accommodations in their existing homes to allow greater independence for themselves or for family members who are disabled. Architects and designers are the primary source of information for these individuals.

PROJECT 1: THREE-BEDROOM RAMBLER

The clients owned a three-bedroom rambler house and had a limited budget from which to draw. They had two adult sons, both of whom used wheelchairs. Their main request was for better access to bedrooms and bathrooms. Figure 8.14 shows the solutions described below.

Access to the sons' bedrooms was improved by repositioning the doors at 90° angles to each other: This change required the installation of a header to replace the bearing wall.

FIGURE 8.11   Natural and recessed lighting display the grab bar in the shower; flexible shower head; note the sky light.

FIGURE 8.12   Grab bar positioned close to the toilet.

FIGURE 8.13   Grab bars are on both sides of the toilet in this bathroom.

The hall bathroom and master bedroom bathroom were combined. The stool remained in the same position, and a large corner shower was constructed in the same area as the master bedroom shower. The door to the master bedroom was retained, providing secondary access for the parents.

The linen closet was removed and a new 3-foot doorway was installed into the bath, while the vanity cabinet and sink were realigned along the hallway wall to provide better access to the stool. Bath linens were stored in the vanity. The sink portion of the vanity was constructed cantilevered so that wheelchair access at the sink could be maintained. Plumbing changes were kept to a minimum, primarily involving the removal of existing fixtures.

FIGURE 8.14   A section of a redesigned three-bedroom rambler house modified for wheelchair accessibility (Architect: Lee Meyer).

PROJECT 2: TOWNHOUSE

The client for the second project was an adult male who used a wheelchair. His main concern was to achieve better access on the second floor, bedroom level, while maintaining an unmodified appearance for future resale purposes. Figure 8.15 shows the solutions described.

The residential elevator was installed in an existing hoistway since the units in the townhouse development were planned with space for residential elevators, connecting all four levels of the structure within each unit.

The door at the bottom of the third-floor stairs was eliminated, and one step was removed and a winder was added. Doors to the office and guest bedroom were realigned in order to provide access to the office. Both doors had been 2 feet wide, and they were replaced with doors of 3-foot widths. The realignment also provided a turnaround space at the end of the corridor.

In the master bathroom, a whirlpool tub and bidet were removed, and a large drive-in shower was constructed. The toilet was repositioned, and the wall between the bath and dressing area was removed. The two-sink vanity was removed, and a single-sink vanity was installed with knee space at the sink and with drawer storage at each side for medical supplies and personal items.

The door to the walk-in closet was removed, and a 4-foot cased opening was installed. Low rods were installed on each side of the closet, with an Elfa basket storage system installed below the window for foldable items. A pocket door was installed in the bathroom, and a 3-foot door was placed at the entrance to the master bedroom.

Changes throughout the balance of the townhouse were minimal: The island in the kitchen was reconstructed to create a lower eating/work surface; the microwave oven was located in a base cabinet, immediately below the countertop, and a pull-out breadboard on roller extensions was installed below the microwave as both a landing surface for hot items and a cutting surface accessible to the chair. The toe-kick and raised bottom of the sink cabinet were eliminated to provide drive-up access to the sink. A bar sink was added at the lower level of the island. Finally,

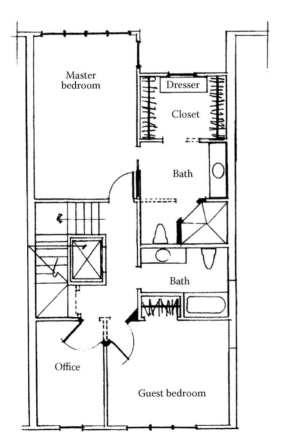

FIGURE 8.15 Floor plan of a redesigned townhouse for wheelchair accessibility. Note the corner door placement (Architect: Lee Meyer).

a washer and dryer were placed in the basement laundry. The front-loading washer and the dryer both featured front controls.

The client remained in the townhouse for three years after the renovation. The new owners had been seeking a luxury townhouse that could be used by an elderly parent with mobility problems.

HILLSIDE HOME WITH ELEVATOR AND OTHER UNIVERSAL DESIGN FEATURES

Designing a hillside home is always a challenge, but when accessibility is added to the design criteria, it becomes a greater problem. This major challenge was faced by Sylvia Sullivan, interior designer and design educator. To meet universal design criteria, the house needed to be accessible and also adaptable to meet the needs of a retired couple and their family.

Given: a 9000+-square-foot boulder and rock-filled upslope lakefront lot with a very small footprint area suitable for a house. The house is 1350 square feet with three bedrooms, three baths, kitchen, dining, and living on three floors, plus garage level (see Figures 8.16 and 8.17). Because of the steep slope and boulders, the driveway location was one determining factor in the decision to install an elevator. The garage elevation was 5 feet lower than the house so the elevator opens into the garage and then proceeds up to the main level and the two additional floors. Alternate access is provided by stairs serving the three floors plus garage. Once the location of the elevator and stairs was determined, other accessibility and circulation issues surfaced. For example, does every area of the house need to be accessible or only the public areas such as the kitchen, living room, one bath, and bedroom? It was decided to make every area wheelchair accessible. In addition, all doors were 36″ wide except the pocket doors, which are 32″ clear. Pocket doors were used in areas such as bathrooms and the laundry room where door swings would be an impediment. Turnarounds of 60″ are incorporated into all bathrooms except the upstairs loft bath. All bathrooms feature tile floors, roll-in showers, and Kohler "Comfort" height toilets. All levels except the loft were made accessible by an elevator, which has five stops. All windowsills were installed low enough so seated persons would have a view of the lake.

Construction of the home started in Summer 2002 and was completed in Fall 2003. Figure 8.17 shows the floor plan of the main level and the elevator location. A picture of the completed home is shown in Figure 8.16. An elevator with five stops and other universal design features add value to this upscale hillside residence that far exceeds the cost of these features.

In Figure 8.18, the Sullivan floor plan is adapted for a family of cooks.

FIGURE 8.16   Hillside home in Big Bear; an elevator provides access to all levels (Sylvia Sullivan, designer and contractor).

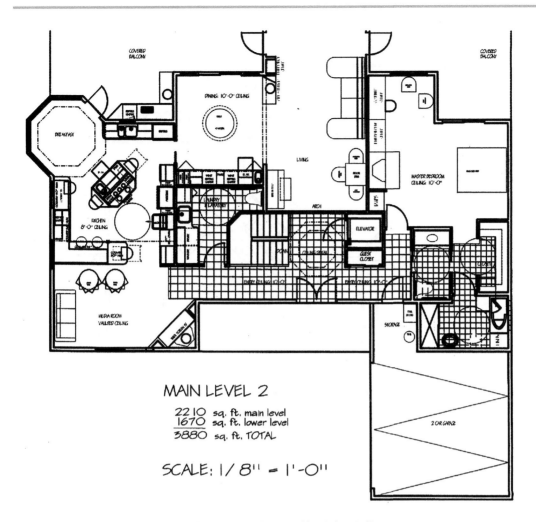

FIGURE 8.17   Floorplan for the hillside home designed by Sylvia Sullivan.

KITCHEN DESIGN FOR A FAMILY OF COOKS: KITCHEN DESIGN STRATEGIES TO INCREASE HOME DINING FOR ALL FAMILY MEMBERS

Today, two-thirds of American adults are obese or overweight. Overeating and its lethal companion, underexercising, are the recognized culprits in this country's rise in obesity rates. As lives grow busier—and waistlines grow larger—a number of nutritionists are calling for Americans to reduce their risk of obesity-related health problems by cooking at home more and eating out less. Many Americans spend more time in traffic these days than in the kitchen. Research shows that people consume 50% more calories, fat, and sodium when they eat out than when they cook at home.

Those of us concerned about the obesity epidemic are battling an entire environment, massive societal change, government policy, and billions of dollars in advertising that all influence family eating habits. Advertising, fast-paced living, convenience foods and treats every time one turns around are keeping us away from healthful eating choices. Americans even eat an average of 30 meals a year in their vehicles and these meals are most likely purchased at fast food restaurants. Individuals and families eat everywhere—an average of five meals a day, counting snacks, with fewer and fewer of these meals "home cooked" (United States, General Services Administration et al., 1984).

The Obesity Summit (June 2004)—*Time* magazine and ABC's attempt to determine the causes of the nation's obesity crisis and its resulting costs in terms of needed health care and workers' lost productivity—brought no "new" solutions. The summary of the findings of studies reported by participants in the Summit listed three major causes of the problem:

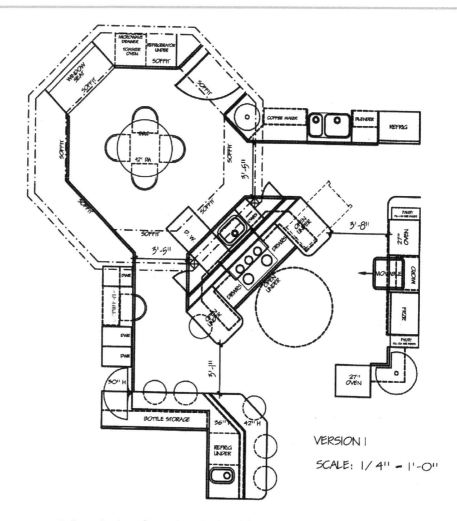

FIGURE 8.18 Sullivan kitchen floor plan; the breakfast nook is in the tower; the food preparation area with range is centrally located with adequate turning radius.

1. Not eating foods high in nutritional content and low in fat and sugar—the so-called Power foods
2. Eating very large servings of these non-nutritious foods
3. Lack of exercise/physical activity

The first two causes seem to be closely related to our nation's increased consumption of "fast foods" and the third is related to an increasingly sedentary lifestyle (hours commuting and sitting in front of the TV, lack of focus on school sports and physical education, more school time spent on "academic" core subjects including hours of homework every night, and the loss of cooking classes/culinary arts and related nutrition education programs in public schools).

In order to reduce our reliance on "fast food," we need to encourage home cooking. To do this, we must make cooking at home an attractive, viable alternative to "eating out" and develop kitchen design strategies and educational programs to increase home cooking and dining for all family members.

In this article, the authors focus on the kitchen design approach to encourage more individuals and families to cook and eat at home. Kitchen design criteria that will contribute to the creation of a supportive environment for a family of "cooks" is described.

The proposed kitchen design will support all family members sharing responsibility for the family's food preparation and enjoyment of family meals. Through these collaborative activities, children learn to appreciate the teamwork contributions of older family members and their parents in addition to learning skills that relate to food preparation.

Several recent research studies have specifically addressed the needs of older people in kitchen planning. In a study titled "An Evaluation of Kitchen and Bathroom Design in Homes Designed for Senior Citizens 2004" (Andes 2004), Glenda Andes found that we are not designing well for seniors and that seniors don't know what supportive design features are appropriate for them of those available. She also found a lack of attention to detail in product selection. (For example, ranges selected for nearly 800 kitchens in a new retirement community in Indiana all had controls located behind the burners at the back of the range [a real safety hazard]. Until this problem was pointed out to the management, they were not aware that this could cause difficulty and no changes were even considered.)

According to Anders (2005), we need to destigmatize supportive kitchen features by attractive appearance and distinguishing between handicap features and supportive features, and then we need to market supportive design features as an asset (value added).

Incorporating universal design features into kitchen design helps us create spaces that are functional and aesthetically pleasing. Using the data collected in a 2003 research study, Dr. Joy Potthoff developed a set of universal design guidelines for kitchens that were supportive of an aging population (Potthoff 2003). These included the following:

1. Two sinks—one, a large two- or three-bowl sink (provides for vegetable preparation), and one, a small single sink possibly located in an island or at the end of a peninsula.
2. Counter space at different heights: 30/36/42 inches.
3. Two good-sized ovens (think 25-pound Thanksgiving turkey size)—not too low
4. Nonslip and easy-clean floors
5. Adequate space for all the cooks to work in the kitchen—minimum aisle width, 44 inches

These same universal design guidelines also ensure good wide traffic/circulation paths for efficiency and safety for the "Family of Cooks" kitchen.

In addition to the creation of a universally designed kitchen, the authors/designers describe special features that accommodate the needs of children, parents, and older family members using the kitchen. Some of these features were highlighted in a kitchen plan designed by Sylvia Sullivan (Sullivan 2005) that incorporates principles of universal design and accommodates many people working together in the kitchen. Arlena Hines related the kitchen design to the National Kitchen & Bath Association (NKBA) guidelines, which include the following (Hines et al. 2010) (see Figure 8.18):

1. Lunch/Snack Prep Center

    Ideal for children and teens for entertaining their friends and serving healthy drinks (smoothies). This versatile addition to the kitchen can be used as an informal eating area for the family, or a buffet serving area, and the bar/counter would be part of an additional food preparation center for the kitchen and would have barstools for seating of those eating or preparing food.
2. Cooking/Food Prep Center

    The major food preparation/cleanup center would require multiple cooking surfaces
    - Multiple microwave ovens placed at different heights and locations in the kitchen (Sharp has a new microwave drawer above a range oven.)
    - Cooking appliances (grills, etc.) can be included in the lunch/snack prep center
    - Trash receptacles located in all centers
3. Cleanup/Prep Work Center*

    The major food preparation/cleanup center would include the following:
    - Dishwasher drawers in all food prep centers
    - Instant hot water dispenser
    - Multiple cleanup stations including sinks

---

* Kitchen Planning Guideline 10—Cleanup/Prep Sink Placement.

4. Refrigerator/Baking Center

This center would include storage for spices, baking pans, and serving dishes

- Pull-down and narrow pull-out storage units, so that all items can be stored at the point of first use
- Bottom freezer refrigerator or side by side for easy access from all heights

5. Storage/Serving Center

The major center for serving would include the following:

- A serving cart on wheels and storage for dishes, glasses, attractive placemats, and napkins
- Filtered cold water dispenser at the sink and on the refrigerator door
- Dishwasher
- Adjacent to dining area for easy serving and cleanup*

6. Planning Center

The planning center desk needed to be large enough to accommodate multiple users (three chairs, two computers + printers, a bulletin board, storage for cookbooks, file information on special diets, grocery ordering and other research activities)

Safety was a major factor to consider in a kitchen with multiple cooks. (Note: An outstanding resource is "Universal Design in Housing" prepared by the Center for Universal Design in Raleigh, North Carolina [www.design.NCSU.edu/cud]. It features detailed design criteria and product characteristics to ensure kitchen safety.)

The overriding principles of a user-friendly kitchen design help families provide a kitchen that (1) supports shared activities, (2) supports all members in food preparation and planning strategies, and (3) enables team decision making and delegation of duties. Family food preparation activities related to the guidelines from the NKBA are as follows:

Selection
Preparation
Serving
Cleanup

The work centers designed for the kitchen to accommodate a family of cooks provide for these activities.

AWARD-WINNING ACCESSIBLE HOME USES UNIVERSAL DESIGN (FAMILY LIVING)

In 1996, the Minnesota chapter of the American Society of Interior Designers awarded this Bloomington, Minnesota home first place in its annual competition as a "Residence Over 3000" square feet. The family knew from past experience that they wanted wide open spaces, on the basis of the contemporary Prairie School (or F.L. Wright) design; one that met the needs of older parents, a young adult son (J) who uses a wheelchair, and their own future needs as aging adults.

The family is also very active in programs for persons with disabilities and hosting meetings and events in their home, so they anticipated the need for visitability. The design/builder team worked closely with the clients and J, addressing safety, egress, accessibility, temperature control, communication, lighting, and visitability. Because J drives, he enters the home through the garage. That entry is ramped. The door entrance in the garage has touch controls to unlock the door, turn on lighted pathways inside the home, and activate the intercom. J needs these devices because he cannot use his hands easily but a person carrying a bag full of groceries or a crying baby would also benefit from these design decisions. J can access his space in the upper level of the house by using the residential elevator (Figures 8.19 and 8.20).

Because J has his own home, he would most often be a visitor to the kitchen and not the regular chef. However, all the appliances are accessible with press or turn action. The refrigerator door has an easy-to-use water and ice dispenser. The kitchen fireplace is lit with a remote control.

---

* Sullivan House.

FIGURE 8.19   Residential elevator hidden behind a full height door lends a feeling of spaciousness to the house (Mary Hackett).

The security system includes an intercom system and a front door camera so that when J is in his lower-level suite, he can hear and see someone at the front door on his TV security channel, use the phone to talk with the visitor, and use remote controls to open doors and light pathways. All outlets in the home are about 3 feet from the floor; accessible to persons of any height whether standing or seated (Sweet's Group 1996).

J's bathroom has a heated floor, which quickly warms the room when the door is closed. The floor slants to the shower area where transfer to shelf seating is easy. The shower controls are on the wall opposite the shower head, so water temperature can be controlled before getting wet. A glass block wall brings in light from the adjacent area. All towel bars are usable grab bars, blending into a very masculine stone and glass decor.

The owners can live on one level when their needs require that they do so. Their master bathroom has a remote-controlled window treatment so that the shade can be drawn up for privacy. The remote control vertical shades in other parts of the home were chosen because easily closed vertical blinds were not available when the house was being designed. The elevator is a standard model.

AWARD-WINNING UNIVERSAL DESIGN HOME (SENIOR HOUSING, CALIFORNIA)

IMPROVING QUALITY OF LIFE THROUGH UNIVERSAL DESIGN

Universal design offers a human-centered approach to creating products and environments that accommodate human differences and the changes people face throughout their lifetime, as well as protecting and enhancing the safety, health, and well-being of all. Universal design unites

FIGURE 8.20 Control panel for the home elevator, a house in Minnesota (Mary Hackett).

occupational therapy, design, manufacturing, and building professionals to create products and environments that add to a client's quality of life. Each professional on the design team brings his or her unique knowledge to the design process and helps increase the ultimate product usability and marketability. An occupational therapist's health care education and experience provide insight into human functioning throughout the life span, how chronic health conditions and injuries affect function, and how universally designed products and environments can improve the performance of daily tasks (NAHB National Research Center 1992).

Occupational Therapist Susan Mack created a universally designed house that received the top award from the Senior Housing Council of NAHB (shown in Figure 8.21).

Universal design improves safety, ergonomics, and work efficiency by taking a proactive approach that enhances safety by removing unnecessary hazards. No-step walkways, entries, and showers reduce the risk of trip/fall injuries; grab bars and slip-resistant flooring reduce the risk of slip/fall injuries.

As our average life expectancy increases, we need to protect our bodies from unnecessary repetitive stress at home as well as at work. Universal design enhances comfort by incorporating the ergonomic principles of joint and back protection to reduce repetitive stress and enhances convenience by incorporating the principles of work simplification. Universal design creates efficient products, work environments, and homes to reduce the time and personal energy required to perform daily tasks, allowing us to participate in other creative and productive activities (Connell et al. 2009).

Twenty percent of the US population has some form of disability. Accessible universal homes welcome people of all ages and abilities; no one is excluded from family gatherings because of unnecessary barriers and hazards. Accessibility in universally designed homes expands their marketability, and well-done universal design is not easily apparent. Universally designed homes are attractive as well as inclusive; they incorporate "best practices" in accessible design, not just minimal code compliance.

Universal design accommodates the tall as well as the small or seated individual. Universal design also takes into consideration how people vary in vision, hearing, strength, size, stamina, coordination, fragility, mobility, and flexibility, and how they use medical, mobility, or adaptive devices.

Universal design is not just a finite list of features but a comprehensive concept, methodology, or way of thinking that can be applied by all professionals who create new products and environments, as well as those involved in redesigns. Not everyone on the team needs to be a universal design expert. But inclusion of a universal design consultant with depth of experience and proven successes helps to ensure the team's success.

UD features of the house designed by Susan Mack are pictured in Figures 8.21 through 8.24.

AWARD-WINNING UNIVERSALLY DESIGNED HOME (SENIOR HOUSING, OHIO)

Another award-winning senior home was built in an Ohio community.

A second UD award winning senior home was built in an Ohio retirement community by Mac Kennedy. It features a step-free entrance (Figure 8.25) and UD interior features (Figures 8.26–8.28).

HOW TO DESIGN FOR AGING IN PLACE—KITCHENS AND BATHS

Kitchen and bath designer, Mary Jo Peterson writes about the problems of aging-in-place for a growing number of elderly persons.

Often, when I have a conversation with other designers regarding design for aging, they talk about parents or grandparents who found that neither they nor their homes were prepared for changes that occurred with age. Yet, many of these changes are predictable, and we should, as designers, be able to incorporate concepts into our designs that accommodate these changes (Peterson and National Kitchen and Bath Association (U.S.) 1998).

FIGURE 8.21   Exterior of award-winning senior housing with step-free entrance (by Susan Mack, OT).

FIGURE 8.22 Accessible cabinets by Susan Mack, OT; cabinet doors are retractable. Note the finished floors under both sink and range cabinets.

FIGURE 8.23 Accessible kitchen in Susan Mack house; multiple counter heights are shown; the side-opening oven provides easy access.

FIGURE 8.24   Step-free shower in the award-winning home by Susan Mack, OT.

FIGURE 8.25   Mac Kennedy, Senior Award-winning house with no-step entrance. Retirement community in Ohio.

FIGURE 8.26 Kennedy house, entrance interior, lever door handle, and flat threshold.

FIGURE 8.27 Kennedy house pocket door into bathroom; the lever handle makes access easier.

FIGURE 8.28 Kennedy house kitchen with side-opening oven in the wall cabinetry. The downdraft cooktop gives an open look to the kitchen.

Ten years ago, my incredible Irish grandmother died at age 99, and her living situation in the last years of her life propelled me to pay closer attention to design as it pertains to aging. Although basically healthy and alert at age 95, her strength and hearing had diminished to a point that she could no longer safely live in the home she'd known since before I was born. This began my commitment to creating environments that would enable older adults to live comfortably in their homes for the duration of their lives.

Personal experiences also prompted New York–based interior designer Rosemary Bakker to focus her efforts on these same considerations. In a *New York Times* article, Bakker related that when her mother returned home from hip surgery, she was faced with trying to maneuver a walker through narrow doorways, over area carpets and raised thresholds, and into a kitchen where she couldn't bend to get food out of the refrigerator or reach pots and pans. Additionally, there was no bathroom on the first floor.

As the article said, "Suddenly, the house that had suited her for 42 years was a time bomb waiting to go off." Out of this experience came Rosemary's book, *Elderdesign*, a resource for designing and furnishing homes for later years.

As kitchen and bath designers, we have an opportunity and a responsibility to design flexibility, access, and support into each project we approach. For the first time in history, there are more people over age 65 than under age 25, and many of the homes we live and work in were not designed for this new longevity.

American Association of Retired People surveys show that more than 80% of the people over age 60 want to remain in their homes. Accessible kitchens and baths are critical to this desire. To this end, designers must address issues of safe movement throughout the home, as well as efficient yet accessible use of the spaces we design.

One response, from the Atlanta-based Concrete Change, is the concept of "visitability." To be "visitable," the organization says, a home must have at least one entry that is accessible, wide enough passage through the main floor, and at least one main floor bathroom that is designed for use by people of varying abilities. For certain projects in Atlanta, parts of Texas and California, and the United Kingdom, visitability is required by law.

As we advance in the aging process, our senses decline, and our flexibility, balance, stamina, and reflexes diminish. These are often compounded by side effects of medications and chronic or injury-related conditions such as arthritis or limited recovery from broken bones. Rather than reacting with denial or depression, we can design to accommodate and support these changes.

Both the kitchen and bath begin with the entry, where the clearance at the opening, maneuvering space around the door swing and threshold must be examined. Sometimes, just reversing a door swing and installing a swing-clear hinge and lever handle to the door will improve the situation. An important break with tradition is to replace the raised threshold at the door with a flush conversion at the entry.

Once in the kitchen or bath, lighting is a critical element to reduce risk. We all realize that generous amounts of task and ambient light are important. In addition, we must avoid glare and use contrast appropriately to guide the way. If we increase the bath lighting, we must also carefully light the path to the bathroom, perhaps with a motion-activated system, as aging eyes will be blinded by a quick change from darkness to bright light, or the reverse.

Criteria for selecting flooring should include slip resistance and some forgiveness for dropped items, or to prevent serious harm in the case of a fall. Pattern or contrast should be gentle and can be used to help guide the eye. Area rugs should be taped to the floor or, better, eliminated. The point at which flooring materials change should be flush.

Clear floor space for maneuvering is relatively easy to accomplish in the kitchen, but often difficult in the bathroom. Pocket doors or reversed door swing help in the bath, as do vanity designs that increase open space below. Particularly in traditional 5 by 8-foot bathrooms, converting from a tub to a roll-in shower will also help.

In the kitchen, planning retractable doors to conceal an open knee space will open up the clear space and provide a storage spot that easily converts to a place to sit while working. A big consideration for storage is that our height decreases as we age, and for many of us, it becomes less comfortable to bend or climb. Design that provides generous storage between 24 and 45 inches off the floor eliminates the need to do either.

This means that the backsplash area in the kitchen becomes valuable for storage, and at least some wall cabinets might be lowered. Rolling storage in either the kitchen or bath can provide flexible clear floor space and storage that moves to the point of use as desired. Open or glass door storage will help accommodate changes in memory.

Support in the form of railings or grab bars is essential as we age, yet this is often distasteful to both clients and designers. With the broad offerings of grab bars today, many coordinated to match accessories, the challenge is minimized.

Says Mary Jo Peterson, if my grandmother's home had been designed to support her, she might have stayed comfortably at home. If Rosemary's mother's home had been originally designed to be supportive, her trauma might have been reduced. A 45-year-old couple and their teenage children might not seem to need "aging in place" design, but their parents or friends might. If we can design beautifully and incorporate solutions respectful of our elders, why wouldn't we?

Universal design and access don't have to be the only focus of our efforts. Rather, we can make them an integral part of every project we design.

HOME DESIGN FOR A LIFETIME

UNIVERSAL DESIGN PRINCIPLES CAN HELP MAKE YOUR HOME MORE FUNCTIONAL AND FASHIONABLE—NOW AND IN YEARS TO COME

When John Salmen and his wife, Ann Scher, were shopping for a new home in the Washington, D.C., area, they sought a house that would grow and change with them. "We wanted a home that would serve us regardless of our stage in life, without sacrificing style for utility," Salmen, an architect, said. He found a circa-1910 Arts and Crafts fixer-upper in a Takoma Park, Maryland, neighborhood. An ambitious three-year renovation converted the historic house so that it could accommodate the couple—and possibly their children or their parents—at any age. They created easy-access stepless entries, widened doorways for ease of movement, and stacked closets so that they could more easily install an elevator. Salmen and Scher are like many other people who want to age in place—comfortably grow old in the home they've spent years enjoying. As a result of that desire, a new design concept has gained popularity over the past few decades. Universal design (UD) is based on the belief that any living space or product should be usable by the widest group of people possible, without complicated plans or features, for the longest time possible. Its genius is its simplicity. When done right, UD is nearly invisible. "The highest compliment we hear is when people say they didn't realize subtle changes, such as wider doorways or a stepless entry, are UD," says Roy Wendt, president of Wendt Builders, in Atlanta.

UD originated with the need for barrier-free access in commercial settings like office buildings, but it later evolved into an all-inclusive philosophy for homes and products (Barrier-Free Environments 1991; Liebrock and Behar 1993). "A common myth about UD is that it's only geared toward the elderly or those with physical challenges," says Richard Duncan, senior project manager with The Center for Universal Design at North Carolina State University, which was founded by Ronald Mace, an architect and wheelchair user who recognized the need for accessibility and user-friendliness in homes. "But UD is really about design that makes life easier, safer, and more comfortable for everyone."

UD is a commonsense approach: For example, if you build a house with a sloped walkway instead of steps, you accommodate everyone from parents pushing a stroller to visitors with impaired mobility. When you replace a faucet that has separate hot and cold knob handles with a lever mechanism, you can nudge it on when your hands are dirty or if you have diminished hand strength. And if you choose a floor plan with a master bed and bath on the ground floor, it may help you—or an aging parent—avoid daily treks up and down the stairs.

"A house with UD features also has enhanced marketability because you can sell your home to people of all ages and abilities," says Susan Mack (2006), with Homes for Easy Living Universal Design Consultants. Whether you're remodeling, building, or working to make your existing home more comfortable and efficient, here are some expert tips to introduce UD to your living space.

If you're remodeling or building new: "UD isn't expensive. The cost is in the design, not the fabrication," says Salmen. If planned from the ground up, you'll spend an estimated 1% to 2% of total costs to introduce UD features.

- *Make entering your home easy.* "At least one entryway should be stepless," Wendt says. If you prefer front steps for aesthetic reasons, plan a barrier-free route elsewhere, such as through the garage (Figure 8.29).
- *Choose wider doors.* "Those that are at least 42 inches wide lend a more spacious feel, plus they offer better maneuverability for tasks such as rearranging furniture or carrying laundry," Duncan says. Pocket doors—panels that slide along a track hidden between walls—are another good choice because they create unobstructed doorways.
- *Create adaptable living spaces.* Subject to local codes, try to position light switches between 36 and 48 inches above the floor and outlets between 18 and 24 inches above the floor to limit bending and allow accessibility by seated individuals and people of all heights.
- *Plan for flexibility in the kitchen.* Install counters at multiple heights to fit various people and tasks. A counter that's 6 to 7 inches lower than the standard 36 inches accommodates a seated person or a child, while a 40- to 42-inch-high section works well as a

FIGURE 8.29 Backyard of John Salmen home with step-free entry and wide doors.

FIGURE 8.30 Kitchen in John Salman's home features a variety of counter heights and front grab bars.

serving center or when you're decorating a The mudroom in the Salman house adds convenience to simple family tasks (Figure 8.31).

- *Rethink appliance layout.* Install modular cooktops, which allow you to customize your cooking area with individual components, such as a grill, pasta cooker, or wok, and place them where they're most accessible. Choose side-opening wall ovens, which are easier to reach into if you're pregnant or have back trouble. Place the microwave at waist level so you don't have to reach overhead for hot foods. Raise dishwashers 6 to 10 inches off the floor, or consider dishwasher drawers to reduce the need for bending and stooping, says Jane K. Langmuir, an architect and UD expert based in Providence, Rhode Island, and Los Angeles.

FIGURE 8.31 Mudroom/bath of John Salmen home; note the open area beneath the counter.

■ *Build in efficiency and safety.* Select full-extension pull-out drawers and racks in lower cabinets so you can reach items stashed in back. Choose C- or D-shaped pulls on cabinets because they're easier to grasp than knobs. "Consider a curbless shower, which has a flat threshold to prevent tripping," Duncan says. Maximize lighting with skylights, recessed lighting under stair treads, and extra lighting in potential danger zones like the head and foot of stairs and landings (see Figures 8.30 and 8.31).

REFERENCES

American National Standards Institute. (1992, Final draft). *American National Standard for Accessible and Usable Buildings (ANSI 117.1-1992).* Falls Church, VA: Council of American Building Officials.

Connell, B., Verschoor, G., & Ricketts, C. (2009). Riojas family #617 [Television series episode]. In *Extreme home make over home edition, (Season 6).* Fresno, CA: ABC Television Network.

Department of Justice (DOJ). (1991). Nondiscrimination on the basis of disability by public accommodations and in commercial facilities; Final rule. *Federal Register, 56*(144), 35544–35961.

Hines, A., Null, R., Potthoff, J., Sullivan, S., & Welch, L. (2010, November). *Kitchen design strategies to increase home dining for all family members.* Portland, OR: Housing Education and Research Association.

Kira, A., Tucker, G., Cederstrom, C., & Cornell University. (1960, January). Housing needs of the aged with a guide to functional planning for the elderly and handicapped. *Rehabilitation Literature, 21,* 370–377.

Kira, A., Tucker, G., Cederstrom, C., & Cornell University. (1973). *Housing requirements of the aged: A study of design criteria.* Ithaca, NY: Center for Urban Development Research, Cornell University.

Koontz, T. A., & Dagwell, C. V. (1994). *Residential kitchen design, a research-based approach.* New York: Van Nostrand Reinhold.

Liebrock, C., & Behar, S. (1993). *Beautiful barrier-free.* New York: Van Nostrand Reinhold.

Mace, R. L.; Barrier-Free Environments, Inc. & National Institute on Disability and Rehabilitation Research (U.S.). (1991). *The accessible housing design file.* New York: John Wiley & Sons.

Mack, S. (2006, March–April). Universal design don't buy, build, or remodel without it. *Ultimate Home Design, 2,* 50–53.

Moore, L. J., & Ostrander, E. R. (1992). *In support of mobility: kitchen design for independent older adults* (Information Bulletin 225). Ithaca, NY: Cornell University, Cornell Cooperative Extension.

NAHB National Research Center. (1992). *Directory of accessible building products* (2nd ed.). Upper Marlboro, MD: NAHB Research Center.

Null, R. L. (1987). *IIA* Universal kitchen design for the low-vision elderly. *Journal of Interior Design Education and Research, 14*(2), 45–50.

Peterson, M. J., & National Kitchen and Bath Association (U.S.). (1998). *The National Kitchen and Bath Association presents universal kitchen and bathroom planning: Design that adapts to people.* New York: McGraw-Hill.

Raschko, B. B. (1982). *Housing interiors for the disabled and elderly.* New York: Van Nostrand Reinhold.

Schumacher, T. L., & Cranz, G. (1975). *The built environment for the elderly: A planning and design study, focusing on independent living for elderly tenants.* Princeton, NJ: Princeton University, School of Architecture and Urban Planning.

Schwab, C. M. (2004, January 1). A stroll through the universal-designed smart home for the 21st century. *The Exceptional Parent, 34,* 24–29.

Smith, E., & Smith, J. (1991). *Entryways: creating attractive, inexpensive no-step entrances to houses.* Atlanta, GA: Concrete Change.

Sweet's Group. (1996). *Sweet's accessible building products catalog file.* New York: Sweet's Group.

United States, General Services Administration. et al. (1984). *Uniform Federal Accessibility Standards.* Washington, DC: General Services Administration, Department of Defense, Department of Housing and Urban Development, U.S. Postal Service.

Marketing Universal Design

9

UNIVERSAL DESIGN AND GREEN DESIGN, AN IMPERATIVE

Many are convinced about the benefits of universal design (UD)—which is design for transportation, buildings, and products that can easily be used for all people, of all ages, abilities, and disabilities. As you already recognize, it is imperative and urgent that we publicize the importance and necessity for UD in this day and age. We cannot rely on a trickle-down theory—just hoping that eventually the health and longer-life benefits of UD will gradually filter down from a few practitioners and advocates. Instead, the UD movement has to become increasingly dynamic and strong by joining forces with like-minded groups.

As John F. Kennedy said not once, but twice, during his presidency, "a rising tide raises all boats"! I am suggesting that the UD movement collaborate and hitch a ride with the green/sustainable movement. Why? Because in the minds of a lot of young, as well as older people, sustainability is powerfully current on people's minds. According to this week's *Newsweek* magazine, "all over America, a post-Katrina future is taking shape under the banner of 'sustainability.' Architects vie to create the most sustainable skyscrapers." It is a hot subject! It is a vital topic! And a recognizably critical issue. A green and sustainable revolution is occurring. By joining together, both movements can help each other. Undoubtedly, the main beliefs underlying sustainable design will take on more universal meaning by embracing UD principles.

A number of features fostered by the sustainable movement dovetail with those of UD. Some of the parallels can be listed as follows:

- Building design can help sustain life from the cradle to the grave.
- Environmentalism is on the upswing and so is UD.
- Attitudes toward green construction have changed from negative to positive and so are attitudes toward building using UD principles.

Both of these movements are about giving people independence in their lives, says San Francisco Bay area architect, Alan Ohashi. For example, UD promotes the freedom to live and move around in a residential environment, where green/sustainable design is attempting to give individuals their independence from big oil and big power companies!

In addition, he mentions, these two movements have inherent efficiencies that generate greater profitability or savings. UD aims to substantially reduce the waste and expense of changing homes because they are not adaptable. Green/sustainable initiatives encourage the use of cheap and renewable energy sources such as the sun and wind. Further, Mr. Ohashi comments, both movements fuel growth in technology and building trades. UD has already created new products aimed specifically at this growing market. Definitely those designers and contractors knowledgeable about UD get a marketing edge. Green/sustainable concepts have also created a slew of new products that give the same edge to technology companies and installers knowledgeable about these concepts (McMillan 1981).

It is important to note that green homes and condos often sell more quickly than their conventional counterparts. Developers and builders should therefore be urged on by the UD design and health community to realize that selling homes designed with both UD improvements and green/sustainable building materials could be highly profitable.

The former CEO of Patagonia, David Olsen, who is one of the corporate pioneers in the development of sustainable materials, buildings, and business practices and who is currently the Director of the Coalition to Advance Sustainable Technology, observes that bringing together the UD and Sustainable Movements is both timely and urgent. He says that sustainability is first and foremost a design problem. Environmental and energy use effects of building and products have essentially been thought of as external to, or only indirectly the responsibility of, the designer, rather than as an integral aspect of design. But now, the ability to design using materials that use fewer resources and energy, last longer, are nontoxic, and can be reused is expanding rapidly. These design challenges can be at the heart of UD.

Directly related to the benefits to be gained from joining together sustainable ideas and universal design principles are health issues. Indoor air pollution can be 10 times worse than LA on a bad day no matter where you live. Off-gassing is a hidden menace. Everyone benefits by choosing nontoxic building materials, good moisture-control detailing, pollution-free, contamination–rejection, and strong ventilation planning. This is especially pertinent for those who suffer from breathing problems and possibly for those with brain-related disorders. Chemical sensitivities are a disability and a growing number of our population suffer from illnesses associated with toxic poisons and off-gassing from construction materials such as formaldehyde, adhesives, and some plastics, among others.

The road to good, sustainable, and universal design remains a bumpy and challenging one. As an example, before coming to Columbus, I queried members of an ASID (American Society of Interior Designers) peer group to receive their comments and ideas for this presentation. Only one member offered a suggestion. The rest said they knew very little about UD. Unfortunately, it is apparent that not many designers, let alone architects, educators, developers, specifiers, health workers, and contractors, are aware or seem to be interested in UD. This mindset has to be changed.

Added to this situation, I find that some of my interior design clients don't even want to hear about incorporating UD materials and design in their homes. For example, I suggested to a client that he install grab bars in his bathroom, not necessarily for now, but for possible future use. His comment, "I don't want to do that. Every time I look at the grab bars, I will feel sick!" However, even though my clients are a bit unsure of exactly what the green/sustainable concept promotes, they more readily approve suggestions for "going green" because they fit in with an acceptable movement. This further argues for UD to work together with the green movement.

We need to constantly keep in mind that everyone deeply desires to live and work in a healing environment from the cradle to the grave. Designers, architects, green planners, occupational therapists, all of us, I believe, need to cooperate in order to raise the tide of healthy design awareness so all the boats that John F. Kennedy referred to will indeed rise together … but those "boats" need to be constructed sustainably with nontoxic adhesives, sealed with no off-gassing paints or stains, made out of reused wood that has not been treated with petroleum-based products, and installed with added grab bars for use when monster waves come their way!

NATIONAL HOME MODIFICATION COALITION

The National Home Modification Action Coalition (NHMAC) was formed in 1992 by professionals representing national, state, and local entities directly and indirectly involved with creating supportive home environments. Participants are committed to exchanges via periodic teleconference meetings, a newsletter and ongoing communication via an Internet List Serve. Coalition members monitor and share information about national, state, and local home modification activity; provide technical assistance to each other and other professionals; and use the Coalition as a forum to discuss strategies and common interests. The Coalition and its members have supported efforts to plan and promote periodic National Conferences and meetings to provide the opportunity for professionals of different disciplines and policy makers to collaborate on a common goal—to increase the availability of home modification for older persons and persons with disabilities. Fundraising efforts have generated corporate and organizational sponsorship for the conferences. To maintain the momentum and successes of the NHMAC, organizations such as AARP, USC's National Resource Center on Supportive Housing and Home Modification, the Center for Universal Design, Adaptive Environments, and the Idea Center (SUNY, Buffalo) have provided ongoing in-kind staff support and leadership. Approximately 1200 professionals participate in the NHMAC with special interest in consumer education, program development, financing mechanisms, and coalition building.

The National Resource Center on Supportive Housing and Home Modification's website, www.homemods.org, is the most comprehensive website on home modification available. The site is designed for professionals interested in home modification, caregivers and consumers of home modification, and policy makers. Also hosted on the site is the National Directory of Home Modification Resources that provides a state-by-state guide to business and services that provide home modification and other support services.

The website provides education and training on home modification through a distance learning program. It serves as a forum for professionals such as occupational therapists, remodelers, and interior designers to share information and to collaborate in problem solving.

The website also serves as a repository of information on home modification and supportive housing. The website features a newsletter that chronicles developments in home modification and is then archived for future reference.

BEYOND TIME, BEYOND SPACE: A HOME FOR ALL AGES: AN INTERACTIVE EXHIBIT

One of the activities taken by the coalition was to create home modification exhibits in easily accessible community focal points. Such exhibits increase involvement and visibility in the community. Exhibits also help legitimize the idea of home modifications to the community, bringing long-term community support for continued programs.

Exhibits were located in donated retail spaces in upscale shopping areas. The representation of aging, disability, and rehabilitation perspectives within the coalition ensured that the exhibits addressed diverse needs and interests.

Over 30 local and national vendors donated home modification and universal design products and materials. They were willing to do so because the community would benefit from their presence and they would receive visibility within mainstream venue for broadening their customer base.

The Pasadena exhibit was open to the public for two months. Over 5000 people of all ages took part in the interactive project, learning of several innovative home modification products. Young children could be seen riding the Stair Glide, middle agers took notes on remodeling kitchens, and older individuals remarked appreciatively on the simple assistive devices they had hitherto been unaware of.

Also, more than 20 groups held their regular meetings at the exhibit site, including the American Institute of Architects. Workshops were conducted for social service and housing professionals.

For example, the Pasadena Neighborhood Housing Services held a meeting of lenders at the exhibit site which focused on financing home modifications and utilizing universal design.

Highlights of the exhibition included the following:

- The GE New Home Essentials Living Center: A fully equipped, universally designed kitchen. (Designed by Mary Jo Peterson for GE.)
- Home Modification Products, such as portable ramps, bathroom fixtures, roll-in showers, grab bars, assistive devices, and so on.
- Photos and presentations of home modification efforts completed by local professionals and social services.
- A Home Resource Center, with extensive listings of local and national resources, which included brochures, checklists, and free manuals.
- Videos of successful home modifications were continually played during the exhibit.

The exhibits brought together diverse groups of people who now work more closely together. These include separate vendors and professionals, as well as service groups and vendors/professionals who are better able to meet client/customer needs.

Over 80% of the visitors indicated that the exhibit greatly increased their knowledge of home modification. And nearly 50% stated that they would be making a change to their own homes or to those of someone they knew as a result of what they had learned at the exhibit.

Numerous other exhibits have highlighted the use of universal design. They utilize kiosks, plans, floor models, photographs, and other techniques to celebrate universal design. Several examples follow.

FIGURE 9.1 Japanese department store gallery highlighted Tripod Design products.

FIGURE 9.2 Japanese store gallery—Tripod Design.

TRIPOD EXHIBIT OF THE DESIGN PROCESS

The Tripod exhibit in a Tokyo department store art gallery showed the process the company used in creating their universally design products (Figures 9.1 and 9.2).

TOYOTA UNIVERSAL DESIGN SHOWCASE

The Toyota Exhibit Hall in Japan was a universal design showcase of over 400 products developed in Japan and abroad (Figures 9.3 through 9.5).

UNIVERSAL DESIGN PAVILIONS AT THE NATIONAL AARP MEETINGS

The Universal Design Pavilions in Chicago, Los Angeles, and Las Vegas in the early 2000s included several complete modular houses and other exhibits.

Larry Weinstein (2007) made presentations and led tours of the homes that highlighted universal design features (Figure 9.6).

FIGURE 9.3   Exterior of the Toyota Exhibit Hall.

FIGURE 9.4   Universal design products from a variety of sources in the Toyota Exhibit Hall.

FIGURE 9.5   Japanese universal design products.

Installation of a platform lift can solve the problem of steps leading to an entryway.

FIGURE 9.6   Modular home from the Universal Design Pavilion at the National AARP meeting (Larry Weinstein).

THE WORLD OF "OFF-SITE" HOMES

At last—a fully accessible,
reasonably priced,
comfortable, beautiful,
energy-efficient
"Home Sweet Home"?

Story and photos by Larry Weinstein

Personal happiness and independence are deeply affected by our ability to perform everyday living tasks and activities … especially within our own homes! Even for perfectly healthy individuals, our ability to successfully accomplish tasks in the home, such as cooking and bathing efficiently and safely, is sometimes seriously impeded by the way most homes have been designed and built through the years.

Physical barriers in our homes prevent us from enjoying independent and safe daily living. Steps leading to the front door create a potential hazard for youngsters and adults and are a major obstacle for anyone with a mobility impairment. Narrow hallways and doorways are hard to get through when our hands are full of groceries, or carrying a young infant, or for someone using a walker or wheelchair. Round doorknobs can be tricky for children's small hands as well as the arthritic hands of adults. Trying to get something from upper-cabinet shelves can be a challenge for anyone who is short or has difficulty stretching and reaching. Lower kitchen and bathroom cabinet shelves are almost impossible to reach into. Poor lighting makes it difficult to see and hinders the performance of simple daily tasks (Bopp 1990; Fisk et al. 1990).

The following are some of the elements of a "Universal Design featured Livable Home":

- A zero-step main entry into the house.
- All hallways are 42 inches minimum width and doors 36 inches wide.
- Lower kitchen and bathroom cabinets with full-extension drawers for ease of use and open finished accessible spaces underneath the kitchen sink, cooktop, and bathroom sinks for comfortable chair and wheelchair use.
- Front-load and front-controls dishwasher and clothes washer/dryer mounted on elevated platforms minimize bending.
- Clear five-foot-diameter open spaces in all rooms.
- A spacious curb-less shower.
- Electrical outlets, light switches, and thermostat mounted at convenient accessible heights.
- Great energy-efficient, high-output fluorescent lighting throughout the house.

(These concepts were illustrated in Chapter 8, Residential Design, with examples of award-winning designs for senior housing; see Susan Mack and Mac Kennedy, Figures 8.22 through 8.29.)

Larry Weinstein continues: I began presenting free workshops on accessible and universal design (UD) applied in the home at the Southern California Abilities Expo in 2001, and at workshops and seminars at other consumer and trade events—and I have been doing so ever since. Through the years, I have often been asked to recommend knowledgeable and experienced reputable architects, designers, and builders who are familiar with residential design and construction for people with disabilities. Unfortunately, they are hard to find.

Recently, I contacted a number of architects and designers throughout the country to inquire who had knowledge and experience in the design and construction of new homes, and the modification of existing homes, to accommodate the needs of people with physical or developmental disabilities. Many of them were familiar and experienced with the Americans with Disabilities Act accessibility details, which relate to commercial environments but do not apply to single-family homes (Wilkoff and Abed 1994). After I reviewed a good number of home designs I received in recent years, it became clear that many architects and designers do not know how to correctly design adaptive or barrier-free living environments.

In 2002, I became the Livable Homes Housing Pavilion consultant for AARP National Events. AARP has more than 39 million members age 50 and above and will celebrate its 50th anniversary next year. Since 2001, AARP's large annual three-day national member event has been in a different major city each year. This happening takes place in a huge convention center exhibit hall filled with numerous pavilions, offers seminars and top-name entertainment, and is attended by thousands of people from all parts of the country.

Installation of a platform lift can solve the problem of steps leading to an entryway (Figure 9.6).

Champion Enterprises, one of the nation's largest "off-site" or factory homebuilding corporations, became one of the AARP 2002 National Event exhibitors. The company wanted to develop a special show home that would meet the needs of people age 50 and above. I offered, at no charge, to help them design and develop an all-new, UD-featured, fully accessible home—if they would agree to build it. They did, and this home was displayed at the AARP National Event in San Diego. This gave me a firsthand opportunity to learn about homes built in a factory instead of outdoors on-site, and thus began my exciting journey into the world of "offsite" factory homebuilding.

Two years later, after designing the Champion Enterprises exhibit at the AARP 2002 National Event, I helped design another new, fully accessible, energy-efficient, reasonably priced UD-featured, 1572-square-foot, three-bedroom/two-bath home for Champion Homes, and a new, beautiful, accessible 674-square-foot, one-bedroom "backyard cottage." These were exhibited at the AARP 2004 National Event in Las Vegas, Nevada, and were toured by thousands of enthusiastic event attendees. The price for this complete home—including all interior finishes, flooring, fixtures, lighting, and appliances—came to less than $100 per square foot excluding land, site-preparation costs, and required permits.

The plans for this home were subsequently given to a few quality on-site homebuilders to determine the home cost using traditional construction methodology. It was almost twice the Champion Homes price. In addition, where it took less than three weeks to build the Champion home from start to finish, it would most certainly take many months or more to construct the same home on-site, exposed to inclement weather, along with numerous delays, contractor no-shows, job-site pilferage, and more.

UD-featured homes offer lower kitchen and bathroom cabinets with open finished accessible spaces under the kitchen sink, cooktop, and bathroom sink (Figure 8.23).

In 2006, I helped design and develop a "Generation Series" of new, fully accessible, UD-featured Champion Homes, one of which made its debut at the AARP 2006 National Event in Anaheim, California, along with an Athens Park Homes beautiful one-bedroom, fully accessible Backyard Cottage. Both were highlight exhibits at the event and were toured by almost 15,000 people.

Working with Manorwood Homes, a division of Commodore Corporation, I helped design and develop the new, fully accessible, UD-featured show home for the AARP 2007 National Event in Boston. On each of the three days, a continuous stream of attendees toured this elegant, reasonably priced home. I had the pleasure of taking the AARP president and a staff member of the US Department of Veterans Affairs' (VA's) national Specially Adapted Housing (SAH) Grant program on private tours through the show home. The VA staff member affirmed the model home on display met all the SAH Program's adaptive features requirements. These accessible homes are available with optional curb-less showers up to five square feet in size.

I have been working in collaboration with the National Manufactured Housing Institute and National Modular Housing Council to motivate their membership of homebuilders throughout the country to offer reasonably priced, fully accessible, UD-featured homes throughout America.

In recent years, unfortunately, things have become progressively worse in the traditional "on-site" homebuilding industry as the availability of quality building materials and experienced reputable subcontractors has continued to decline. I believe that factory-built home construction is really the best way to go.

Manufactured and modular homes are completed in the factory in a fraction of the time it takes to build a traditional home. Because these homes are built in factories, manufacturers are able to use sophisticated tools, jigs, and technologies that ensure all walls, floors, and ceilings are square and plumb. Major components such as these are not only nailed together, as in site-built homes, but are also bonded with special adhesives. They are engineered for trouble-free structural durability and are built stronger and with more lumber than most site-built homes because they must endure the rigors of transportation or being lifted and set on site, while maintaining excellent structural integrity.

The factory environment can be very efficient, eliminating the possibility of on-site construction-related problems, including delays owing to bad weather. Furthermore, factory-built homes are seldom subject to delays owing to back-ordered materials or slipshod work by unreliable subcontractors. Once the home is delivered, it is closed and sealed within days, and the remaining on-site work is completed within weeks. This ensures early occupancy and eliminates damage caused by poor weather and vandals.

HOMES FOR VETS

A few years ago, I began developing a new program or initiative to get a number of national off-site homebuilding corporations to commit to building fully accessible and reasonably priced, energy-efficient homes to meet the needs of servicemen and -women with severe injuries and disabilities along with some of the more than 54 million Americans of all ages with some form of permanent disability. I have worked closely with the VA SAH Program to ensure that the designs and specifications I give to national off-site homebuilding corporations fully comply with all criteria set forth by the program.

Visit www.livablehomes.org and go to the "Who Builds" section to link to the Manorwood Universal Lifestyle Series Homes and the Champion Generation Series Homes websites.

If you have an active military service–incurred severe disability and have been honorably discharged or are still in active service, you may be eligible to receive a VA new adaptive home grant up to a maximum of $50,000. Modular, fully accessible homes built by any homebuilding corporation or company may be eligible for purchase using SAH grant funds. Each home must incorporate certain adaptive features to address veterans' specific needs and then reviewed by VA and approved on an individual basis.

CHICAGO UNIVERSAL DESIGN ARCHITECTURE COMPETITION

As part of an architecture competition that featured universally designed homes, the winning entries were actually built. Pictures of the winning designs were displayed on Kiosks as part of the UD pavilion at the AARP meeting in Chicago. A major exhibit was created to summarize and conclude the project (see Figures 9.7 and 9.8).

FIGURE 9.7   Intergen Chicago Housing Competition Exhibit.

FIGURE 9.8   Intergen 1 Chicago Housing Competition Exhibit.

ART MUSEUM EXHIBIT IN SEOUL, KOREA

Universal design seminars with international participation accompanied the museum exhibit in Seoul, Korea, in 2006 (see Figures 9.9 and 9.10) (Chung-a 2007).

UNIVERSAL DESIGN EXHIBIT: IGM GALLERY, USC MEDICAL CAMPUS

The sponsors of this exhibit solicited competitors from interior design students. Advertising through Interior Design Educator's Listserv resulted in contributions from Georgia State University, Lansing Community College (Michigan), and Bolling Green State University (Ohio). Invitation-only events

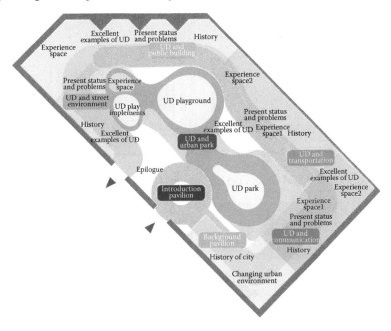

FIGURE 9.9 Floor plan of the 2005 Art Museum Exhibit in Seoul, Korea.

FIGURE 9.10 Portion of a poster from the exhibit in Seoul with many universal design products.

provided in-depth opportunities for selected groups. Some presentations emphasized universal design for aging. Exhibits were open for a three-month period (beginning March 2006) (see Figures 9.11 through 9.16).

Programming included several seminars. Dr. Yeun Sook Lee from Korea brought a group of her graduate students to California to participate in one of the seminars. Two representatives from the Toto Company in Japan visited the exhibit and photographed all of the items displayed.

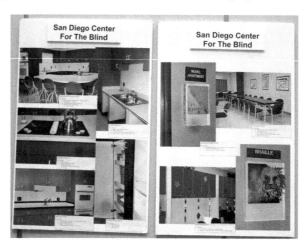

FIGURE 9.11   Poster in the IGM Gallery on USC medical center campus.

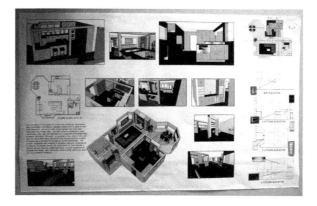

FIGURE 9.12   Poster in the IGM Gallery on USC medical center campus (student contest winner from Georgia State University in Atlanta).

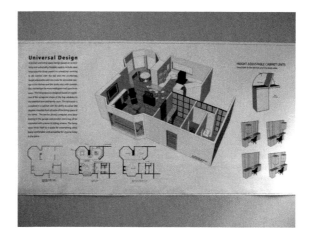

FIGURE 9.13   Poster in the IGM Gallery on USC medical center campus; a student contest winner, Kitchen Design. Georgia State University, Atlanta.

FIGURE 9.14   IGM Gallery (USC Keck Medical Center Campus) display of Baci wall mirrors created for the exhibit.

FIGURE 9.15   IGM Gallery. GE movable sink in exhibit.

NATIONAL KITCHEN AND BATH ASSOCIATION (NKBA)

Student Competitions Room Layout and Floor Plan (see Figures 9.17 and 9.18).

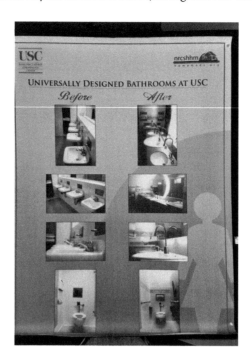

FIGURE 9.16   IGM Gallery poster of USC bathroom remodel.

FIGURE 9.17   NKBA student competition winner's kitchen sketch (Sanchali Srivastava).

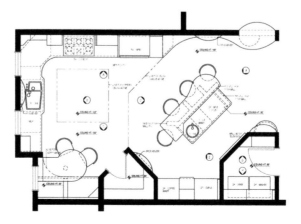

FIGURE 9.18   Student competition entry exhibits by Sanchali Srivastava.

FIGURE 9.19   NARI award-winning kitchen.

HOME REMODELING ASSOCIATION (NARI) AWARD-WINNING KITCHEN

(See Figure 9.19.)

TEACHING UNIVERSAL DESIGN THROUGH COMMUNITY BUILDING SERVICES LEARNING TECHNIQUES

Teachers of interior design have been extolling the benefits of universal design in the classroom for years. Instructors also have found that because universal design features are not yet standard in many homes, students need actual experiences with features and clients to understand its benefits. Experiential learning techniques can actually change student attitudes and project solutions. While classroom instruction helps students to incorporate universal design features into their design solutions, it is also clear that if true change is going to occur in the housing/building industry, builders, remodelers, and consumers must also be educated about its benefits.

There is clearly a need to take information about universal design into the community. In addition, many university administrators have been encouraging or even mandating that faculty be engaged in outreach, fostering ways to increase community relationships. Universities can reach out to communities in several ways. Service-learning courses are a viable way for universities to serve the community and also provide valuable experiential learning for their students. In addition, Land Grant Universities use the Cooperative Extension Service to educate thousands of state constituents daily through workshops, mass communication, publications, and computer programs. Finally, faculty often work with industries in cooperative research and consulting. Susan Zavotka and Meg Teaford describe the successful efforts used to accomplish this at The Ohio State University.

COMMUNITY BUILDING

There are two major ways that service can be provided to a community. One is based on the needs of a community; the other is asset based. In an asset-based scenario, the university works with members of the community to determine what they already possess that could be used to address a problem. It allows the valuable tools of the community to be used in solving community problems. The end result is empowerment of all participants.

Use of the asset-based approach is critical in developing a Universal Design Education Program. The community already possesses retail industries that provide products and in-store professional consultants for builders and homeowners. Furthermore, government and social service agencies provide in-home consulting for safety and assistive devices. Interior and kitchen designers and other design professionals offer design services.

To develop a universal design educational project that would have direct impact on decision makers in the building industry, a collaborative effort was needed that involved several The Ohio State University (OSU) departments, extension, social service agencies, and retail operations where home building products are sold.

METHOD

Initially, faculty members from OSU Interior Design, Family Sciences, and Occupational Therapy met with state cooperative extension specialists to develop grant proposals for the educational program. As the team explored the possibilities for dissemination of information in the community, the asset-based approach revealed that a social service agency (State Department of Aging) and a representative from a retail home improvement store (Lowe's Home Improvements and Warehouse) were beneficial.

The end result of the educational program was a service-learning course where students (interior design and occupational therapy) spent the first four weeks in the classroom learning about universal design, available products, and how to conduct a workshop. Students then teamed up with state cooperative extension agents to help teach workshops at either a Lowe's Home Improvement Store or a public center identified by the extension agent. Students next conducted home assessments for workshop participants during a two-week period after the workshop. Finally, students would spend two weeks back in the classroom evaluating and revising the workshop and assessment process.

Prior to the student workshops at Lowe's, project team members met with key store employees and the Lowe's sales associates for initial training sessions about universal design. Lowe's employees contacted vendors for additional products, produced a booklet of universal products available in the store, and devoted two end-cap displays for products featuring universal design. University faculty and extension agents provided posters, information publications, product identification cards, a tabletop display in the store, and persons to sit at the table the day of the workshops to answer questions. Faculty and extension services also provided extensive publicity through press releases and flyers sent to news media, senior centers, and other social service agencies.

On the days of the workshops, team members conducted one-hour workshops for the general public providing participants with a packet of educational information (Figure 9.20). On the same day, Lowe's Pros conducted mini-workshops on installation procedures for various products throughout the store.

This outreach activity is an outstanding example of community education. It was funded by multiple Outreach and Engagement grants beginning with an OSU CARES grant and course support from the College of Human Ecology Learn and Serve Initiative; this project involves OSU Extension Family and Consumer Sciences staff as well as Dr. Susan Zavotka, Associate Professor, Consumer and Textile Sciences, and Dr. Meg Teaford, Assistant Professor, Allied Medical Professionals. The teaching laboratory was made possible by the partnership with Lowe's Home Improvement Warehouse and Dave Fox Remodeling. Universal design concepts, such as use of door levers for handles rather than knobs, improve ease of access for all populations (Price et al. 2004).

Finally, a universally designed kitchen and bath have been built in partnership with OSU, Lowe's, and Dave Fox Remodeling in the McCormick Building at the Farm Science Review site in London, Ohio. The OSU facility is unique because it is permanent and is available for tours, teaching, and research (most UD demonstration facilities are only open temporarily because of the cost to maintain them). See the pictures from the partnership in Chapter 3, Figures 3.33–3.38.

FIGURE 9.20   Lowe's employee using the UD information table setup by OSU students and faculty.

ASID DESIGNER SHOW HOUSE IN ATLANTA, GEORGIA

In the Atlanta show house each room was designed by a different ASID (American Society of Interior Designers) designer. The kitchen is featured in Figure 9.21. For a UD laundry with ample space to maneuver, see Figure 9.22.

A residential elevator was also included in the house. Unfortunately, this show house was open for only a limited time because it was to be sold. However, many people were able to visit it and photographs of its interior are still available for education purposes (elevator, Figure 9.23).

FIGURE 9.21   ASID designer show house kitchen.

FIGURE 9.22   UD laundry in the Atlanta UD show house.

FIGURE 9.23   Residential elevator in the UD designer show home.

FIGURE 9.24   Extreme Home Makeover kitchen design in sketchup by Mark Johnson.

FIGURE 9.25   Extreme Home Makeover laundry, sketchup computer rendering by Mark Johnson.

3-D COMPUTER DESIGN APPLICATION

Mark Johnson of Kraftmaid cabinets worked with the Extreme Home Makeover TV program to produce UD designs for a handicapped homeowner/client. Several sketchup designs produced by computer software are shown in Figures 9.24 and 9.25. These can be produced quickly and easily by a trained technician and enforce design criteria.

"ALL FOR ONE," UNIVERSAL DESIGN CONCEPTS FOR EVERYONE

As defined by founder Ron Mace, "Universal Design is a strategy that aims to make the design and composition of different environments and products usable for everyone. It attempts to do this in the most independent and natural manner possible, without the need for adaptation or specialized design solutions. The intent of the Universal Design concept is to simplify life for everyone by making the built environment, products and communications equally accessible, usable, and understandable at little or no extra cost. The concept further emphasizes user-centered design by following a holistic approach to accommodate the needs of people of all ages, sizes, and abilities." Consequently, universal design is becoming an integral part of the architecture, design, and planning of the built environment.

UNIVERSAL DESIGN AND AGING

We have all become familiar with the baby boomer bulge, a highly diverse group whose members are poised to enter the last third of their lives. In addition to the special needs facing this group, add responsibility for aging parents and raising grandchildren. Designers are faced with a multiple-needs population, and the challenge is to create supportive living environments for everyone. Universal design is the best strategy for accomplishing this goal.

Achieving universal design is a complex process that has been further complicated by a lack of recognition of need by an increasingly aging population. Older people must first get past the stigma of being considered "disabled" after a lifetime of relative independence and competence. Few are willing to admit that their bodies are changing and that their demands on their environment are also changing, despite the availability of design information and assistive devices.

Marketing to a specialized "elderly" population is like trying to reach a nonexistent group. After all, we generally consider "elderly" to be 15 years older than we are, whether we are 15, 55, or 85 years old. No one wants to be labeled as "old" or "disabled," so any products or services that are targeted for this group are sure to be met with a lack of interest.

Universal design is a general approach that provides maximum appeal and benefit for all age groups, rather than to a niche market such as the frail elderly or disabled/wheelchair users. We need to take a universal design approach to ensure safe, comfortable, convenient, and accessible dwellings for everyone, not just the elderly.

FINDING EXAMPLES

Because universal design is invisible and inclusive, it will meet the needs of an aging population (and all others). This is not only an advantage but also a challenge. It means that we need to show good examples of universally designed products and environments and explain why and how they incorporate universal design principles. We also need to be able to evaluate whether products and environments can, to the greatest extent possible, be used by everyone. Some basic questions to use are related to the four universal design ideas originally developed by the concept's founder, Ron Mace. They are as follows:

- Is it easy to use and take care of? An example would be a single-control faucet with a non-shiny surface.
- Is it adaptable? Can it be used in a variety of ways to accommodate a variety of users? For example, dishwasher drawers that can be used separately or together and that come with multiple wash programs are good alternatives.
- Is it accessible? Side-by-side refrigerators/freezers provide access to both refrigerator and freezer space from any height. Double-door refrigerators with a bottom freezer also provide easy access.
- Is it safe? This has increasingly become a concern among manufacturers of kitchen and bath appliances. Look for marketing terminology on product literature and packaging.

A good example is the relatively new decorative grab bar from Rohl. The literature for the grab bars states, "Rohl grab bars are designed to give added security in and around the bathroom and provide years of trouble-free service. These grab bars provide one of the easiest and most efficient ways to make any bathroom area safe."

Asking these basic questions will help you to identify good examples of universal design and help you to avoid problems. Rather than talking about good design for aging in place, look to the promotional terms used by manufacturers. These include *ergonomic, easy to use, easy care, easy living, supportive, comfortable, comfort height, relaxed lifestyle,* among others. For instance, Kohler's Highline Pressure Lite Toilet literature mentions that it is one of Kohler's Comfort Height models, which are often shown next to a regular chair and are not mentioned in a disabled context, thus sparing the ego of the potential user.

DESIGN SOLUTIONS

It is also important to be able to identify safety issues with appliance design and placement. Ranges with the controls placed behind the burners and microwave ovens placed over the burners are both major safety hazards. Electric ranges with staggered burners and controls in the front can be a better option. If you are not using dishwasher drawers, dishwashers should be raised 6 inches to 8 inches off the floor and placed so that they are accessible from either side. Even refrigerators now are available with comfort height drawers. They provide flexibility in placement (Figure 9.26). Many of the new storage options are good examples of universal design because they provide easy access to all areas in the cabinets and can be adapted to meet specific storage needs. Think about including pullout cutting boards and roll-out shelves in your projects. D-shaped handles on cabinet doors also provide a better grabbing surface as opposed to knobs or cutouts. In addition, one also needs to think of creative ways to provide inexpensive and attractive alternatives to expensive special designs, such as pricey, electronic, adjustable-height sink cabinetry. You also might achieve flexibility by planning a variety of counter heights in the kitchen and have sinks and cooktops at several different levels. An eating area in the kitchen provides easy wheelchair accessibility or a place to sit in a regular chair while preparing food.

Along these lines, doors should be wide, with at least a 32-inch clear opening (36 inches is recommended, and since this is the basic size of an exterior door, it shouldn't be much more costly; it does create a look of spaciousness throughout the house) and have lever handles instead of knobs. Light switches, thermostats, and other controls should be mounted between 36 inches and 48 inches from the floor for easy reach. Rocker-style switches are easiest to use.

In the bathroom, low mirrors, nonslip floors, faucets with antiscald temperature controls, and doors that open outward (if someone falls, they won't block the entryway) are but a few of the design strategies that can make the space more useful for everyone in the house.

In the end, designers wishing to survive in the 21st century will need to rethink their marketing strategies, products, and services to meet the needs of a rapidly changing population. Universal design is going to be the strategy of choice.

FIGURE 9.26   Refrigerator drawers from Kitchen Aid in display kitchen at the Kitchen & Bath Expo.

REFERENCES

Bopp, K. D. (1990, January 1). How patients evaluate the quality of ambulatory medical encounters: A marketing perspective. *Journal of Health Care Marketing, 10*(1), 6–15.

Chung-a, P. (2007, September 11). Rolling rains report: Universal design in Korea. *Design shows way to comfort*. Retrieved June 11, 2013, from www.rollingrains.com/archives/000655.html.

Fisk, T. A., Brown, C. J., Cannuzzari, K. G., & Naftak, B. (1990, January). Creating patient satisfaction and loyalty. *Journal of Health Care Marketing, 10*(2), 5–15.

Gorman, J. (1992). Critical Condition. *Interiors, 151*(12), 28, 32, 36 passim. Retrieved June 11, 2013, from MEDLINE®/PubMed®, a database of the U.S. National Library of Medicine.

McMillan, N. H. (1981). *Marketing your hospital, a strategy for survival* (p. 77). Chicago, IL: American Hospital Association.

Price, C., Zavotka, S., Teaford, M., & Holmes, P. (2004). Universal design: An interdisciplinary partnership to promote ease of living. *Natural Resources and Environmental Issues, 11*(1, Article 8), 6.

Weintsein, L. (2007, December). The world of 'off-site' homes. *PN/Paraplegia News, 61*, 32-38.

Wilkoff, W. L., & Abed, L. W. (1994). *Practicing universal design: An interpretation of the ADA*. New York: Van Nostrand Reinhold.

Sustainability

HOW UNIVERSAL CAN IT BE IF IT'S NOT GREEN?
HOW GREEN CAN IT BE IF IT'S NOT UNIVERSAL?

Connie Barker of the Environmental Health Network has been a long time advocate for sustainable designs, including both green and Universal Design. Green, or more accurately, sustainable design, has increasingly become an important concern in many realms, but particularly so in the design of the built environment. Over the last 15 years, standards and codes for designing and maintaining buildings, and for conducting building operations in ways that optimize energy efficiency, conserve resources, utilize recycled materials, improve indoor environmental quality (IEQ), and lessen the carbon or ecological "footprint" of those living or working in them have proliferated. Some of the better known standards include the US Green Building Council's LEED (Leadership in Energy and Environmental Design) program, Canada's Green Globes Program, The Green Guide for Hospitals and Healthcare Program (embraced by Kaiser Permanente and other large health care providers), and Build It Green's Green Points Rated Program. The reasons for such growth are obvious. Concern with global warming, constantly rising energy costs, reports of "sick buildings," and the negative impacts that some modern built environments have on the health of their occupants, along with general concerns regarding dwindling resources, growing landfills, and models of growth and resource consumption that are clearly not sustainable at current rates, have all combined to create widespread concern that the design and maintenance of the built environment be brought into alignment with growing ecological knowledge and concerns. But very little has been done relating the growing body of knowledge regarding such sustainable design to basic principles of universal design.

This chapter offers a brief overview of some of the access challenges presented by the shift to "Green Building," as well as some examples of green projects with an access focus; information regarding IEQ as a neglected aspect of Universal Design; discussion of same IEQ aspects of the most widely used Green Building Standards (e.g., LEED); and information on several IEQ-focused guidelines and standards that can be helpful both in understanding IEQ as a part of access responsibilities, accommodating IEQ-based access needs, and bringing basic principles of universal, inclusive, and accessible design to bear on the creation of sustainable and green certified buildings and spaces.

In this chapter, we then go to one of the classic lessons on sustainability, "Cradle to Cradle: Remaking the Way We Make Things" for a look at some of the major problems facing our society today.

CRADLE TO CRADLE: REMAKING THE WAY WE MAKE THINGS

A QUESTION OF DESIGN

The post-Industrial Revolution society

- Puts billions of pounds of toxic materials into the air, water, and ground every year
- Results in gigantic amounts of waste
- Puts valuable material into holes all over where they can never be retrieved (landfills)
- Requires thousands of complex regulations
- Measures productivity by how few people are working
- Creates prosperity by digging up or cutting down natural resources and then burning them or burying them
- Erodes the diversity of species and cultural practices

Allergies, asthma, and "sick building syndrome" are on the rise. Yet legislation establishing mandatory standards for indoor air quality is practically nonexistent.

Rather than a readily identifiable illness, some people develop an allergy, or multiple chemical sensitivity syndrome, or asthma, or they just do not feel well, without knowing exactly why. (See Connie Barker's experience in the Ecology House case study.) Even if we experience no immediate ill effects, coming into constant contact with carcinogens like benzene and vinyl chloride may be unwise.

Think of it this way. Everyone's body is subjected to stress, from both internal and external sources. These stresses may take the form of cancer cells that are naturally produced by the body (by some accounts, as many as 12 cells a day), exposure to heavy metals and other pathogens, and so on. The immune system is capable of handling a certain amount of stress. Simplistically speaking, you could picture those stressors as balls your immune system is juggling. Ordinarily, the juggler is skillful enough to keep those balls in the air. That is, the immune system catches and destroys those 10 or 12 cells. But the more balls in the air—the more the body is besieged by all kinds of environmental toxins, for example—the greater the probability that it will drop the ball, that a replicating cell will make a mistake. It would be very hard to say which molecule or factor was the one that pushed a person's system over the edge. But why not remove negative stressors, especially since people don't want or need them?

Some industrial chemicals produce a second effect, more insidious than causing stress: they weaken the immune system. This is like tying one of the juggler's hands behind his back, which makes it much harder for him to catch the cancer cells before they cause problems. The deadliest chemicals both destroy the immune system *and* damage cells. Now you have a one-handed juggler struggling to keep an increasing number of balls in the air. Will he continue to perform with accuracy and grace? Why take the risk that he won't? Why not look for opportunities to strengthen the immune system, not challenge it?

The resulting problems are environmental illness including multiple chemical sensitivities (MCS).

Today's industrial infrastructure is designed to chase economic growth. It does so at the expense of other vital concerns, particularly human and ecological health, cultural and natural richness, and even enjoyment and delight. Except for a few generally known positive side effects, most industrial methods and materials are unintentionally depletive. Yet just as industrialists, engineers, designers, and developers of the past did not intend to bring about such devastating effects, those who perpetuate these paradigms today surely do not intend to damage the world. The waste, pollution, crude products, and other negative effects that we have described are not the result of corporations doing something morally wrong. They are the consequence of outdated and unintelligent design.

Green design and universal design really need to be studied and marketed together. In the beginning of Chapter 9, "Marketing UD" designer Mary Lou D'Auray eloquently stated the importance of combining the two concepts. In a recent project, "Re Green," the American Society of Interior Designers, and the Green Building Council worked together to develop a major research and teaching agenda that will be described later. Because universal design has not found the widespread recognition experienced by "green design," it just makes good sense for each to build on the strengths of the other (Barket et al. 2009).

This chapter continues with a case study of a project that was started over 20 years ago—resulting in a housing complex that has been very successful because of its universal design and sustainability.

ECOLOGY HOUSE

Ecology House is an 11-unit affordable apartment development in San Rafael, California (Figure 10.1) that houses persons with environmental illness/multiple chemical sensitivities (MCS). This tiny apartment complex has attracted national media attention. Many misstatements of fact have been reported. Ecology House, Inc., the developer and owner of Ecology House, produced this report in an attempt to provide accurate information regarding building materials, costs, and results. We hope the report will provide useful information to others building housing for persons with MCS. The report reflects input from the architect, contractor, development consultant, members of the design team, current residents, onsite property manager, and the Ecology House, Inc. Board of Directors. However, the information and conclusions in this report are the responsibility of Ecology House, Inc.

In November of 1994, residents moved in to the first affordable housing apartment building in the United States constructed and maintained with materials that can be safer for persons with MCS.

What is it? MCS is also called environmental illness. It is an acquired condition in which people become intolerant to normal levels of everyday chemical and biological substances that cause no acute problems to most people. When the accumulation of toxins in the body becomes greater than the ability to get rid of them, even minute amounts of any substance may trigger symptoms.

FIGURE 10.1 Ecology House—an 11-unit affordable apartment development in San Rafael, California, designed for persons with multiple chemical sensitivities (MCS).

The person often develops sensitivities to new substances. Common toxic substances are everywhere—some are new building construction and remodeling materials like fresh paint, new carpeting and adhesives, cleaning products, office machines, automotive exhaust, tobacco smoke, molds, pesticides, fragrances in cosmetics, body products, air fresheners, and food additives.

Because it is a hidden disability, MCS has a somewhat different impact on social functioning than disabilities that are apparent because of appearance or use of equipment (i.e., canes or walkers). A list of criteria for hidden disabilities includes impaired functioning and restricted lifestyle, especially in social and work realms; a daily focus on pain or fatigue; feeling defensive and having to explain or defend your condition to others; facing stigma, blame, or other's trivializing one's symptoms; and having one's behavior misinterpreted or labeled, for example, the person whose memory loss is "flaky," the person in pain who gains weight is "lazy." There are three invisibility issues common to people with MCS that influence social relationships and access:

1. Fluctuations in day-to-day functioning. Although fluctuation is common in disabling chronic illnesses, it can be extreme in people with MCS. In a work setting, a person who is capable and productive in a protected environment, like an office in their low-toxic home, or a workplace that has accommodated his/her needs, may suddenly experience neurological and cognitive impairment in a less toxin-controlled environment. In his description of home offices in universally designed houses (Chapter 6), Charles Schwab talked about how the home office can be a supportive work environment for persons with chronic disabilities. If workers are required to perform in a less safe environment, or if their "safe" environment becomes temporarily compromised, they might experience memory loss, difficulty reasoning, severe breathing problems, or mobility problems like difficulty walking or standing. This extreme variability complicates and exacerbates the issue of invisibility.

2. Invisible and fluctuating barriers. Barriers to access for persons with MCS usually fall into the general category of poor indoor environmental quality (IEQ), which includes both poor indoor air quality and large numbers of electromagnetic fields (EMFs). Both air quality and EMFs are, of course, inherently invisible. And both can fluctuate widely on any given day as well as over a longer period. A room that is safe enough for a person with MCS one day may be treated with pesticides overnight, or repainted, or have its carpet cleaned and be unsafe the next day.

   The combination of invisible barriers and their fluctuating nature creates an unusual set of access challenges. As was previously mentioned, invisibility exacerbates the tendency to believe that such barriers are unreal and their removal unnecessary. This fluctuation creates two barrier issues: First, variability means that CREATING AND MAINTAINING ACCESS, rather than performing a one-time design accommodation (like building a ramp), becomes an ongoing process. Second, precisely because it is the ongoing, recurring activities of a great many other individuals that create the barriers, only the consistent cooperation and assent of such people as building maintenance personnel, grounds upkeep staff, and so on, can effectively remove or mitigate most of the barriers.

3. The advantages and perils of passing. One property of all hidden disabilities, like other socially problematic non-obvious traits, is that their hidden-ness creates the possibility of "passing"—not letting others know that you share a socially stigmatized trait or behavior. Avoiding being treated as lazy, mentally ill, or malingering can easily seem like a matter of psychic and emotional survival to people who routinely experience belittling and non-belief regarding their disability from all corners of their social world.

Not surprisingly, this constellation of situations results in profound stress, frequent unwillingness to ask for accommodation, and serious shortcomings in what accommodations are requested.

A partial solution to MCS access is therefore the addition of IEQ to the concept of universal design. Changes to routine building construction, maintenance, and occupancy policies are much more likely to provide consistent access than individual requests for accommodation. Universal design is really the process of codifying building and design elements that provide better access for one group, but ultimately also benefit many other groups. Access for people with MCS could be improved by expanding the concept of universal design to include the use of less toxic building materials, building and grounds maintenance techniques, occupancy guidelines, and other techniques or procedures that enhance and improve IEQ. Since IEQ guidelines are increasingly being included as part of various green building guidelines and are used in some certified green products, it is now more feasible than ever to begin codifying such standards for universal design. As a building constructed to accommodate disabled people using universal design (all units are fully wheelchair accessible; see Figure 10.2), and as a building that has been given special recognition as an example of multifamily green design, a study of Ecology House is an excellent place to begin this process.

FIGURE 10.2   The second floor of Ecology House features less toxic building materials such as the stainless steel elevator interior.

The relevance of MCS or environmental illness for the design community is that many of the materials used in building construction, interior finishes and furnishings, and products for installations contain toxic chemicals that can trigger MCS and are being implicated in other conditions such as asthma, chronic obstructive pulmonary disease, attention deficit disorder, and so on. Therefore, it is essential that designers specify the safest, healthiest materials possible. By building in a way that keeps the most vulnerable population (people with MCS) safe, the general population is protected as well. This is a precautionary approach to design that emphasizes IEQ as an important feature of universal design. It is also a "green" approach to design that highlights the overlap between "green" building and universal design. One of the best examples of a building that features both green design and universal design is Ecology House.

Ecology House was designed and built in the early 1990s using US Department of Housing and Urban Development (HUD) funding—Section 811—PRAC program. The development team had difficulty meshing the widely varying needs of persons with MCS with code requirements and funding limitations. Katie Crecelius, a development consultant, said, "We felt like we were walking on Jell-O in designing and planning for Ecology House." Therefore, universal design features were incorporated to set the design framework and special IEQ features were included in the construction.

Ecology House was built to create a supportive environment for persons with environmental illness and multiple chemical sensitivities (MCS). Because persons with MCS often display symptoms similar to those of mental illness, their problems are frequently misdiagnosed. Because of this, it is difficult to clearly define environmental illness or MCS. The authors of Cradle to Cradle have compared the body to a juggler. The effects are different for different people, but this comparison helps explain why a person can develop asthma or other allergies overnight. As the number of toxins we encounter increases, the number of people with allergies also increases.

Ecology House planners recognized the often conflicting needs of people suffering with MCS. Therefore, planners understood that many, but not all, persons with MCS could be helped in a single building.

The success of Ecology House may be measured through occupancy data. Ecology House's 11 units are currently occupied by 11 residents with MCS, 8 of whom moved in immediately after the building was completed in November 1994. During the first year of occupancy, 3 residents moved out because they could not satisfactorily tolerate some of the materials used in their apartments. Three others have since moved in.

Residents seem to tolerate most materials and design techniques used. Some residents report that their health has substantially improved by living in a safer environment. Connie Barker, one of the original residents, is an example of this. Before she got sick, Connie Barker was a math major at UC Berkeley. After she got sick, she couldn't even add a column of numbers. When she applied for an apartment at Ecology House, the resident committee was hesitant to admit her because they didn't think she would live until the end of the year. Connie thrived in the new environment and has become an articulate advocate for safer environments for persons suffering from environmental illness or MCS.

However, others have not been able to tolerate some of the original materials selected in the early 1990s. Some residents have had difficulty with the interior wall finish, a hard gypsum plaster with no paint or sealer. Others liked the plaster. Many residents have not tolerated the paint smell from the factory-baked enamel finish on the steel cabinets. Part of these problems could be related to a shortage of curing time for the baked enamel finish. Because some residents were so desperate for housing, they might have said that the environment worked when it really did not. (Preference was given to those who were homeless and spending over half their income on shelter.) Different people suffering from MCS have different problems. Connie Barker has difficulty with paint smells. Several years ago, the city was painting curbs in her neighborhood and had not contacted residents. She had a window open and started feeling sick. She left the area, escaping in her safe RV vehicle, and then felt better as she got further away from Ecology House.

FUNDING

Six years of work by volunteers, who were later the incorporators of Ecology House, Inc., occurred before residents moved in. In 1988, Marin Homes for Independent Living (MHIL), a Marin

County, California, nonprofit housing corporation, began the process of developing new afford-able and safer housing for persons with MCS. Later, MHIL asked Ecumenical Association for Housing to serve as cosponsor for Ecology House. Ecology House, Inc. obtained grants from nine public and private sources, including the HUD, Marin County, the San Rafael Redevelopment Agency, Marin Community Foundation, and the San Francisco Foundation. Major funding for Ecology House came from HUD's Section 811 program. Therefore, Ecology House was developed according to HUD Section 811 regulations.

Perhaps the most important funding component for Ecology House is the ongoing HUD rental assistance reserved for the development. Ecology House residents pay only 30% of their income for rent. HUD rental assistance pays the difference between 30% of the resident's income and the actual rent.

SITE SELECTION

HUD Section 811 funds were reserved in late 1990 with the requirement that construction begin by September 30, 1992. Their goal was to find a sunny location in rural Marin or Sonoma County. Developers searched these counties for two years for a low-density, rural site. They found many wonderful sites. Unfortunately, local land use restrictions or water and sewer moratoria pre-cluded using any of these sites.

Finally, in late 1992, with HUD in the process of canceling the Section 811 fund reservation because of the delay in locating a site, Ecology House, Inc. acquired an urban site already zoned for apartments. Although the site benefits from good air circulation, it is in a high-density loca-tion with traffic fumes and neighbors using herbicides. Additionally, the site is close to power lines. Gauss meter readings for the EMF levels from those lines are relatively low. However, those with extreme EMF sensitivity may have difficulty tolerating the site. The decision to go ahead with a less-than-perfect site was made because Ecology House, Inc. determined that it was better to benefit some with MCS than to let the Section 811 funds go back to the Federal Treasury and give up the idea of developing affordable housing for persons with MCS.

DESIGN AND CONSTRUCTION

Ecology House, Inc. board members, including individuals with environmental illness/MCS, participated in a complex design process, coordinated by Development Consultant Katherine Crecelius. The architect, Kodama Associates, researched published materials about nontoxic building. However, the primary research used to design the building and select materials was the personal experience of persons with environmental illness/MCS and volunteer consul-tants. Every building component was discussed. On occasion a clear favorite would emerge; other times not. Many alternate materials were analyzed. Some materials could be made avail-able to be "sniff tested" by a team of up to 4–5 persons with MCS before selected for use. On the whole, those that were thoroughly tested and approved are those that have proven the most tolerable.

Some participants in the design process were frustrated by the lack of time for design and the lack of funds for paid safe-building consultants. HUD regulations limited developers to a three-year development period after HUD Section 811 funds were reserved. Since it took 2.5 years to find a site and then get local preliminary design approval, the time for detailed design and selec-tion of materials was short.

Ecology House, Inc. used a Bay Area architect with a track record designing HUD-funded, low-cost housing for persons with disabilities (see Figure 10.1). The architect selected had not previously designed housing for persons with MCS. At the time, they found no Bay Area archi-tects with safe-building design experience in affordable housing. Unfortunately, time and cost constraints precluded using paid safe-building consultants to work with their architect.

The General Contractor, Joseph DiGiorgio and Sons, used great care in construction tech-niques to keep the apartments clean and safe. No smoking was allowed on site at any time during construction. Only nontoxic cleaners were used.

Ecology House, Inc. based its choice of materials on input from persons with MCS; cost constraints and a multitude of HUD regulations made the total process difficult. Others who build multiunit housing with safer materials may make different choices, based on the initial evaluation.

COST

There are three general categories of multifamily development costs—land, soft costs (architect, engineering, fees, legal, title, etc.), and construction costs. Ecology House land, architectural, and other soft costs are the same as for standard HUD-financed apartments on this site. The data relevant to unusual costs are the increment between Ecology House construction cost and the construction cost for a standard HUD-financed, 100% wheelchair-accessible apartment complex. The total extra construction cost increment for the special materials was approximately $200,000, or approximately 20% more than the construction cost (or 11% of the total development cost) for these units if they were to be occupied by disabled persons without MCS. The cost difference was made up by local community development block grants. IEQ features included the following:

- Radiant heated baseboard units eliminated the need for forced air vents (see Figure 10.3)
- Materials that did not need painting
- Zero-VOC (volatile organic compound) interior paint
- Tile and concrete flooring instead of carpeting (Figure 10.4)
- All-metal cabinets, shelves, and countertops (Figure 10.5)
- Whole house water filtration system
- Vapor barriers in walls
- Powerful attic fans on a thermostat
- Formaldehyde free insulation
- Styrofoam moisture barrier between foundation and building (to discourage mold)
- Untreated concrete throughout (Figure 10.6)
- Concrete rather than asphalt parking lot
- Parking lot access away from building
- Pea-gravel barriers between building and landscaping
- Low-maintenance, low-pollen, fragrance-free landscaping (Figure 10.7)

The total of the differences in cost between concrete and asphalt, tile and carpet, steel cabinets and particle board cabinets, and so on was $200,000 for 11 units or about $18,000 per unit. Total development cost for Ecology House (including land, soft costs, and construction cost) was $1.8

FIGURE 10.3 Radiant heated baseboard units eliminated the need for forced air vents in Ecology House (the generator is in another building).

FIGURE 10.4   Tile flooring in the reception area of Ecology House looks as good today as it did over 20 years ago when it was new.

FIGURE 10.5   In kitchens, metal cabinets feature a baked-on surface and stainless countertops, sinks, and so on.

million. When comparing Ecology House per-unit costs with per-unit development costs for other buildings, several factors should be considered:

- High development costs characterize all new construction in the San Francisco Bay Area. Bay Area land costs and city and utility fees are extremely high.
- Extra costs are associated with HUD funding. For instance, HUD required that all apartments be 100% wheelchair accessible, not just ground floor apartments. Therefore, an elevator was required.
- All small developments have much higher per-unit costs than larger projects because common costs such as site work, elevator, community room, office, and utility lines are spread over a small number of units.

FIGURE 10.6 Untreated concrete is used throughout the Ecology House exteriors and in the parking lot.

FIGURE 10.7 Low-maintenance, low-pollen, fragrance-free landscaping.

In addition to considering the impact on air quality, materials were selected on the basis of durability and a maintenance requirement to minimize introduction of irritation chemicals in the future. It's not just the materials used to construct Ecology House that make it a healthy environment for residents, but how it is used and maintained. The addition of a venting room (Figure 10.8) has helped the atmosphere of Ecology House living quarters.

FIGURE 10.8   A venting room off the hallway allows residents to air out garments and so on before moving them into their apartments.

Green design note: the initial cost of tile and concrete flooring was determined by the heavy shipping weight of quarry tile and its costly manufacture, but now with 20 years of evidence, its selection has become a good green design decision because the tile looks like the day it was installed. Furthermore, it requires no harsh cleaning materials; vinegar and water suffice. If the HUD requirement for carpeting had been chosen, through traffic wear and soiling, it would have needed to be replaced at least three times in this period. Several years ago, Connie Barker gave a presentation at the Coverings conference in Florida. Her presentation was well received by the stone fabricators. Unfortunately, although there were many different stone vendors represented in the exhibit floor, Connie could not go through the demonstration area/hall because of off-gassing from the instant carpet covering added to beautify the exhibits.

RESIDENTS

The 11 rental apartments in Ecology House were developed to be occupied by 11 very low income individuals with MCS. During the first 2.5 years, 3 residents moved out because they could not satisfactorily tolerate the building. A few others have declined to move in after staying in the apartment for a test period of 1–5 days.

Ecology House, Inc. anticipated turnover because residents would find after they moved in that they could not tolerate the unit. Therefore, during the initial rent-up process, Ecology House, Inc. encouraged prospective residents to stay temporarily in their unit to determine if they could tolerate the unit before moving in. Had all the initial 11 residents made the decision to move in after a 1- to 5-day stay in their potential units, the 3 who ultimately moved out might have determined unit intolerability prior to moving in.

When Ecology House was initially occupied, HUD regulations required that preference be given to applicants who were homeless, displaced, paying more than 50% of their income for rent, or living in substandard housing. Partially because of this regulation, some of the initial residents came to Ecology House from desperate circumstances. They were sick. They had not been able to locate safe housing. They had little or no support or understanding from family and very little income. When offered a new, very affordable apartment, it may have been impossible to turn it down, even if they had difficulty tolerating the apartment during the test stay.

The 11 apartments at Ecology House have been occupied since opening, and there is a lengthy wait list of people eager to live in this healthy environment. Ecology House is a true demonstration project from which other affordable housing developers, interior designers, and builders can learn.

Another leader in the sustainability effort that has established healthy home standards is Bau-Biologie & Ecology, Inc. (IBE), a nonprofit educational organization dedicated to bringing together the technical expertise, biological understanding, and ecological sensitivity to create healthy homes and workplaces (Figures 10.9 through 10.10).

Access to buildings and services for people with chemical sensitivities

Extract from *Access to buildings and services: Guidelines and information*
Australian Human Rights Commission

Use of chemicals and materials

A growing number of people report being affected by sensitivity to chemicals used in the building, maintenance and operation of premises. This can mean that premises are effectively inaccessible to people with chemical sensitivity. People who own, lease, operate and manage premises should consider the following issues to eliminate or minimise chemical sensitivity reactions in users:

- the selection of building, cleaning and maintenance chemicals and materials (see *Note* below);
- the provision of adequate ventilation and ensuring all fresh air intakes are clear of possible sources of pollution such as exhaust fumes from garages;
- minimising use of air fresheners and pesticides;
- the provision of early notification of events such as painting, pesticide applications or carpet shampooing by way of signs, memos or e-mail.

For more information on ways to eliminate or minimise chemical and fragrance sensitivity reactions look at http://www.jan.wvu.edu/media/MCS.html and http://www.jan.wvu.edu/media/fragrance.html
Note: There are a number of relevant environmental and occupational health and safety regulations and established standards, however, as is currently the case with other standards referenced in building law, compliance with those standards may not necessarily ensure compliance with the DDA.
http://www.humanrights.gov.au/disability_rights/buildings/guidelines.htm

© Australian Human Rights Commission July 2007, updated April 2008

Australian Human Rights Commission www.humanrights.gov.au
Telephone: (02) 9284 9600 TTY: 1800 620 241 General enquiries: 1300 369 711
Disability Rights: disabdis@humanrights.gov.au

This poster was produced by:
Allergy and Environmental Sensitivity Support and Research Association Inc.
Reg. No. A0006141S ABN 32 386 589 943
P.O. Box 298. Ringwood. Vic 3134 Phone: (03) 9888 1382 www.aessra.org

FIGURE 10.9 This educational poster from the Australian Human Rights Commission stresses the importance of understanding chemical sensitivities and how you can promote healthy living.

FIGURE 10.10   Housing units designed for persons with multiple chemical sensitivities by pH Living feature modular construction.

HEALTHY HOME STANDARD

WHAT IS BUILDING BIOLOGY?

The American headquarters for the International Institute for Bau-Biologie & Ecology (IBE) was founded in Clearwater, Florida, in 1987. IBE is a 501-C(3) nonprofit educational organization dedicated to bringing together the technical expertise, biological understanding, and ecological sensitivity to create healthy homes and workplaces. The principles of Building Biology are based on the premise that what is healthy for the occupants will be healthy for the environment (ecologically sustainable). IBE holds nature as the golden principle and yardstick in terms of what is healthy.

Bau-Biologie (Building Biology) emerged in Germany because of problems with post-war housing construction. Much of the criteria and values used in the verification testing sections of the Healthy Home Standard were taken from the Supplement to the Standard of Building Biology Testing Methods SBM-2008.

In addition to working on public information campaigns, IBE offers training and seminars on how to perform indoor environmental assessments, the building sciences, and natural building methods. IBE focuses on providing the following kinds of services:

- To endorse or teach courses, workshops, and seminars covering the field of healthier and more natural building and lifestyle
- To advise and provide support and networking for those who are committed to a healthier and more natural building industry in the products and services that they provide
- To make information on healthier and more natural building materials and services available to the public and to the building industry
- To advise and cooperate with other relevant people, including environmental, research, health, community, local, and central government organizations to encourage a healthier and more natural-built environments and lifestyles

THE 25 PRINCIPLES OF BAU-BIOLOGIE

The following list of 25 principles was developed by Anton Schneider, PhD, founder of the Institut fur Baubiologie and Oekologie. These principles can be used while planning the construction of a natural and ecologically friendly home or while remodeling an existing one.

1. Make sure the building site is geologically undisturbed.
2. Place dwellings away from industrial centers and major traffic roads.
3. Place dwellings well apart from each other in spaciously planned developments amidst green areas.
4. Plan homes and developments individually taking into consideration the human aspect and the needs of family life and nature.
5. Use natural and unadulterated building materials.
6. Use wall, floor, and ceiling materials that allow the diffusion of moisture.
7. Allow natural self-regulation of indoor air humidity using hygroscopic materials.

8. Consider sorption of building materials and plants (inside and outside), which allow filtration and neutralization of toxic airborne substances.

9. Design for a balance between heat storage and thermal insulation in living spaces.

10. Plan for optimal surface and air temperature.

11. Use thermal radiation for heating buildings, employing solar energy as much as possible.

12. Promote low humidity and rapid desiccation in new buildings.

13. Utilize building materials that have neutral or pleasant natural scents and that do not emit toxic vapors.

14. Provide for natural light and use illumination and color in accordance with nature.

15. Provide adequate protection from noise and infrasonic vibration or sound conducted through solids.

16. Use building materials that do not have elevated radioactivity levels.

17. Preserve the natural (DC) air electrical field and physiologically beneficial ion balance in space.

18. Preserve the natural (DC) magnetic field.

19. Minimize technical (AC) electric and (AC) magnetic fields.

20. Minimize the alteration of vital cosmic and terrestrial radiation.

21. Utilize physiological knowledge in furniture and space design.

22. Consider proportion, harmonic orders, and shapes in design.

23. Use building materials that do not contribute to environmental problems and high energy cost in the production process.

24. Do not support products or building materials that overuse limited and irreplaceable raw materials.

25. Support building activities and production of materials that do not have adverse side effects of any kind and that promote health and social well-being.

THE HEALTHY HOME STANDARD FOR CONVENTIONAL CONSTRUCTION

The Institute for Building Biology & Ecology (IBE) has developed the Healthy Home Standard (HHS), an assessment that results in a letter grade (A, B, C, D, F) given to a home in terms of how well it is likely to support occupant health. Like other standards, it has checklists. It is unique in that it also requires a visual inspection and actual verification testing using test equipment and laboratory analysis in three categories: indoor air quality, electromagnetic radiation (EMR), and water quality.

Concerning specific building materials and new construction, it is assumed that all materials and jobsite activities conform to applicable US Green Building Council, Environmental Protection Agency Indoor Air Quality, American Lung Association Healthy Home, and regional building codes.

INFORMATION SUMMARY SHEETS

The following information is presented to give some background about the parameters in the HHS and the intention of IBE in including these parameters in the HHS. This information is not meant to be a substitute for training and knowledge about the subject matter. The HHS should only be used by a qualified, indoor environmental professional with knowledge and field experience. For more information, courses, and training seminars, visit www.buildingbiology.net.

As stated, the Home Health Standard document includes checklists for the environmental variables most likely to affect a person's health; where applicable, it also contains procedures for Verification Testing and descriptions of Test Conditions and Mitigation Options. The following selections from the IBE Health Home Standard are examples of the information available there.

LEAD PAINT

Lead (Pb), is a chemical element. It is a heavy metal found in paint, pipes, air solder, lead crystal decanters (e.g., brandy stored 5 years in crystal decanter—20,000 g/L, 1300 g/L is allowable), cookware glazes, bullets, fishing sinkers, cosmetics, printing ink, paints on toys, gasoline, and so on.

ASBESTOS

Asbestos is a generic term used to describe a number of fibrous materials found in various concentrations across the Earth's surface. It is used for its strength, flexibility, and fire resistance and is placed in building materials.

Asbestos was recognized as a hazardous building material by the EPA2 in 1972 and can still be found in older homes and buildings. There are about 75 product sources of asbestos. Most uses were banned in 1978; that for floor tile was banned in 1989.

MITIGATION OPTIONS

If the asbestos-containing material is in good shape and nonfriable (unable to be crushed by normal hand pressure) or if it is encapsulated (www.safeencasement.com) or enclosed, and not subject to physical damage, it probably can be left alone and monitored. If it is in a friable or damaged state, or subject to damage, then it should be encapsulated or removed by a licensed contractor.

INDOOR AIR QUALITY CHECKLIST

The IBE Health Home Standard includes detailed checklists with the following structure:

Number	Assessment Element	Element Value	Applicable Value for This House or Write N/A	Points Awarded for This House
B.1	Structure			
B.1.n				
B.2	**Heating, Cooling, and Ventilation** Note: It is possible to build a house in a mixed climate without using air-conditioning by use of proper building orientation, eve overhangs, thermal mass, and building design. In this case, an air conditioner would not be required. Building Biology considers the ideal type of heat to be radiant heat. In this case, a central heating system is not required. However, a ventilation system is still required to provide fresh air and filtration. Duct work may still be needed for ventilation purposes if radiant heat is used in order to meet ASRAE 62.2-2003. Building Biology encourages design and construction techniques and materials that result in natural ventilation wherever possible.			
B.2.1	The heat is radiant type—radiator, baseboard, in floor, walls, and ceilings.	1		
B.n.n				

FORMALDEHYDE AND VOCS

Air sampling for volatile organic compounds (VOCs) and formaldehyde relies on the use of special pumps, receptacles, and an extended period. Always follow specific directions provided by the analyzing laboratory.

MOLD

Samples should be taken to accommodate at least one sampling location per floor and HVAC (heating, ventilation, and air-conditioning) system. Sampling location(s) should be focused on critical areas such as occupied bedrooms, high occupancy areas, or any area of concern determined during the pre-qualification and initial checklist.

CARBON MONOXIDE

The most significant carbon monoxide (CO) sources like gas cars and gas ovens cannot be corrected, only better ventilated. Exposure to increases in 8-hour average ambient (outdoor) CO of just 1 to 2 ppm has been shown to be life threatening for both people with asthma and people with heart disease, who show a statistically significant increase in emergency room visits whenever this happens.

GUIDELINES

- CO from starting a car in the garage and then leaving gradually migrates into the house, commonly causing levels in rooms adjacent and above to exceed 100 ppm (Aerotech Tech Tips #106 1/12/04).
- The National Institute for Occupational Safety and Health stipulates 200 ppm as the level immediately dangerous to life and health.
- Occupational Safety and Health Administration maximum exposure level is 50 ppm average over 8 hours.
- Building Biology: no allowable sustained increase of indoor CO over outside level.

VENTILATION

Proper ventilation mitigates some of the above problems.
A filtered supply of fresh air should be ensured for the following reasons:

- To provide sufficient oxygen (outdoor air: 21% oxygen)
- To avoid increased levels of carbon dioxide and other air pollutants
- To regulate indoor air humidity (except as noted in a hot, humid climate that requires conditioned OA)
- To supply naturally occurring negatively charged air ions

In general, the air exchange rate is a good indicator of the overall indoor air quality.
For information regarding types of ventilation, see the online Indoor Climate course (IBE 204.1). Remember that the Building Biology Way is through natural ventilation. In building biology, the building envelope is regarded as a third skin that is able to breathe. This is not achievable via conventional construction used in North America.

ELECTROMAGNETIC RADIATION

In the course of evolution, all living organisms have adapted themselves to this unique radiation climate prevalent on planet Earth. This natural balance is being threatened now because over the last 100 years, humans have been very busy adding their own versions of electromagnetic energies.

WHAT IS EMR?

The types of EMR are related to each other by the rate at which each vibrates. The rate of vibration is termed frequency in cycles per second or Hertz. The electromagnetic spectrum visually relates each type of radiation to the others by the frequency of vibration. The spectrum is shown below. These energies include the following:

Those emanating from a building's electric power distribution system (wiring):

- AC (alternating current [60 cycles/second]) electric fields (also called ELF electric fields)
- AC magnetic fields (also called ELF magnetic fields)
- Dirty electricity

Those associated with communications radiation (radio frequency fields) and found in the air. These are produced by information carrying radio waves such as:

- Cordless telephones
- Cell phones
- Pagers
- Wireless Internet
- Bluetooth
- Broadcast TV and digital broadcast TV
- AM and FM radio
- Emergency and military communications

AC electric fields are present at all times when wiring is energized. They are emitted by the wiring in the walls, floors, and ceilings and by the cords to electricity-using devices. EMR cannot be sensed by most people; hence, measurements must be made to find and reduce these fields.

Many years ago before the discovery of solid-state devices now found in all electronic equipment, the electricity that powers our homes, buildings, and factories was like a meadow in the country, gently varying, quiet, harmonious, and clean. Today, our electricity is no longer harmonious, quiet, gently varying, and clean. It is filled with abrupt changes in character (voltage) described technically as electromagnetic interference. The vernacular terms are *dirty electricity* and *dirty power*. This dirt is much like noisy static you might hear on a radio playing wonderful classical music from a station far away. The underlying music is there and it could be beautiful and relaxing, but all that static is irritating, so you change the station or turn off the radio.

Unfortunately, with dirty electricity, turning it off is not possible. Dirty electricity is everywhere in our environment. You are generating this dirt in your very own house, as are your neighbors and the office down the block and the factory in the industrial part of town.

The dirt is produced by the workings of all of our electronic equipment like computers, TVs, radios, microwaves, compact fluorescent lights, light dimmers, digital clocks, and cell phone chargers.

The dirt emanates into air space in the immediate vicinity of the generating device. The wires in our buildings transmit this dirt signal around the entire structure. There are some electrically sensitive people who seem to feel that this signal inside the entire dwelling emanated into space around the wires. So far, we have not been able to measure these energies in rooms, only near the wires and in the vicinity of the generating device.

The common pollutant and health hazard not described here is water quality. The IBE Health Home Standard contains an additional material on this subject.

pH LIVING

Another group that has become a major factor in the sustainability movement is pH Living, which is a design-build group and think tank specializing in healthy, sustainable, transparent, and affordable housing solutions worldwide. They design, factory manufacture, and install custom homes that are certified by an independent third-party assessor as meeting the Healthy Home Standard developed by the International Institute for the Building Biology and Ecology

(IBE). Their team of designers, scientists, and visionaries are dedicated to building healthier homes and a healthier planet. Their goal is to become the leading home provider to people suffering with chemical sensitivity and environmental allergies and for folks who just want a healthier home for themselves and their children. They offer an array of services and products that range in scale from self-sufficient guest cottages to multifamily housing and emergency housing solutions for disaster relief zones.

pH Living recognizes the growing need for people to have access to homes consciously produced without harmful toxins so they may enjoy healthier lives. Their site represents the first effort to meet the needs of people with electrical sensitivites in addition to chemical and other better known allergies. (See the marketing drawing in Figure 10.10).

TAKE IT FOR A TEST DRIVE

The initial offering for pH Living was the Sanctuary testing unit built for chemically sensitive clients to evaluate prior to purchasing. pH Living recognized that within the CS community, there was considerable individual variability in the pollutant concentrations and specific pollutants that causes reactivity. In view of this, the group realized up front that their product would not satisfy the needs of every person with chemical and electrical sensitivities. The Sanctuary represented their best efforts of bringing together state-of-the-art technology and unique building techniques at an affordable price. They urged potential customers to arrange to test drive the Sanctuary unit to evaluate the home's ability to meet their specific needs and sensitivities.

The test unit is located in Bucks County, Pennsylvania, near Quakertown (see Figure 10.11), situated on 16 beautiful acres of land. The site was selected for its proximity to PHL international and their east coast manufacturing facility. The unit is 5 minutes away from a preeminent multiple chemical sensitivities doctor. A short drive into town, you will find a health food store, a compound pharmacy, and a variety of food choices, including organic farm produce.

FIGURE 10.11   The Sanctuary test unit in Pennsylvania was created for persons with multiple chemical sensitivities who are encouraged to "test drive" the unit to see if it works for them. Note: placing the test unit on concrete pillars helped reduce the incidence of mold, but the structure would not be accessible for persons with mobility problems.

WHAT COMES STANDARD IN OUR pH LIVING pureHOME?

Our homes are built to support your health and that is our standard benchmark. Quality, sustainability, transparency, and affordability have been included at no additional cost to you. Here is our short list:

- Forest Stewardship Council–certified framing and NAUF (no added urea formaldehyde) plywood substrates and sheathing
- Cement board siding in a variety of styles
- Solar reflecting and EMR reducing aluminum roofing
- Military-grade scrim reinforced electromagnetic radiation barrier
- Formaldehyde-free pure fiberglass insulation
- Double-glazed, anodized aluminum, LowE EMR reducing windows and doors
- Energy Star appliances
- Metal-clad electrical cabling and stainless steel cover plates for EMR reduction
- 25 SEER split HVAC system with on-demand air purification and dehumidification control
- ERV or HRV (energy or heat recovery ventilator) for continual fresh air
- Merv 12 particulate and activated coconut charcoal filtration
- Electric water heater
- Recessed incandescent lighting or porcelain tile flooring
- Solid wood or aluminum kitchen shelving with aluminum supports
- Quartz slab countertops
- Dual-flush, low-flow toilets
- Low flow kitchen and bath faucets
- All-copper supply lines and ABS waste lines
- Carbon-based whole house water filter
- Zero-VOC (volatile organic compound) paints, stains, and sealers
- Healthiness certified to the Healthy Home Standard by IBE

A UNIVERSALLY DESIGNED ACADEMIA

Another demonstration project that included indoor air quality concerns in the design was a new classroom building created several decades ago by interior design professor Louise Jones. In the case study, she described the process she used in creating a multifunction building on a university campus (Jones 2008).

The campus of Eastern Michigan University features a building that exemplifies the concept of "good design meets the needs of the people who use the space." Dr. Louise Jones, interior designer and interim associate dean, was asked to work with architects and the general contractor to ensure that the new home for the College of Health and Human Services addressed four goals:

- Use universal design to meet the needs of everyone who uses the building regardless of their physical ability or stature
- Provide leading-edge technology for teaching and learning
- Provide a healthy environment by using green/sustainable finishes and furnishings
- Foster a sense of community

The outcome is a building that houses classrooms, laboratories, offices, and meeting places in a model for academic buildings in the 21st century. Classrooms feature leading-edge instructional technology that is easily controlled from a touch panel at the podium. Faculty and students can move from computer-enhanced presentations, to 35 mm slides, to DVD/CD files, just by touching the screen. Awkward equipment carts and complicated instructional manuals are unnecessary.

Sound and lighting can also be adjusted from the touch panel to accommodate diverse users' needs.

The furnishings in the classrooms are adjustable to accommodate people of different physical stature or those with disabilities. The custom podium has two surface areas to accommodate either standing or seated speakers. Student tables and chairs are on casters so that the room can be easily arranged to accommodate different teaching and learning preferences, for example, lecture, small group, large group, or role-play. Lecture notes from the electronic whiteboards can be downloaded to students' computers to facilitate discussions during class time.

State-of-the-art laboratory spaces include a Home Care unit that resembles an efficiency apartment. Students in interior design, nursing, or occupational therapy can mock-up different environments to simulate clients' homes for delivery of home health care or to design supportive environments for people who have disabilities or are frail. There are 1000+ network connections in the building to enable students with laptop computers to easily communicate with peers or professors and link to Internet resources. Two student commons and a café in a two-story atrium space provide comfortable environments for studying or group work. The design is inclusive, going beyond ADA and barrier-free codes, to ensure that everyone's needs are considered so that no one has to ask for special accommodation.

The building also serves as a demonstration site for environmentally responsible design. Furnishings and finishes were selected to "do no harm" to the Earth's environment. Indoor air quality was protected to ensure a healthier environment for everyone who uses the building, but particularly for those with allergies, asthma, or multiple chemical sensitivities. Tests done three months after occupancy showed no measureable quantities of volatile organic compounds (VOCs), chlorofluorocarbons, formaldehyde, mold, or fungus. Indoor air tested better than outdoor air—an unusual condition in today's sealed buildings.

In selecting finishes and furnishings for the building, Dr. Jones used a "cradle to cradle" approach, as described by William McDonough (McDonough and Braungart 2002). Dr. Jones considered the source of the raw materials, the manufacturing process, delivery, and installation, as well as maintenance and ultimate disposal. The goal was to use recycled or sustainable materials that could be manufactured without harming the environment, installed and maintained without degrading the indoor air quality, and recycled at the end of useful life. For example:

- Water-based paints were chosen not only for durability but also for low-odor and no-VOC off-gassing.
- Linoleum, selected as the primary flooring material, was made of natural materials that do not off-gas formaldehyde or VOCs and did not require waxing, thus protecting the indoor air quality from the chemicals in both waxes and strippers.
- Soda pop bottles were recycled into the fabric used to upholster the office panels and into vinyl chair components.
- Recycled steel and aluminum in the tables and chairs was finished with a powder-coat paint in order to protect the environment and provide a hard, durable finish.
- When necessary for acoustical control, the carpeting was woven of a recyclable polymer with an impermeable layer between the front and recycled backing material, allowing it to be vacuumed for routine maintenance or wet mopped to clean up spills.
- Hot water extraction removed stains to protect indoor air quality from harsh cleaning chemicals. At the end-of-useful life, the carpet tiles could be recycled into new carpet tiles—a "closed loop" process that used the worn-out product as the raw material for the next manufacturing cycle.
- Wooden products were finished with a water-based stain and an ultraviolet-cured, water-based polyurethane finish to prevent off-gassing of oil-based products.
- The upholstery fabric was made of sustainable materials manufactured using only chemicals that are safe for people and the environment instead of the more than 7500 potentially harmful chemicals typically used.
- Tables used a substrate of recycled materials sealed to prevent off-gassing.
- Signage throughout the building alerted users to the green/sustainable features in order to increase awareness of the beauty and practicality of selecting products that protect indoor air quality and are environmentally responsible.

The College wanted to create an environment that fosters a sense of community and belonging for all students, faculty, staff, and visitors. Dr. Jones' design for the Everett L. Marshall Building successfully supported this goal by considering the needs of everyone who had an opportunity to spend time there while protecting the birthrights of future generations.

COMMON CHEMICAL CONCERNS

Problem	Solution
Wood products: plywood, particle, chip, and hardboard; paneling (formaldehyde outgassing)	Use solid wood or exterior grade plywood that's been sealed with a vapor barrier; steel beam; nontoxic tilt-up concrete slabs; or rammed earth construction.
Subfloor: particle board or plywood made with formaldehyde	Use formaldehyde-free subfloor (Wonderboard, Homasote 440, or Carpetboard); or seal with vapor barrier.
Floors and coverings: carpet and rugs (made of toxic synthetics and treated with toxic substances for stain protection)	Use carpet sealers, untreated natural fiber carpet (fibers grown without pesticides), wood flooring or tile; nontoxic cement; or natural linoleum (Forbo).
Vinyl floor covering: petrochemicals and plasticizers	Use natural or ceramic floor tiles (commercial grade recommended for large areas), and nontoxic cement.
Wood floor finishes: solvents and sealers	Use natural finishes and provide adequate ventilation and drying time.
Wall surfaces and coverings: vinyl wallpaper (plasticizer outgassing); glues (formaldehyde, fungicide, or mildewcide); drywall joint compound (preservatives, formaldehyde, and mildewcides)	Use wall coverings of paper, linen, or metallic foil; use low-toxic glues or wheatpaste glue; or use low-toxic joint compounds.
Adhesives (used in flooring installation): formaldehyde, hydrocarbon solvents, and reactive chemicals	Use water-based or natural adhesives.
Tile set and group: preservatives	Use Portland thinset and grout without preservatives.
Paints, stains, and finishes: oil-based and alkyd products contain hydrocarbon solvents; water-based paints may contain biocides and solvents	Use low- or nontoxic products that are water based and don't contain mildewcides and fungicides.
Wood preservatives: fungicides and mildewcides	Use low-toxic or natural wood preservatives and naturally rot-resistant woods (redwood, cedar, cypress).
Adhesives and glues: toxic hydrocarbon solvents	Use natural and nontoxic glues.
Furniture: plywood and particle board with formaldehyde	Use solid wood furniture, metal, or natural materials (bamboo, wicker, rattan, etc.).
Fabrics: formaldehyde, plasticizers, fungicide, and so on, for stain and wrinkle resistance	Use untreated natural fiber fabrics. Washable fabrics can be soaked in hot water and laundered several times to remove the stain-resistant finish and biocides.

CALIFORNIA 01350 AND INDOOR AIR QUALITY

Emerging from years of study and controversy in the environmentally conscious west is the applied research tool, California Section 01350 (Lent 2009). It is being used to identify design directives used in building and maintaining a facility. This special environmental requirement standard developed by the State of California to address the impact of building materials on indoor environmental quality in buildings, as well as other environmental performance issues, is the only health-based building material specification at this time. Technology has helped to produce its sophisticated measuring technique. It can identify pollutants in the air and also calculate their exact percentages (concentrations).

This protocol and associated specification was initially developed for a model office-building project in Sacramento, California (the Capitol East End Project). It quickly proved its usefulness in this project, providing manufacturers with a consistent protocol to test their products, and has already resulted in product changes to reduce emissions of troublesome chemicals. In the new office building, there was no measurable problem with pollution. Then, people moved in. Applying CA 1350, engineers identified the two new pollutants found and the percentage of each. They found pollutants related to the chemicals present in certain deodorants and also in dry cleaning fluid. Having this information provided evidence that could then be used to mitigate the problem.

CA 1350 was originally developed in 2000 for California's Modular Office Furniture Specification and has since been modified. Section 01350 received wide acceptance from building materials manufacturers because of its flexibility and relative low cost and because it is the only health-based building material specification. CA 1350 is an important scientific tool for screening primarily major interior finishes on the basis of:

- Emissions testing protocol
- Hazardous content screening
- Avoiding mold and mildew from construction practices

The Section 01350 (www.ciwmb.ca.gov/greenbuilding/specs/section01350) specification language is now being integrated into other broader specification programs, including the following:

- Scientific Certification Systems (Indoor Advantage Indoor Air Quality Performance Environmental Certification Program SCS-EC10-2004, their EPP carpet specification and the Resilient Floor Covering Institute FloorScore (www.scscertified.com/iaq)
- Green Guide for Health Care (www.gghc.org)
- US Green Building Council LEED (www.usgbc.org)
- Collaborative for High Performance Schools Best Practices (www.CHPS.net)
- Institute for Market Transformation to Sustainability (MTS) SMART standards (MTS.sustainableproducts.com)
- Carpet & Rug Institute's Green Label *Plus* Carpet Testing Program (www.carpet-rug.com)
- California's Reference Specifications for use in all major state construction (www.ciwmb.ca.gov/greenbuilding/Specs/Section01350)

MATERIALS SCREENING COMPONENTS

The 01350 standard has three key components related to materials screening for indoor air quality:

1. Screening based on emission testing for exposure
2. Construction adhesives component screening
3. Reporting (by the contractor) on compounds measured

Emissions testing protocol is coupled with modeling of a planned building to predict the concentration of each of the chemicals in the finished project's indoor environment.

But another advantage of the protocol is that it can be applied to a building after it is occupied. A structure can pass inspection when it is empty, yet experience elevated measures of pollutants once it is occupied. People do introduce changes to air quality as they unpack their belongings and begin their daily routines—whether it is a commercial building or a residence. Once a problem is identified, mitigating steps can be taken.

CA 1350 is an important new tool that can assist us in moving toward healthier and more fully accessible buildings and homes. Working with this research strategy can contribute to green and healthy specifications in homes and commercial establishments.

HEALTHY INDOOR AIR FOR AMERICA'S HOMES

INDOOR AIR QUALITY HEALTH EFFECTS

Major Indoor Pollutants

Kills Quick	Kills Many	Serious Impairments	Irritation, Discomfort
Carbon Monoxide • 500 deaths per year, US residential	**Tobacco** • 430,000 deaths per year **Radon** • 15,000 to 20,000 deaths per year	**Lead** • 3,000,000 mild elevated levels • 250,000 serious elevated levels **Dust Mites** • Account for 1/3 of 14 million doctor visits per year **Mold** • Allergens, toxic particles, VOCs	**Formaldehyde** • Strong irritant **Mold, Mildew** • Allergens, toxic particles, VOCs **VOCs** • Irritants, possible or known carcinogens

Source: USEPA, CDC Indoor Air Quality Health Effects.

Note: A good publication with more information on this topic is called "Indoor Air Pollution: An Introduction for Health Professionals," found on USEPA's web site.

Effects of Poor Indoor Air Quality on Health

Eye Findings
- Irritant or allergic conjunctivitis (burning, sensation of dryness, redness)

Nasal Manifestations
- Rhinorrhea, nasal obstruction
- Irritant rhinitis
- Allergic rhinitis, chronic sinusitis

Respiratory Manifestations
- Chest tightness, cough with or without fever, shortness of breath with exertion
- Nonspecific abnormalities (no chest x-ray or lung function abnormalities can be documented)
- Asthma
- Hypersensitivity pneumonitis
- Infectious pneumonia
  - *Legionella* pneumonia
  - *Aspergillus* pneumonia (in immunosuppressed persons)
  - Tuberculosis
  - Others (most common in immunosuppressed persons)

Oropharyngeal Manifestations
- Dryness, irritation of the throat

Lung Cancer

General Symptoms
- Headaches, lethargy, fatigue, poor concentration
- Nonspecific complaints
- Systemic effects of hypersensitivity pneumonitis
- Variant of organic dust toxic syndrome (humidifier fever)
- Carbon monoxide poisoning

Skin Manifestations
- Dryness, irritation, rash

Note: One must always consider other causes for these problems in addition to looking at factors in the home environment. Also, the person with underlying lung disease likely will be more sensitive to dust, gases, and fumes of all types.

Physical Factors That Affect Health

Temperature

Ideal temperature is between 20°C and 23°C (68°F and 73°F).

Relative Humidity

Should remain between 35% and 55%. Lower humidity leads to skin drying, irritation. High humidity leads to growth of molds (*Aspergillus*, *Stachybotrys*, etc.) and bacteria that cause other illness.

Ventilation

Poor ventilation allows carbon dioxide to accumulate as well as other gases. This can be a problem in modern, airtight buildings.

Lighting

UV rays from fluorescent lighting cause photochemical reactions, which lead to formation of "smog" that can irritate eyes.

Odors

Odors, such as those of swine confinement buildings, may seep into the home from the outside. Approximately 30% of persons may respond with headache, malaise, and so on. The mechanism by which this occurs is a point of much debate.

Secondhand Smoke

Short-term effects include mucous membrane irritation. Long-term effects include lung cancer from inhalation of carcinogens within the smoke. There is also good evidence that passive cigarette smoke in the home contributes to the development of asthma in children and causes increased risk for respiratory infections.

Asbestos

Inhalation of fibers causes inflammation and tumors. This is manifested as pleural plaques (benign scar-like changes seen on chest x-ray or during surgery or autopsy), fibrosis (asbestosis), and cancers in the lung. It can also cause cancer of the lining of the lung (pleura) and in the abdominal and abdominal cavity (mesothelioma).

Asbestos fibers are ubiquitous in the environment (from insulation, brake linings, etc.) but more common in urban areas and are often found in low numbers in lungs of healthy persons. There is NO strong evidence that asbestos insulation causes risk to those who live in buildings that have asbestos insulation.

Because of present-day Occupational Safety and Health Administration standards, low exposures from living and working in buildings are estimated at 1 million times less than exposures around the time of WWII, which caused recognition of asbestos-related cancers.

Carbon Monoxide

Carbon monoxide exposure occurs through inhalation. This substance binds to the hemoglobin molecule with a much greater affinity (200 times) than oxygen. Thus, oxygen is displaced (the amount dissolved in the blood is not enough to sustain life) and leads to tissue hypoxia. Tissues most dependent on large amounts of oxygen suffer most (brain, heart).

Severe exposures can cause brain damage, evident after recovery from the acute effects. It can also cause heart attacks. The most common source of carbon monoxide poisoning in the home is from incorrectly vented furnaces and heaters. However, one must remember that carbon monoxide poisoning can also occur after use of methylene chloride containing varnish removers in a poorly ventilated setting.

Noise

Low-frequency noise (20–100 Hz), such as from machines, can cause mental irritation and tiredness.

Nitrogen Dioxide (NO$_2$)

Concentrations are high at times in homes that use gas stoves and kerosene heaters. There is some evidence that high NO$_2$ levels are linked to a greater prevalence of respiratory symptoms in schoolchildren. Persons with underlying lung disease may also be sensitive to NO$_2$.

Wood Smoke

This can cause exacerbation of asthma because it acts as an irritant. It may be a cause of chronic bronchitis. Respiratory symptoms may be more common in homes where there is a wood burning stove. Modern airtight stoves with a well-functioning chimney or flue reduce carbon monoxide and particle levels in the home.

Radon

Radon forms in the soil from radium and uranium. Breakdown products of radon (radon daughters) in the lung after radon inhalation may cause lung cancer in some persons. This effect is best understood in miners exposed to radon gas underground. This can occur in nonsmokers, but is more likely if the person does smoke.

At this time, it does not appear that the risk from radon exposure in the home is a large one. Some authors argue that radon exposure in the home is too low to cause lung cancer.

Animal Dander

Dander from cats and dogs that live indoors as family pets is a very important cause of allergic symptoms in sensitized people. These people can suffer from asthma, allergic rhinitis, and allergic conjunctivitis. Dander can persist in the environment long after the pet no longer lives in the home.

Mites (*Dermatophagoides pteronyssinus* and *Dermatophagoides farinae*)

Mites live in carpet, bedding, and so on, and consume skin particles from humans. They are a common source of allergy (asthma and allergic rhinitis) and are the offending substance in "house dust." Management of this problem includes keeping humidity low, good ventilation, frequent vacuuming, removing carpets, and covering pillows and mattresses.

Cockroaches

Cockroach antigen is associated with allergic asthma. This may be one cause of the increased prevalence of asthma in inner city populations.

Mold Spores

These can cause an allergic reaction (rhinitis or asthma). Hypersensitivity pneumonitis is another type of allergic reaction that consists of fever, chills, dry cough, and a flu-like feeling, all or some of which happen 4–6 hours after repeated mold spore inhalation (like from a humidifier) in susceptible persons. It consists of an inflammatory reaction in the lung that causes release of mediators of inflammation from the lung. These circulate and cause fever and other systemic effects. If persons are immunocompromised (e.g., from cancer), fungi can cause infection in the lung.

This information may not be inclusive of all IAQ Health Effects.

Susanna Von Essen, MD
University of Nebraska Medical Center
600 S. 42nd Street, Omaha, NE 68198-5300
Phone: (402) 559-7397
Fax: (402) 559-8210
E-mail: svonesse@unmc.edu
November 1996, Revised October 1999

Most people are familiar with the department store perfume and cosmetics counters. While many women cluster around them for free samples and attention, others take deep breaths and hurry by trying not to inhale. While localized ventilation may or may not be an issue, the concentration of odors overwhelms the environment. The following article sheds some light on what may be in the air.

Canada has been a leader in exposing the health risks of secret chemicals in fragrance and in an article titled "Not So Sexy" showed how their environmental groups are fighting the problem.

NOT SO SEXY: THE HEALTH RISKS OF SECRET CHEMICALS IN FRAGRANCE

When sprayed or applied on the skin, many chemicals from perfumes, cosmetics, and personal care products are inhaled. Others are absorbed through the skin. Either way, many of these chemicals can accumulate in the body. As a result, the bodies of most Americans and Canadians are polluted with multiple cosmetics ingredients. This pollution begins in the womb and continues through life.

Most unfortunately, widespread exposure and a long-standing culture of secrecy within the fragrance industry continue to put countless people at risk of contact sensitization to fragrances with poorly tested and intentionally unlabeled ingredients (Schnuch 2007).

Product tests initiated by the Campaign for Safe Cosmetics and subsequent analyses, detailed in this report, reveal that widely recognized brand-name perfumes and colognes contain secret chemicals, sensitizers, potential hormone disruptors, and chemicals not assessed for safety (Sarantis 2010).

In addition to the secret chemicals found via testing, some chemicals that are disclosed on the labels of the products in this report also raise safety concerns. They include sunscreen and ultraviolet-protector chemicals associated with hormone disruption (Schlumpf 2004).

ABOUT THE ENVIRONMENTAL WORKING GROUP

Environmental Working Group (EWG) is a nonprofit research and advocacy organization based in Washington, DC, and founded in 1993. The team of scientists, engineers, policy experts, lawyers, and computer programmers pores over government data, legal documents, scientific studies, and their own laboratory tests to expose threats to health and the environment and to find solutions. The mission of the EWG is to use the power of public information to protect public health and the environment. EWG specializes in providing useful resources (like Skin Deep and the Shoppers' Guide to Pesticides in Produce) to consumers while simultaneously pushing for national policy change.

ABOUT ENVIRONMENTAL DEFENCE CANADA

Environmental Defence protects the environment and human health. They research. They educate. They go to court when they have to. All in order to ensure clean air, clean water, and thriving ecosystems nationwide, and to bring a halt to Canada's contribution to climate change. Nationwide. www.environmentaldefence.

EXECUTIVE SUMMARY

A rose may be a rose. But that rose-like fragrance in your perfume may be something else entirely, concocted from any number of the fragrance industry's 3100 stock chemical ingredients, the blend of which is almost always kept hidden from the consumer.

People have the right to know which chemicals they are being exposed to. They have the right to expect the government to protect people, especially vulnerable populations, from hazardous chemicals. In addition to required safety assessments of ingredients in cosmetics, the laws must be changed to require the chemicals in fragrance to be fully disclosed and publicly accessible on ingredient labels.

SECRET CHEMICALS

Avoiding questionable fragrance ingredients in personal care products, under current laws, is nearly impossible. Numerous products used daily, such as shampoos, lotions, bath products, cleaning sprays, air fresheners, and laundry and dishwashing detergents, contain strongly scented, volatile ingredients that are hidden behind the word "parfum" or "fragrance."

Increasingly, personal care products have claims like "natural fragrance," "pure fragrance," or "organic fragrance." None of these terms has an enforceable legal definition. All can be misleading. One study found that 82% of perfumes based on "natural ingredients" contained synthetic fragrances (Rastogi 1996).

SENSITIZERS

During the last 20 years, fragrance contact allergy has become a major global health problem (Scheinman 2002). Many scientists attribute this phenomenon to a steady increase in the use of fragrance in cosmetics and household products (Johansen 2000; Karlberg 2008). Fragrance is now considered among the top five allergens in North America and in European countries (de Groot 1997; Jansson 2001) and is associated with a wide range of skin, eye, and respiratory reactions.

Many of the sensitizing chemicals in perfumes and colognes are also found in a wide range of other products, increasing a consumer's total exposures and overall risk for developing allergies. For example, limonene is a fragrance chemical that is commonly used as a solvent in cleaning products and degreasers where it may be listed as "citrus oil."

HORMONE DISRUPTORS

A significant number of industrial chemicals, including some in fragrances, can act as hormone disruptors by interfering with the production, release, transport, metabolism, and binding of hormones to their targets in the body (Gray 2009; Rudel 2007).

Recent research has clearly demonstrated that even at low doses, exposure to hormonal disruptors during susceptible periods can have drastic consequences for health later in life. Scientists are especially concerned about the impact of hormone-disrupting chemicals during critical times, such as fetal development (Breast Cancer Fund 2008).

THE SELF-POLICING FRAGRANCE INDUSTRY

The International Fragrance Association (IFRA) sets voluntary standards for fragrance houses and the manufacturers of fragrance ingredients. The compliance program, initiated in 2007, tests fragrance samples for prohibited ingredients (the program historically has only looked at prohibited ingredients and is now beginning to look at restricted ingredients as well). If there are violations, the supplier's name is posted on IFRA's website as not complying with the IFRA Code of Practice. IFRA has banned or restricted approximately 150 ingredients from fragrance (IFRA 2010).

IFRA's recommendations are based on research conducted by the Research Institute for Fragrance Materials (RIFM). IFRA members are given access to a database generated by RIFM that houses safety information—and testing gaps—on the more than 3100 fragrance ingredients used by IFRA members.

SAFER PRODUCTS AND SMARTER LAWS

Products we put on our bodies should not contain chemicals that could damage our health. Yet, because of gaping holes in federal law, it is perfectly legal for perfumes, colognes, body lotions, shampoos, and other cosmetics and personal care products to contain sensitizers, hormone disruptors, reproductive toxicants, carcinogens, and other toxic chemicals linked to harmful health effects.

The lack of full disclosure regarding the ingredients that make up fragrance is only one of the problems associated with the cosmetics industry. While the Government of Canada has a list of restricted and prohibited ingredients in Canadian cosmetics that helps manufacturers make sure that they are not selling products that will cause harm (Health Canada 2009a), the legal authority of this list is unclear and any prohibitions do not pertain to impurities (or by-products). Furthermore, there are more than 1000 chemicals, including carcinogens, mutagens, and reproductive toxicants, that are legally banned in European cosmetics (European Parliament and Council Directive 2003/15/EC and Cosing 2009), many of which are not on the Canadian Cosmetic Ingredient Hotlist.

As our test results show, short of sending your favorite perfume to a laboratory for testing, shoppers have no way of knowing exactly which of the 3100 fragrance ingredients may be hiding in their beauty products or even in their child's baby shampoo.

BE JUST BEAUTIFUL

Environmental Defence Canada makes a plea to the public. One-time use of fragrances highlighted in this report may not cause harm. But cosmetics and personal care products are used repeatedly and in combination with other consumer products that can also contain hazardous chemicals. Research by government agencies, academia, and independent organizations finds widespread human exposure to multiple chemicals (CDC 2009); we are all regularly exposed to various toxic chemicals from air, water, food, and household products. People can also be exposed to the same chemical from multiple sources. Here's what you can do to protect yourself, your loved ones, and future generations from unnecessary exposure to toxic chemicals in personal care products.

1. *Choose products with no added fragrance.* By choosing products without fragrance, you can reduce toxic chemical exposures for yourself and your family. It is important to read ingredient labels, because even products advertised as "fragrance free" may contain a masking fragrance. Visit the website, www.environmentaldefence.ca, for tips and resources to help find safer products and to link to EWG's Skin Deep: www.safecosmetics.org.
2. *Less is better.* If you are very attached to your fragrance, consider eliminating other fragranced products from your routine and using fragrance less often.
3. *Help pass smarter, health-protective laws.* Buying safer, fragrance-free products is a great start, but we can't just shop our way out of this problem. In order for safer products to be widely available and affordable for everyone, we must pass laws that shift the entire industry to nontoxic ingredients and safer production.
4. *Demand that cosmetics companies fully disclose ingredients and support those that do.* Tell cosmetics companies that you want them to fully disclose the ingredients in the products

they make—including the chemicals that are hiding under the term "fragrance." You can find companies' toll-free customer hotlines on product packages and online, and calling them only takes a moment. We've provided some helpful talking points on our fragrance report fact sheet, which you can find online at www.environmentaldefence.ca. Companies need to hear from you, the potential customer—you have the power to vote with your dollars! In the meantime, support companies that fully disclose ingredients.

For additional information, visit the following web sites:
www.environmentaldefence.ca
www.SafeCosmetics.org
www.CosmeticDatabase.com

ACHIEVING CLEAN INDOOR AIR

The original concept of the plans in the "smart home" is the idea of combining universal design and environmentally friendly building features. This addresses the entire home environment and its building systems.

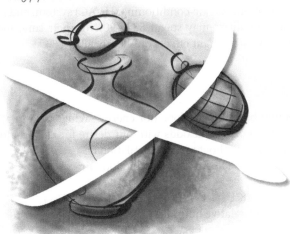

www.naaf.no

Unfortunately, some people get unwell from products that smell nice to others...

THANK YOU FOR NOT USING

PERFUME

Applies also to perfumed hygiene products
(detergents, fabric softener, deodorant, shampoo and hair spray)

Perfume free air is fresh air for everyone

www.ud2012.no

FIGURE 10.12   The poster "Thank You for NOT Using Perfume" was featured in the registration area for the Oslo Universal Design conference in Norway in June 2012 (UD 1012).

Now included as a universal design feature, clean indoor air is the common denominator that connects traditional universal design and what is known as "green housing." Green building is also called sustainable building.

Sustainability basically means building and designing with materials and methods that do not use up natural resources, creating a burden for future generations. The two terms are interconnected but not truly the same.

Universal design in housing is clearly sustainable as far as the actual space planning goes as it will put less of a burden on family caregivers, and people are able to stay in their homes longer and more independently since the residences are created for general accessibility.

Achieving clean indoor air was encouraged at the Oslo UD Conference in June 2012 by posters that asked attendees not to use fragrances (Figure 10.12).

POOR AIR AT HOME

Respiratory disabilities in the United States have the highest rate among children under 18 and the fourth highest among adults. Senior citizens and people with physical disabilities may also have weakened immune systems and could be more vulnerable to mold. For this reason, moisture and humidity levels must be controlled.

The air quality in our homes can be a cause for concern. The Environmental Protection Agency (EPA) estimates that we are exposed to two to five times more pollution indoors than outdoors. A few of the largest indoor air problems are odors or contaminants; problems with a poorly designed heating, ventilation, and air-conditioning (HVAC) system; and people themselves. Since people with physical disabilities may spend 80%–90% of their time indoors, air pollutants in sealed environments have significant effects on their health.

WHAT MAKES GOOD AIR?

The concept of a tighter building envelope (for energy savings, and in an effort to keep outdoor air contaminants out) can also be at odds with our desire for clean indoor air. It may also keep contaminants in the house, without proper ventilation design.

The following are the characteristics of good indoor air:

- Temperature: 72°F–78°F
- Humidity: 40%–60% relative humidity
- Air velocity: 20–30 fpm in ducts
- Dilution ventilation: 20 cfm per person

Some of the shower and toilet areas in my home designs are called "wet rooms." These have floor drains in the accessible shower and in the area between the shower and toilet.

Wheelchairs will naturally carry water out of the shower in these areas, and people using wheelchairs will probably be in the area longer than an ambulatory person. Proper ventilation and product selection are vital to mitigate high rates of humidity and eliminate mold growth.

THE RIGHT TOOLS

The bathroom should have a minimum of 8 air changes per hour and maybe even 10 in the wet room area. It is important to consult a qualified mechanical/plumbing engineer or contractor at all times. A kitchen should have a minimum of 15 air changes per hour; laundry and basements should have a minimum of 6 air changes per hour.

Another tool for controlling indoor air quality is a well-designed HVAC system. I like to specify radiant in-floor heating/cooling systems as they do not move air or dust and other irritants. This has major benefits for those with asthma or who have environmental hypersensitivity by providing clean indoor air and heat/cooling that can be connected to a geothermal heat pump or solar hot water heater.

If a forced-air system is used, the filters should always be high quality and well filtered to eliminate gaps that may allow unfiltered air flow. It is also a good idea to include a radon emissions detail that can be obtained from EPA.

SAFE ROOMS

Radon is at high levels throughout the country, especially in the Midwest. An odorless, colorless gas, it can increase the chances of lung cancer. It radiates up through foundation walls and through cracks at the edge of slabs. Radon is classified as a class A carcinogen known to cause cancer in humans. Some other class A carcinogens are arsenic and asbestos.

People with extreme indoor air hypersensitivity may consider installing an "Andair" bacteria and gas filtering system. This is also a good system for a storm or fallout safe room.

These units are often specified by the US Army for biological and nuclear fallout protection. They come from Switzerland. You may be surprised how many people are building safe rooms. Larger accessible bathrooms double nicely as safe rooms since they are larger by nature because of their wheelchair-use design (see "Riding the Storm Out," December 2010).

THE OBJECTIVE

The goal is to build universal design homes for use by people of all abilities while including clean indoor air and an overall energy efficient environment.

The case studies and other materials included in the first part of this chapter tell success stories of what happens when universal design and green design have been combined to face the problems created by increases in toxins and flawed environmental choices. As we face similar problems, where can we as design professionals and students go to get help for our projects? Many answers can be found in the second edition of their document: "REGREEN: Residential Remodeling Guidelines." The cover of that document is shown in Figure 10.13.

REGREEN is really a major initiative put together by two organizations, the American Society of Interior Designers (ASID) and the US Green Building Council (USGBC). This is a reference to provide assistance to designers and design students for integrating data from a variety of sources.

FIGURE 10.13   The cover photo for the "REGREEN Guidelines" California resource put together by ASID and the USGBC.

REGREEN RESIDENTIAL REMODELING GUIDELINES 2008

Professional Resources

www.regreenprogram.org
www.regreenprogram.org/resources

Contact the American Society of Interior Designers

  ASID
  608 Massachusetts Ave., NE
  Washington, DC 20002-6006
  Tel: (202) 546-3480
  Fax: (202) 546-3240
  E-mail: asid@asid.org

Contact the US Green Building Council

  USGBC
  1800 Massachusetts Ave., NW
  Suite 300
  Washington, DC 20036
  Tel: (202) 828-7422
  Fax: (202) 828-5110
  E-mail: REGREENprogram@usgbc.org

ASID

The ASID is a community of people—designers, industry representatives, educators, and students—committed to interior design. Through education, knowledge sharing, advocacy, community building, and outreach, the Society strives to advance the interior design profession and, in the process, to demonstrate and celebrate the power of design to positively change people's lives.

ASID endorses the following principles of environmental stewardship:

- Advocacy for safe products and services: Interior designers should advocate with their clients and employers the development of buildings, spaces, and products that are environmentally benign, produced in a socially just manner, and safe for all living things.
- Protection of the biosphere: Interior designers should eliminate the use of any product or process that is known to pollute air, water, or earth.
- Sustainable use of natural resources: Interior designers should make use of renewable natural resources, including the protection of vegetation, wildlife habitats, open spaces, and wilderness.
- Waste reduction: Interior designers should minimize waste through the reduction, reuse, or recycling of products and encourage the development and use of reclaimed, salvaged, and recycled products.
- Wise use of energy: Interior designers should reduce energy use, adopt energy-conserving strategies, and choose renewable energy sources.
- Reduction of risk: Interior designers should eliminate the environmental risk to the health of the end users of their designs.

ASID believes that, whenever feasible, interior designers should endeavor to practice sustainable design. Interior designers should meet present-day needs without compromising the ability to meet the needs of future generations.

Of the Society's 20,000 practicing interior designers, 6500 practice primarily in the commercial field and 4000 practice primarily as residential designers. The remaining 9500 work in both commercial and residential design. Professional members of ASID must pass rigorous acceptance standards: they must have a combination of accredited design education and full-time work experience and pass a two-day accreditation examination administered by the National Council for Interior Design Qualification.

ASID Industry Partners include nearly 3000 member firms with more than 8000 individual representatives, uniting the professional designer with manufacturers of design-related products and services.

The Society's membership also includes more than 12,000 students of interior design. ASID has more than 300 student chapters at colleges, universities, and design schools with two-year and four-year programs throughout the United States and a "virtual" chapter through Rhodec International.

US GREEN BUILDING COUNCIL

The built environment has a profound impact on our natural environment, economy, health, and productivity. Breakthroughs in building science, technology, and operations are now available to designers, builders, operators, and owners who want to build green and maximize both economic and environmental performance.

The USGBC is coordinating the establishment and evolution of a national consensus effort to provide the industry with the tools necessary to design, build, and operate buildings that deliver high performance inside and out. Council members work together to develop industry standards, design, and construction practices as well as guidelines, operating practices and guidelines, policy positions, and educational tools that support the adoption of sustainable design and building practices. Members also forge strategic alliances with industry and research organizations, federal government agencies, and state and local governments to transform the built environment. As the leading organization that represents the entire building industry on environmental building matters, USGBC's unique perspective and collective power enable members to effect change in the way buildings are designed, built, operated, and maintained.

USGBC's greatest strength is the diversity of its membership. USGBC is a balanced, consensus nonprofit organization representing the entire building industry, comprising more than 12,000 companies and organizations. Since its inception in 1993, USGBC has played a vital role in providing a leadership forum and a unique integrating force for the building industry. USGBC programs are distinguished by several features:

- Committee based—The heart of this effective coalition is the committee structure, in which volunteer members design strategies that are implemented by staff and expert consultants. USGBC committees provide a forum for members to resolve differences, build alliances, and forge cooperative solutions for influencing change in all sectors of the building industry.
- Member driven—Membership is open and balanced and provides a comprehensive platform for carrying out important programs and activities. USGBC targets the issues identified by its members as the highest priority. In annual reviews of achievements, USGBC sets policy, revises strategies, and devises work plans on the basis of members' needs.
- Consensus focused—USGBC members work together to promote green buildings and, in doing so, help foster greater economic vitality and environmental health at lower costs. The various industry segments bridge ideological gaps to develop balanced policies that benefit the entire industry.

Development of the REGREEN Residential Remodeling Guidelines was managed and implemented by ASID and USGBC staff and included review and suggestions by many technical advisors.

BACKGROUND

Green remodeling is the design and construction of projects that reduce the environmental impacts of remodeling, including energy, water, and materials consumption; waste generation; and harmful emissions, both indoors and out.

Although the principles that govern green residential remodeling are shared with all other design and construction projects, more than a few aspects are unique to green remodeling:

- Range of projects. Residential remodeling covers everything from painting a room to refitting a kitchen to gutting a whole house down to the framing and then rebuilding.
- Existing conditions. In new home construction, we generally have just the site to consider, but in residential remodeling, there are existing conditions that range from room configuration to hazards such as mold, lead, and asbestos.

- Custom work. Whether in design or construction, just about every residential remodeling project is custom, with very little opportunity for the sorts of economies of scale that occur in production building.
- Professional–client relationship. Remodeling professionals almost always have a client; "spec" remodeling projects are rare. The closer relationship with homeowners requires skills and perspectives not generally required for new construction projects.
- Occupants. Remodeling professionals must plan and often conduct their work on the basis of the health, safety, and schedules of real people. Even if the work requires that the home be vacated, the timing and duration of such a period must be very carefully orchestrated.
- Sequenced or staged projects. Many residential remodeling projects are phased or sequenced projects ("we want to do the bathroom and then the kitchen"), and this can make for challenging orders of operation in both design and construction, often requiring innovation and improvisation by remodeling professionals.
- Integration. A cornerstone of all types of green building is systems integration, and residential remodeling adds a new dimension to this integration: integrating the old or existing with the new.

Those unique aspects of residential remodeling mean that a best-practices guide, rather than a rating program, is appropriate.

This program and best-practices guide to green residential remodeling have been developed through a partnership between the ASID Foundation and the USGBC.

WHOLE-HOUSE, SYSTEMS-THINKING APPROACH

It is easy and tempting to boil down green building to simply product selections and glide over or even ignore the challenges of green building as a process. The REGREEN Program and the REGREEN Residential Remodeling Guidelines are about products and process, about synergies and unintended consequences. In green building, it is rarely a single product or building component or a collection of attributes that results in a building's being labeled "green." Green building is almost always about how systems work together to reduce environmental impacts. In the REGREEN guidelines, systems thinking and integration are encouraged by the cross-listing of strategies by project and environmental category, as well as by the "potential issues" section. In the electronic version of the REGREEN guidelines, electronic links emphasize the systems nature of green residential remodeling.

A FOCUS ON PROFESSIONAL INTEGRATION

Let's be frank: getting interior designers, architects, engineers, builders, and trade contractors all on the same page is not easy and is not an everyday occurrence. Yet, that is exactly what the REGREEN Program and guidelines do. Content, resources, and case studies tie together best-practice design and construction, and all building professionals are included. Green residential remodeling does not just suggest professional integration at all levels and across all disciplines; it requires it.

And although the target audience of the REGREEN guidelines is building professionals—interior designers, architects, remodelers, and the trades—it should come as no surprise that savvy homeowners and do-it-yourselfers have already shown strong interest in this resource. They want green remodeling integration, too.

GREEN VERSUS "GOOD" DESIGN

One of the challenges in developing resources for green building is deciding how to address what constitutes green design and construction versus what constitutes good design and construction.

What is the relationship between the two? The REGREEN Program and guidelines work from this perspective: you can have a quality project that is not a green project, but you cannot have a green project that is not also a quality project. Good design and construction are the foundation of green design and construction. For example, you can't have just efficient lighting; it must also be effective lighting. Similarly, beauty is an integral part of green design and construction; the beauty of a building or project is the starting point for durability, one of the most important attributes in green building.

DEALING WITH CLIMATE AND SITE

A very significant aspect of building green is designing and constructing for the climate and site. The REGREEN guidelines handle this aspect of green residential remodeling in three ways:

- Strategies. Certain strategies suggest varying degrees of implementation, depending on climate—for example, additional insulation or better-performance windows in colder climates.
- References and Resources. Many of the sources of additional information yield climate- and site-specific guidance.
- Case Studies. Although by no means a comprehensive approach to climate and site, the case studies provide examples of how green residential remodeling can be expressed in types of projects in particular climates and on particular sites.

WHAT THE REGREEN GUIDELINES ARE, AND WHAT THEY ARE NOT

The REGREEN Guidelines are comprehensive, but not stand-alone. The Guidelines depend heavily on vetted links to additional information on specific topics. Given the nature of green building and the depth and breadth of residential remodeling in particular, the Guidelines have to depend—as do you, the user—on connecting with the best resources for more extensive coverage of techniques, strategies, and materials.

The REGREEN Guidelines are PDF (portable document format)-based resources, for now. It was clear to the developers of the Guidelines that electronic resources would offer some significant advantages in covering green residential remodeling. Such a resource would allow for different avenues of initial approach to green residential remodeling, making systems integration and systems thinking natural and easy—and would accommodate evolution and constant improvement to the REGREEN Program. But although the goal is to publish an electronic resource with learning programs to support the use of the guidelines, today it is only available in PDF form. Refer to www.regreenprogram.org for new program updates.

The Guidelines can be accessed electronically to help with specific design decisions. This comprehensive set of guidelines was designed to be the best resource available with a strategy library, case studies, building assessment guidelines, and specific project descriptions.

The REGREEN Guidelines is project based, not project specific. These guidelines can provide guidance on green remodeling for a variety of projects (10 to date), but it cannot give definitive guidance on a specific project. If you are unsure how a particular method, material, or design feature fits into your project, use the principles of green design and construction in these guidelines. The REGREEN Guidelines is primarily single-attribute product selection guidance. We simply do not have comprehensive, multi-attribute tools today to compare and weigh recycled content and recyclability, locally sourced and low-emitting materials, and manufacturing and maintenance environmental impacts. In these guidelines, we have used different proxies for reduced environmental impact rather than a full life-cycle analysis; those products meeting the various criteria are referred to as "environmentally preferable." You will need to accomplish your own balance of various product attributes in weighing the value of one production selection strategy versus another. Look at the product considerations resources on www.regreenprogram.org to augment the product strategies in the REGREEN Guidelines Strategy Library.

The REGREEN Guidelines is not a rating system. The developers of these guidelines have dove-tailed as much as possible with the content and resources of the LEED for Homes Rating System (www.usgbc.org), but not to the extent that any sort of rating or certification can be applied to green residential remodeling projects completed using these Guidelines. Where applicable, we have referenced standards and certifications used by the LEED for Homes Rating System, including the following:

- Energy ratings—HERS, EPA Energy Star
- Water efficiency criteria—EPA WaterSense
- Material selections—Forest Stewardship Council wood certification, GREENGUARD Children and Schools, Green Seal
- Indoor air quality—ASHRAE 62.2

PROJECT CASE STUDY

Many in the building profession understand and internalize principles and practices best when they are expressed through an actual project. For each of the 10 project types, a representative green remodeling project is captured through photos; comments from the designer, builder and client; and discussion of key elements of design and construction.

But be careful how much you read into each of the case studies. They cannot capture all the relevant strategies. They do not necessarily represent the final word or "platinum" perspective; they are guidance, not guided. You are likely to get as much from the "Lessons Learned" as you are from the "Project Features" section of each case study. The case studies were selected on the basis of a long list of attributes; they can only be improved by having lots of company as the REGREEN Program grows and evolves.

FOLLOW INDIVIDUAL STRATEGIES INTO THE STRATEGY LIBRARY

There are nearly 200 write-ups of green remodeling strategies in the Strategy Library. The Strategy Library is organized by environmental topic, so while you are following up on one strategy, say, on water efficiency, you will see other water efficiency strategies that may be relevant to this or another remodeling project you are involved with.

Users of the guidelines can, of course, simply browse the Strategy Library using its table of contents. Many experienced green remodeling professionals will use the guidelines in this way as a sort of brainstorming checklist to remind them of opportunities to consider.

THE WATER SUPPLY

Another option to maintain an ongoing healthy home requires attention to the water supply. Filtering drinking and cooking water is easily achieved with a good plumbing system and special appliances. Even the bathroom sink can be equipped so that you can drink, brush, and rinse with fresh filtered water.

Using newer products, filtration happens right in the faucet and eliminates the need for an under-the-counter system. Enjoy hot, cold, and filtered water while saving money and protecting the environment. For the budget minded, a trip to the drugstore to purchase a paired pitcher and filter achieves the same healthful benefits.

REFERENCES

Barker, C., Hines, A., & Null, R. (2009, March 25–28). Where green design meets universal design: a tour of ecology house. *Interior Design, Educators Council, 2009 Annual Conference, The Spirit of Exploration: The Gateway to New Frontiers-Present, Future, Past.* St. Louis, MO.

Jones, L. (2008). *Environmentally responsible design: Green and sustainable design for interior designers.* Hoboken, NJ: John Wiley & Sons.

Lent, T. (2009, August) Improving indoor air quality with the California 01350 specification. *www.healthy-building.net. Healthy Building Network.* Retrieved June 11, 2013, from www.healthybuilding.net/healthcare/CHPS_1350_summary.pdf.

McDonough, W., & Braungart, M. (2002). *Cradle to cradle: remaking the way we make things.* New York: North Point Press.

Sarantis, H., Environmental Defense (Organization), Campaign for Safe Cosmetics. Environmental Working Group (Washington, DC) & Gibson Library Connections, Inc. (2010). *Not so sexy: The health risks of secret chemicals in fragrance.* Toronto, ON, Canada: Environmental Defense.

What Is the Americans with Disabilities Act?

CONTENTS

WHAT IS THE AMERICANS WITH DISABILITIES ACT?

The most often-cited number of Americans with disabilities is 43 million. This represents over 17% of the total population that is directly affected by the passage of the Americans with Disabilities Act (ADA). But that number may be vastly underestimated because the statistic is drawn from government information on disability-related payments (and the latest census now reports the number at 49 million), and it does not take into account individuals who are secure enough financially to seek aid or those who have disabilities that they learn to live with without seeking assistance from government agencies (U.S. Department of Labor 2013).

Medical equipment manufacturers and suppliers estimate the number to be more than double the official figure, as did Ronald Mace, the president of the Center for Universal Design (formerly Barrier Free Environments, Inc.) in Raleigh, North Carolina, although for slightly different reasons. Mr. Mace's estimate was based on a broader definition of what constitutes a disability (including the fact that each day millions of people find themselves temporarily disabled for varying lengths of time from accidental injuries). The medical community based its figure on actual equipment demand.

A reasonable assumption would be that the number of Americans with disabilities that interfere with their day-to-day functioning is 80 million or more (or about 32% of the total population). There are some 37 million Americans with arthritis alone, most of whom could benefit from the universal application of lever door handles (as discussed in Chapter 2), and many of whom qualify for accommodation under the ADA.

Disability in America

43 Million, according to the ADA
- Hearing impaired: 22 million (including 2 million deaf)
- Totally blind: 120,000
- Legally blind: 60,000
- Epileptic: 2 million
- Partially or completely paralyzed: 1.2 million
- Wheelchair users: 1.4 million
- Developmentally disabled (cerebral palsy, for example): 9.2 million
- Speech impaired: 2.1 million
- Mentally retarded: 2.5 million
- HIV infected: 1 million (estimated)

Source: Office of Special Education and Rehabilitation Services, Centers for Disease Control

---

The ADA (Public Law 101-336) was essentially an extension of the Civil Rights Act of 1964. In much the same manner as the Civil Rights Act established protection on the basis of race, color, national origin, sex, and religion, the ADA provided protection against discrimination on the basis of disability in the areas of employment, public accommodation, state and local government services, and telecommunication services.

The ADA, however, differed from the Civil Rights Act in a very important way: It required that accommodation be made to remove any barriers to full participation by people with physical or mental impairments. It thus reflected a changing view about the nature of a disability. Underlying the act was the belief that a disability is brought about by an environmental or attitudinal barrier, not by the person with a physical or mental impairment. The ADA exhorted Americans to find ways to remove those barriers, thereby minimizing the effects of the disability and providing equal opportunity for a historically isolated and ignored group.

The ADA was passed by the House of Representatives on July 12, 1990, by a margin of 377 to 28. The Senate followed the next day by a margin of 91 to 6. President George Bush signed the Act into law on July 26, 1990 (see Figure 1.1).

A Blessing for the Presidential Signing of the Americans with Disabilities Act

The White House
July 26, 1990
Rev. Dr. Harold H. Wilke

The reverend's address blessed the assembly with the following prayer:

Today we celebrate the breaking of the chains which have held back millions of Americans with disabilities.

Today we celebrate the granting to them of full citizenship and access to the Promised Land of work, service, and community.

Bless this gathering this joyous celebration.

Bless our President as he signs the Americans with Disabilities Act and strengthen our resolve as we take up the task, knowing that our work has just begun.

Bless the American people and move them to discard those old beliefs and attitudes that limit and diminish those among us with disabilities.

THE 2010 REGULATIONS

In July, 2010, Attorney General Eric Holder signed final regulations revising the Department's ADA regulations, including its ADA Standards for Accessible Design.

The revised regulations amend the Department's 1991 title II regulation (State and local governments), 28 CFR Part 35, and the 1991 title III regulation (public accommodations), 28 CFR Part 36. Appendix A to each regulation includes a section-by-section analysis of the rule and responses to public comments.

Detailed information on the law and regulations and assistance materials are available online at www.ADA.gov.

RELATED LEGISLATION

Several other federal acts either directly influenced or strongly affected the enactment of the ADA, and they continue to play a supportive role in creating equal rights for Americans with disabilities. The three most significant of these were the Equal Education for All Handicapped Children Act of 1975, the Architectural Barriers Act of 1968, and the Rehabilitation Act of 1973.

THE EQUAL EDUCATION FOR ALL HANDICAPPED CHILDREN ACT OF 1975

The Equal Education Act directed all public school systems to offer free, equivalent education to children who had been segregated, isolated, or ignored—in short, children who were being discriminated against on the basis of a disability that kept them out of consideration for a mainstream education.

The "mainstreaming" that occurred as a result of this legislation brought about two important events. First, it introduced thousands of Americans to populations they had had little or no contact with, and as people began to interact more fully with children who happened to have disabilities, many preconceived notions about capabilities and "limitations" began to change. The particular child began to take precedence over the particular disability.

Second, a generation of Americans with various physical and mental impairments has now been formally educated in a mainstream setting. And these individuals expected equal opportunity in employment and services. The ADA provided that opportunity.

THE ARCHITECTURAL BARRIERS ACT OF 1968

The Architectural Barriers Act (ABA) (PL90-480) mandated that all facilities funded partially or wholly by federal funds and intended for public use should be designed and constructed in an accessible manner. The act applied to new construction, renovations, and leased facilities.

The ABA had little application to general public spaces, such as shops, theaters, restaurants, and private offices, and offered no consequences for noncompliance. It also did not cover those parts of leased buildings not occupied by federal programs. Thus, a federal office might be located on the third floor of an inaccessible building, making its programs unavailable to many people with disabilities.

THE REHABILITATION ACT OF 1973

While the ABA was directed at accessibility, the Rehabilitation Act of 1973 was most concerned with discrimination. Not only did this act prohibit discrimination against those with disabilities in all federal agencies or programs receiving federal financial assistance, it also mandated affirmative action for people with disabilities.

Section 504 of the act specifically required that federal programs be conducted in barrier-free environments, but allowed (as did the ADA) services to be provided in alternate sites. Thus, the act did not specifically require retrofitting of an environment since its primary aim was to provide equal access to programs.

The Rehabilitation Act was in many ways the model for Title I of the ADA, which deals with employment. For all federal agencies, all recipients of federal financial assistance, and any contractors or subcontractors who received more than $2500 in federal contracts, many of the ADA provisions of nondiscrimination on the basis of a disability have been provided since the mid-1970s. The Rehabilitation Act is not superseded by the ADA for covered entities: The ADA does not exempt a federal contractor from following the affirmative action provisions of the Rehabilitation Act even though the ADA does not specifically set similar requirements for its covered entities.

The Rehabilitation Act has provided some important lessons. Access to the work world was gained for the first time by many individuals who had found it difficult, at best, to find supportive employment. The history of their employment has shown that people with disabilities are capable of working and are willing to work and that accommodations that assist them in their work were available and relatively inexpensive. The basic groundwork for newly covered entities under the ADA had already been laid.

THE REASONING BEHIND THE ADA

The text of the ADA listed two reasons for its enactment. First, discrimination against people with disabilities on the basis of historical isolation, misunderstanding, and stereotype was unjust and counter to the spirit of the Constitution. Second, such discrimination was extremely costly: More than 60 billion dollars were spent each year on maintaining a population that wanted to be independent. The ADA sought to integrate these individuals into the mainstream work world and into society as a whole.

The ADA addressed each of these concerns in its five titles, setting forth guidelines for employment and access to goods and services while defining who was covered under each of its provisions and establishing means of redress against discrimination. The ADA also presented some important defenses for entities that were unable to reasonably accommodate individuals covered by the law. Each of these titles are examined in the sections that follow, with special consideration given to Titles I and III, since these will have the most significance.

WHAT IS A DISABILITY?

Under the ADA, a disability was

  i. A physical or mental impairment that substantially limits one or more major life activities
 ii. A record of such an impairment
iii. Being regarded as having such an impairment

The first definition had some important qualities that needed to be examined in detail:

A. Impairment meant
   ■ Any physiological disorder, cosmetic, disfigurement
   ■ Anatomical loss
   ■ A recognized mental or psychological disorder
      Physiological disorders included speech, hearing, and vision impairments, as well as mobility or dexterity loss (e.g., from arthritis). They also included medical conditions such as cancer and diabetes. Mental or psychological disorders include conditions listed in *The Diagnostic and Statistical Manual* (the encyclopedia of conditions used in the mental health professions) as well as learning disabilities such as dyslexia, identifiable stress disorders, and mental retardation.
B. *Substantially limits* meant that for an impairment to be considered a disability, it must cause the individual to be unable to perform, or be significantly limited in performing, one or more of the major life activities.
C. Major life activities are walking, talking, seeing, breathing, hearing, learning, caring for self, working, and so on.

A person who has a history or record of an impairment is protected under the ADA even if the disability is no longer "active" (such as when cancer is in remission).

A person who is rumored or thought to have an impairment is protected under the ADA from discrimination to the same extent as he or she would be if there was an actual disability. Thus, a person who is rumored to be HIV positive cannot be discriminated against on the basis of that belief.

TITLE I: EMPLOYMENT

The first section of the ADA established guidelines to be followed by private employers, state and local governments, employment agencies, labor organizations, and joint labor-management committees. All employers of 15 or more are covered by the law. This number includes part-time employees who have worked for 20 or more weeks in the current or previous year. Also, foreign-based US companies, their subsidiaries, and firms controlled by Americans must comply with the ADA in their employment of US citizens unless such compliance would violate a foreign country's laws.

WHO IS A QUALIFIED INDIVIDUAL WITH A DISABILITY?

There are two related definitions given in the ADA that describe who is to be considered qualified for protection under the act. One deals with employment and the other deals with accommodation and services.

TITLE I: EMPLOYMENT

Under the Employment provisions of the ADA, a qualified individual with a disability was one who could perform the *essential functions* of a job with or without reasonable accommodation.

An essential function was any part of a job that must be done by the individual holding that position. Since the ADA listed reassigning nonessential job tasks as one way to reasonably accommodate an employee, the employer must be able to define the essential functions of any job.

Under this definition, a qualified person with a disability would be one who met the educational and skill requirements and who could perform the major (i.e., essential) tasks of the job with or without one or more reasonable accommodations.

TITLE III: PUBLIC ACCOMMODATIONS AND SERVICES

Under Title III, a qualified individual with a disability is anyone who could take advantage of a public accommodation or service if he or she did not have a disability. The entity providing the service must modify its operation and facility (removing architectural, communication, and transportation barriers and providing auxiliary aides and services) to make the accommodation/service usable by the individual with a disability.

NOTE: See the sections "What Is a Reasonable Accommodation?" and "Defenses against Charges of Discrimination" in this chapter for related information.

Employers of fewer than 15 workers are exempt from ADA compliance (although many may be covered by the Rehabilitation Act of 1973), as is the Executive Branch of the federal government (covered by the Rehabilitation Act of 1973), corporations fully owned by the US government, Native American nations, and bona fide private membership clubs (except labor organizations) if they are exempt from taxation.

Religious organizations are not exempt from Title I of the ADA, but they are allowed to give preference to members of their own religion.

MANAGING EMPLOYEES UNDER THE ADA: AN EXAMPLE

You are the human resource manager for Common Places, Inc., a mid-sized (35 employees) architectural and design firm that has hitherto been exempt from federal employment regulations. You have a new position opening for a computer-aided design specialist in your corporate offices.

With the ADA in place, you realize that you need to reconsider how you will handle filling open and new positions. Your first concern is the job description itself. One of the more important distinctions called for by the ADA was that an individual must be able to perform the essential functions of a job. Consequently, your task is to define what those functions are, and you will probably want to state them in the job listing and certainly in any longer job description you create. This prepares you for identifying whether an individual with a disability will be able to perform the job with or without some reasonable accommodation.

The position had historically included some minor job tasks, such as holding training sessions for interested designers and architects on new releases of the software used within a classroom setting. One of your applicants has a speech disability that makes it difficult for some people to understand her. You might then consider whether the training is central to the position you're offering, whether other specialists can conduct the training (an equivalent task that the applicant could manage would then be traded in kind), or whether the training could be done in a different way by a person with such an impairment. You cannot simply refuse to consider this applicant on the basis of a belief that she will not be able to teach in the manner hitherto used. However, you can ask her to describe how she would be able to manage this task if you do decide it is part of the essential functions of the job. She might suggest creating a self-guided multimedia training package, solving the need for dedicated training personnel.

In considering the "essential functions" requirement, you need to also consider the "reasonable accommodation" requirement, realizing that such accommodation may call for some creativity and communication on your part and on the applicant's part (see "What Is a Reasonable Accommodation?").

You've written the description of the job, listing the essential functions as they are currently conducted. If you are going to post the position both company-wide and in the local paper, you'll want to consider offering large-print copies for any present employees who may have a visual impairment. You should also consider the possibility that you'll need an assistive device or qualified translator if a person who is hearing impaired decides to apply for the opening. (If the position requires that the applicant take a test to determine skill level, you'll want to consider similar alternatives to the test itself and the test environment.) Note, however, that it is the prospective applicant's responsibility to inform you of any accommodation needed at this stage. You are not expected to plan for every disability.

You receive more than a hundred resumes and decide to follow up with several people who look promising. Thus, you send each an application to fill out prior to your meeting, asking for more specific information about their education, the computer software packages they're proficient in, and their prior employment. Note that you cannot ask for any information that might reveal a disability. Your next step is the initial phone contacts.

The first applicant you call informs you that he uses a wheelchair and asks if your office is accessible. Fortunately, you're located on the ground floor of a multistory building that has automatic doors opening into its lobby, and your hallways and office doors are wide enough for wheelchair access. However, another applicant informs you through a Telecommunication Device for the Deaf (TDD) operator that she is deaf, and although she reads lips, she would be much more comfortable with a sign-language translator.

You panic for a moment, wondering how you'll locate a translator, how you'll explain to your boss about the added expense, and whether this is a reasonable request (i.e., accommodation). But, lucky for you, the applicant informs you she has contacts through a local support agency that provides such services free of charge. So you set up an interview date and continue making calls.

In the event that the second applicant had not made the offer of assistance, you would have needed to consider whether your company had the resources to supply such an accommodation. The ADA did not establish specific "equations" to determine whether any given company is capable of, and thus liable for, reasonable accommodations; the federal government has, however, provided several examples of what it considers to be a reasonable accommodation. And it also allowed an "undue hardship" defense to any claim of discrimination (see "Defenses against Charges of Discrimination" for a detailed explanation). The validity of that defense would depend on the total resources of your company, and any claims of discrimination would be decided on a case-by-case basis. In this example, it would be reasonable to expect that you could provide such a service if it was requested of you. It is reasonable to expect you as a human resource manager to know of the services available in your community, especially given the requirements of the ADA. Also, had you not been so lucky as to work out of an accessible facility, you would have had to set up an interview with the person using the wheelchair in an accessible space, even if you knew the person would not be able to access his or her workspace on a day-to-day basis given the present layout of the building.

You set up 10 interviews, and one of your applicants seems to you to have fairly severe arthritis. You think to ask about this, but then catch yourself. Under the ADA, you cannot inquire about or even refer to a real or perceived disability. You could, however, ask the individual to explain or demonstrate how he or she would perform an essential job function. Then it would be the applicant's responsibility to point out his or her capabilities with or without a reasonable accommodation.

After the interviews, you thought you knew the best person for the job, who turned out to be none of the applicants just mentioned. However, in a conversation you had with her former employer, you were told about some drug problems, and you're concerned. You'd like to ask her straight out or have her submit to a medical screening before you offer her the job, especially since the second-best applicant is the man who uses a wheelchair, which you feel will be only a minor obstacle to overcome. What can you do?

There are some things you cannot do. First, you cannot require a medical test before you make a conditional offer of employment. Second, you cannot require a test at all if it is not required of all employees in the same position. Again, lucky for you, your company has had a prescreening policy in place for over a year, so you make the offer conditional on the medical exam.

The exam turns up two things. First, while taking down her medical history, the physician learns that she did have a drug (cocaine) problem, for which she received treatment. You cannot, however, now withdraw your offer on the basis of that information. Under the ADA, a past drug addiction (but not mere casual use) that has been successfully treated qualifies as a disability. The exam also revealed through testing that she has recently used amphetamines (without a prescription). Consequently, because you have a policy in place requiring drug testing because you do not wish to hire a person currently using illegal drugs, and because the ADA does not cover individuals who currently use illegal drugs, you withdraw the offer.

You call your second choice, make him an offer, and he passes the medical exam without any problems. Although he had a disability, he was well qualified for the position and you were able to make reasonable accommodations to ensure that he works efficiently and happily. These accommodations included purchasing an adjustable workstation (or modifying an existing one); ensuring a clear path of travel to his work area and to common areas such as restrooms, copy rooms, meeting rooms, and the cafeteria; and ensuring that each of these areas is accessible and usable by the person with the disability.

WHO WAS PROTECTED?

The ADA provided protection from discrimination in employment for any qualified individual with a disability. (See "What Is a Disability?" for a discussion of what constitutes a disability.) According to the ADA, a qualified individual with a disability is one who "meets the skill, experience, education, and other job-related requirements of a position held or desired, and who, with or without reasonable accommodation, can perform the essential functions of a job." There were two important qualifying definitions: *reasonable accommodation* and *essential function*. (See "Who Is a Qualified Individual?" and "What Is a Reasonable Accommodation?" for clarification of these terms.)

Also protected are those people who have a record of having a disability and those who are regarded as having a disability. In the first case, an individual who had a history of drug addiction but who has been successfully rehabilitated and has remained drug free is protected from discrimination under this provision of the ADA.

In the case of an individual being regarded as having a disability (e.g., a young man is rumored to have a multiple personality disorder and does not), the same protection applies as would be given for a true impairment.

WHO IS NOT PROTECTED?

The ADA specifically notes certain individuals as not being protected by its regulations. These include anyone who currently uses drugs illegally and people with the following sexual and behavioral disorders:

- Transvestitism, transsexualism, pedophilia, exhibitionism, and voyeurism
- Compulsive gambling, kleptomania, or pyromania
- Psychoactive substance use disorders resulting from current use of drugs

The ADA did not prohibit the use of drug testing in the workplace. Protection was not extended to those individuals who currently use illegal drugs or to those who drank on the job when policies existed against such activity.

EMPLOYMENT PRACTICES COVERED BY THE ADA

An employer could not discriminate against a person with a disability at any stage of his or her employment, and that coverage extended to individuals who became disabled while they are employed. The law required that reasonable accommodations be made to provide equal opportunity for all employees at every stage of their employment. The list of protected employment practices included the application process, testing, hiring, assignments, evaluation, discipline, training, promotion, medical examinations, layoff/recall, termination, compensation, leave, and benefits.

The law applied to a present employee to the same extent it applied to a new applicant. An employee with a disability was also entitled to equal access to all activities offered to other employees, including parties given at a private home or public club, sporting events, picnics, retreats, and so on. This aspect of the ADA revealed the far-reaching impact of the law on everyone's lives. It was important to be aware of just how many barriers exist in a world of motion, sound, and visual acuity that go largely unnoticed. Steps needed to be taken to dismantle those barriers and to start building environments that accommodate everyone.

ACCOMMODATING EMPLOYEES WITH DISABILITIES

The most constructive approach an employer could take to complying with the ADA was to talk to employees about what they needed to manage their work environment. Once their suggestions had been obtained by the employer or designer, the particular accommodation could be planned for and implemented.

Most people with disabilities had plenty of experience accommodating themselves to the world. They knew the tools and tricks of creating a workable environment for themselves. (See Karen Hirsch's article on personal history interviews in Chapter 3 for more on this.) In fact, the ADA did not allow an employer to provide an accommodation that was not acceptable to the individual with the disability. A solution must be offered that was mutually agreeable. The case study offered by Owen Cooks on Purdue University's ADA compliance plan in Chapter 5 pointed out that soliciting input from individuals with disabilities before changes are implemented could actually save money and might show that the employer had made a "good-faith" effort to make reasonable accommodations (see "Defenses against Charges of Discrimination" for more on the good-faith defense).

In some cases, employers would need to be creative, imagining themselves in the place of the employees, seeing through their eyes. The government has set up several programs to assist employers: The Job Accommodation Network, for example, offers toll-free assistance to help create workable solutions to accommodating any number of disabilities. Their services are available by phone, TDD, and modem.

WHAT IS A REASONABLE ACCOMMODATION?

In general, an accommodation would fall into one of three categories:

- Making the workplace/facility accessible
- Changing the way a job/service was normally handled/provided
- Providing or modifying equipment or assistive devices

If a person who uses a wheelchair could perform a programming job given an adjustable workstation, horizontal files, and clear access to a workspace and common areas, then it would be a reasonable accommodation for most companies to make the requisite changes to allow such a person to hold that position. A person with a hearing impairment could reasonably expect to have an assistive device available at a theater open to the general public, and he or she should be able to work, live, and relax in a well-designed acoustic environment that could be adapted to specifically accommodate his or her impairment. People with visual impairments could reasonably expect to encounter environments that were free of hazards and that offered clear signals for navigation. They should also be provided with large-print, software enhancements that enlarge print on a computer screen, or even a reader both on the job and at places of public and private service (e.g., a library).

Other examples of reasonable accommodation were the following:

- Installing an entry ramp
- Enlarging doors and making at least one restroom accessible
- Installing redundant alarms that alert by visual, auditory, and tactile clues
- Using sound-absorbing materials on office dividers to cut down on background noise

Reasonable accommodation could also include offering a flexible work schedule, part-time work, and reassignment to a vacant position (in a case where an employee became disabled and could no longer perform the essential functions of his or her current job).

TITLE II: PUBLIC SERVICES

Title II had two subsections: The first dealt with general state and local government services, and the second dealt with transportation concerns.

In most cases, state and local governments have been covered by the Rehabilitation Act of 1973, because most receive some measure of federal funding. Consequently, in employment, they are in many ways required to meet stricter standards in hiring people with disabilities than those required by the ADA. Virtually all state and local governments are bound by the same employment provisions for private entities in Title I.

DEFENSES AGAINST CHARGES OF DISCRIMINATION

The ADA sets up several specific defenses for any entity accused of discriminatory practices against an individual with a disability. There might be times when compliance with the ADA seems unattainable, and the law has established several categories from which to draw a defense.

I. *Undue hardship/burden*: This defense could be used in cases of employment and accommodations. If an entity could show that making an accommodation (such as installing a ramp, providing a translator, and so on) would involve a substantial difficulty or expense, it might be granted an undue hardship exemption. Note that there are no set guidelines for determining what constitutes undue hardship. Cases were decided individually.

II. *Direct threat*: This defense was used in cases where an entity considered the inclusion of an individual with a disability to pose a substantial risk of harm to the safety and health of herself or others. It must be based on valid medical knowledge or objective evidence, not on speculation. The requirement must be the same for all other individuals, and the entity was required to attempt to eliminate or reduce the risk with a reasonable accommodation before it could claim this defense.

III. *Readily achievable*: This defense was used mainly in reference to removing architectural barriers. Such removal must be both easily accomplishable and able to be done without great expense. Installing an entry ramp may in some instances be unreasonable (the slope of land may be too great, or the company's resources may be too small). Note that even in such cases where physical removal of a barrier is unreasonable, the entity may still be required to provide other means of service (e.g., home delivery/pickup).

IV. *Fundamental alteration to nature of goods and services*: This defense was used in Title III to protect entities for whom making certain types of accommodations would fundamentally change the way they do business or the goods being offered.

V. *Good-faith effort*: If an entity has made several attempts to accommodate a person with a disability but has been unable to do so, this defense was taken into account when deciding what remedies would be required, if discrimination is found.

SUBTITLE A: GENERAL SERVICES

Subtitle A of Title II guaranteed qualified people with disabilities equal access to all services and programs or activities provided by a public entity. This included all public meetings, schools, recreational facilities, and libraries, for example. Virtually any activity open to the public at large that was the province of a state or local government must be accessible to individuals with disabilities. Accessibility in this case might require removing architectural barriers; making facilities fully accessible; and providing assistive devices, alternative means of presentation, or readers and translators (see Figure 11.1).

New construction and alteration to existing facilities must meet the standards set forth by the Americans with Disabilities Act Accessibility Guidelines (ADAAG). These guidelines were similar—and sometimes identical—to the earlier guidelines set by the American National Standards Institute (ANSI) A117.1. ANSI guidelines are voluntary suggestions (although they were widely adopted as code by states and municipalities); ADAAG were minimum requirements that must be met. An extensive presentation on Purdue University's plan for their stadium remodel is shown in Chapter 4, Facility Management. In addition, the case study by Gilbert G. Sanjoy provides coverage of court cases related to ADA compliance.

The ADAAG contained general design standards for building and site elements, such as parking, accessible routes, ramps, stairs, elevators, doors, drinking fountains, bathrooms, alarms, signs, telephones, and automated teller machines. They also contained specific technical standards for restaurants, medical care facilities, libraries, and transient lodgings. Selections from these guidelines appeared throughout this book, but the designer should be familiar with the complete set while remaining aware that these represent minimum standards. Selections throughout the book stress the importance of moving beyond any mere set of guidelines for creating universal design.

In a case study later in this section, Susan Zavotka describes a restaurant remodeling project that was changed to put it into ADA Compliance (see Figure 11.1).

SUBTITLE B: TRANSPORTATION

PART I

Subtitle B covers public transportation and was further divided into Parts I and II. Part I concerns public transportation by bus, light and rapid rail, or any other vehicle that operates on a fixed-route schedule. The purpose of this part was to ensure that such systems be made available to individuals with disabilities as rapidly as possible.

Any new vehicle ordered after August 25, 1990, must be usable by individuals with disabilities, including those who use wheelchairs. Accessible buses (similar to those now available in major US cities) are shown in the new buses available in Kumamoto, Japan (see Chapter 7). (Temporary relief might be granted for this requirement if the public entity makes an effort to purchase an accessible vehicle but finds that none is available.) Also, any vehicle that is remanufactured and can be expected to be in service for five or more years must be made accessible during that remanufacturing. (Used vehicles were also covered by this rule, unless accessible vehicles can be shown to be unavailable.) An exception to the rule existed for historical vehicles, if making them accessible would fundamentally alter the character of the vehicle (see "Defenses against Charges of Discrimination" for a definition of this and other defenses).

Public entities were also required to provide a paratransit (equivalent and accessible) system for individuals with disabilities if they offer a fixed-route system. The paratransit system must be comparable to the regular system (i.e., pickup time must be similar). However, public entities that can show that implementing such a system would result in an undue financial burden could petition for an exception.

If the public entity offered a demand-responsive system (such as a senior center bus), then that system must meet the same accessibility standards for new vehicles detailed above, unless they offered an alternative system that provided an equivalent service for those with disabilities.

All new transportation facilities must be accessible as outlined in the ADAAG. Also, any alterations must follow these guidelines to the maximum extent feasible. Section 227 specifically mandated that all key transportation stations after January 26, 1995, must have become accessible; however, extensions of up to 30 years might be granted in cases that present extraordinary difficulties.

In general, public entities were bound by the same rules that applied to Title III (Public Accommodations).

PART II

This subsection of Title II dealt with intercity and commuter rail transportation. The same rules governing Part I applied here. Some compliance dates, however, differed. Accessibility standards must have been met by July 26, 1995, and all key stations must have been made accessible by July 26, 1993 (again, extensions are available in some cases). In a Chapter 2 case study, Pattie Moor's students designed a light rail system in which the cars and stations were designed to meet all the requirements of the ADA.

TITLE III: PUBLIC ACCOMMODATIONS AND SERVICES OPERATED BY PRIVATE ENTITIES

Title III affects nearly every business imaginable: every grocery, restaurant, theater, zoo, museum, gas station, and laundromat. If a service is offered to the public at large, it is covered by the ADA.

There are 12 general categories listed in the ADA as public accommodations:

1. Places of lodging
2. Establishment of serving food or drink
3. Places of exhibition or entertainment
4. Places of public gathering
5. Sales or rental establishments
6. Service establishments
7. Stations used for specific transportation
8. Places of public display or collection
9. Places of recreation
10. Places of education
11. Social service center establishments
12. Places of exercise and recreation

There are two specific exceptions to coverage given in the ADA—religious organizations and private clubs (as defined by the Civil Rights Act of 1964). Even though religious organizations were exempt from compliance, voluntary compliance was urged by many church leaders. All other private entities that have dealings with the general public and whose operations affect commerce were prohibited from discriminating against an individual on the basis of a disability. They must attempt to accommodate the needs of all individuals by allowing full access to and participation in the services offered. A private entity was required to remove existing architectural and communication barriers in existing facilities and transportation barriers in existing vehicles, where such removal is readily achievable. If the means were not readily achievable, that entity must provide alternative service that is equal in kind.

Furthermore, such access and participation must be offered on an equal basis in an integrated setting. Thus, a person with a hearing impairment had the right to an assistive device when attending a movie and should not have to sit in a special section apart from the general public to make use of that device. Likewise, a person using a wheelchair had the right to enjoy a film while sitting in a central part of the theater, not just in a special section removed from the general audience.

TRANSPORTATION

The transportation guidelines under this title are similar to those in Title II. However, the standards were different for a private entity that provided some transportation services (e.g., a hotel shuttle service) but that is not primarily in the business of transportation. An exception is available for historical vehicles whose character would be fundamentally altered if made accessible.

NEW CONSTRUCTION AND ALTERATION

All new construction of commercial facilities must be accessible except where it is structurally impracticable to do so (note that this exception does not apply to the Title II regulations for public, i.e., government, facilities). Consequently, such construction must at minimum meet the standards set forth in the ADAAG. All alterations must be accessible to the maximum extent feasible, to include parking, entry ways, restrooms, water fountains, and telephones (TDDs). The ADA specifically exempts the requirement for elevators in buildings of fewer than three stories with under 3000 square feet of space per story, unless that facility is a shopping mall or center or contains health care providers (the Attorney General has the authority to add to this list).

EXAMINATIONS AND COURSES

All facilities where examinations and courses related to applications, licensing, certification, or credentialing for secondary or postsecondary schooling, or for education, professional, or trade purposes were given must be accessible or made available at an alternate site. All material must be available in a variety of accessible formats to accommodate specific disabilities (e.g., on tape for the visually impaired).

TITLE IV: TELECOMMUNICATIONS

Title IV amends the Communications Act of 1934 in two significant ways, affecting both individuals with speech and hearing impairments, and public and private entities faced with providing alternative communication devices (such as TDDs) as reasonable accommodations and auxiliary aids.

TAX INCENTIVES FOR BUSINESSES

DISABLED ACCESS TAX CREDIT

Internal Revenue Code Section 44 allowed an eligible small business to elect a tax credit equal to 50% of eligible access expenditures between $250 and $10,250 for a credit of up to $5000 in any given year.

An *eligible small business* was one that has gross receipts for the preceding year of less than $1 million or had no more than 30 full-time employees. Eligible access expenditures include any amounts spent to remove barriers, purchase assistive devices, provide readers, and so on (i.e., any reasonable accommodation expense).

This credit could be elected in more than one tax year.

TAX DEDUCTION TO REMOVE BARRIERS

Internal Revenue Code Section 190 allowed any entity that must remove a barrier to accommodate a person with a disability (including transportation barriers) to take up to a $15,000 deduction each year for such alteration. However, the alteration must meet the standards set by the Architectural and Transportation Barriers Compliance Board.

Expenses for new construction were not eligible.

The ADA amends Title II of the Communications Act by requiring the implementation of telecommunication relay services both within and between states. These services allow communication by print over existing phone lines to a special operator who verbally relays the message to the recipient and then sends the reply back to the originator by print.

Title IV also amends Section 711 of the Communications Act, requiring that all public service broadcasts produced or funded in whole or in part by the federal government be closed captioned. The producer is responsible for this provision; however, if an individual television station receives the captioning but fails to broadcast it, then that station is liable.

TITLE V: MISCELLANEOUS PROVISIONS

Title V addressed several concerns that were pertinent to the rest of the act but that may require further clarification. These were brief sections that are important but somewhat disconnected.

SECTION 501

The first section of Title V declared that no part of the act should be interpreted as allowing lesser standards to be applied than were otherwise required by existing federal, state, and local laws (the Rehabilitation Act of 1973 and pursuant regulations are noted). This section also pointed out that the ADA did not affect rules about smoking in places of employment, public accommodation, or transportation. Nor does the ADA place restrictions on insurance companies, except to specify that insurance plans cannot be used to discriminate against a person with a disability.

This section repeats the rule noted elsewhere in this chapter and in the ADA that individuals with disabilities were not required to accept accommodations, aids, services, opportunities, or benefits that they do not want. The belief that the entity and the individual must work together to find a mutually agreeable solution to any problems of discrimination was underscored throughout the ADA.

OTHER SECTIONS

Section 502 waives a state's immunity from prosecution (as established in the 11th amendment to the Constitution), allowing a state to be sued for discrimination under the ADA.

Section 503 protects an individual who was filing a complaint or suit of discrimination from any form of harassment or retaliation.

Section 504 set the deadline for accessibility guidelines to be issued (ADAAG) by the Architectural and Transportation Barriers Compliance Board.

Section 505 allowed for attorney fees to be collected for the prevailing party in any judgment (federal government is excluded).

Section 506 set guidelines for the creation of technical assistance manuals and programs, including disseminating information to appropriate parties through publications and training programs. This section also established that failure to receive technical assistance cannot be used as a defense against a discrimination charge.

Section 507 required a study of the policies of the Federal Wilderness Areas to be conducted to see if they inhibit the ability of individuals with disabilities from full access and enjoyment. It also reaffirmed that wilderness areas could not prohibit the use of wheelchairs.

Section 508 specifically noted that transvestitism is not considered to be a disability; thus, individuals are not covered by the act solely on the basis of their being transvestites.

Section 509 extended the coverage of the ADA in all its provisions to the House of Representatives and the Senate.

Section 510 underscored earlier statements in the act that current illegal use of drugs is not protected under the ADA, and it specifically stated that nothing in the act was to be construed as affecting current policies on drug testing.

Section 511 defined some categories of individuals or conditions that are not considered to be disabilities as defined by the ADA: homosexuality and bisexuality; transvestitism, transsexualism, pedophilia, exhibitionism, voyeurism, gender identity disorders that are not the result of physical impairments, and other sexual behavior disorders; compulsive gambling, kleptomania, or pyromania, and psychoactive substance use disorders resulting from current use of drugs.

Section 512 amended the Rehabilitation Act of 1973 to include the ADA definitions of disabilities in terms of alcohol and drug addiction.

Section 513 encouraged the use of alternative means of dispute resolution.

Section 514 allowed individual provisions of the act that are found unconstitutional to be removed without affecting the remainder of the ADA.

REMEDIES FOR DISCRIMINATION

The ADA provided for various remedies and courses of action to be taken in cases of discrimination. Specific steps were outlined for each of the act's titles.

TITLE I

The employment section of the ADA is enforced by the Equal Employment Opportunity Commission (EEOC) using the same procedures used to enforce the Civil Rights Act. To report any discrimination, individuals must contact their local EEOC office (on their own behalf or on behalf of another; individuals are protected from retaliation for reporting). The EEOC will investigate and try to resolve any discrimination found. If no resolution is possible, they will either file suit or issue a "right to sue" letter.

Remedies include hiring, reinstatement, promotion, back pay, front pay, restored benefits, reasonable accommodation, attorneys' fees, expert witness fees, and court costs. Punitive and compensatory damages are also available in cases of intentional discrimination to where no good-faith effort has been made, limited to the following amounts:

No. of Employees	Maximum Damages
15–100	$50,000
101–200	$100,000
201–500	$200,000
500+	$300,000

Note: Punitive damages are not available against state and local governments.

TITLE II

Lawsuits can be brought by private parties, or complaints can be filed with appropriate federal agencies (Agriculture, Education, Health and Human Services, etc.). Complaints filed with the Department of Justice would be referred to the appropriate agency. Remedies under this title are the same as for Section 504 of the Rehabilitation Act of 1973. Attorneys' fees are awarded to the prevailing party.

TITLE III

Private parties can bring lawsuits to stop discrimination. In these cases, attorneys' fees may be awarded. Suits can also be brought by the Attorney General in cases of gross abuse. In these cases, monetary damages (but not punitive damages) and civil penalties can be awarded: up to $50,000 for a first offense and up to $100,000 for subsequent violations.

WHAT THE ADA MEANS FOR THE FUTURE

A person who has a disability may see the ADA as a welcome relief to a system of neglect and disregard but is possibly a bit skeptical that sweeping changes will be seen within the first several years. Given the history of other civil rights movements and legislation, this skepticism is probably justified. However, the nature of this act—especially its requirement that employers and business and governments work with individuals to find reasonable accommodations—will bring about some necessary changes. A person with a disability also can find some support in the legal means of redress available for the first time (see "Remedies for Discrimination").

The majority of Americans should look at the ADA as an opportunity for enriching their experience of others and for finding competent and willing workers and consumers. For designers, the ADA and the growing interest in universal design presents an opportunity for creatively changing the built world. The ADA and subsequent information generated about meeting the needs of people with disabilities will change every environment for the better, providing spaces and products that support people throughout their life spans.

Case Study: One Restaurant's Efforts at ADA Compliance

The owners of the Monk Restaurant in Bexley, Ohio (a suburb of Columbus), had completed a remodeling project during the year before the ADA became effective. They had attempted to make the restaurant accessible for people with disabilities and were interested in finding out whether they were in compliance with the ADA guidelines, even though the remodeling was not covered by the law.

The design team conducted a two-day analysis of the site, and Figure 11.1 graphically shows areas of compliance and noncompliance. Areas in gray on the floor plan are in compliance and included the following:

- *Accessible routes*: Both exterior and interior access routes are within the required guidelines. Sloped curbs are provided where needed. Doors meet minimum width clearance and have appropriate hardware.
- *Main dining room*: Ground-floor dining meets the 5% of total seating requirement. However, the restaurant needs a management plan to insure that accessible tables are available for those who need them. Because of limited space, construction of ramps to two other dining areas on separate levels was determined to be unreasonable.
- *Bar/lounge*: Tables in the bar areas are accessible and meet the 5% of total seating requirement.
- *Bathroom area*: Areas on the floor plan that are shaded and numbered to match the following key were found to be not in compliance with the ADA and included the following:
  1. Bathroom doors need the International Symbol for Accessibility.
  2. Flush controls for the urinals in the men's bathroom are 52 inches from the floor. This is 8 inches above the maximum allowable height. *Recommendation*: Move urinals to a lower height. Estimated cost = $300.
  3. Paper towel dispensers are 57 inches above the floor. The required height is 40 inches. *Recommendation*: Move existing dispensers to required height. This could be accomplished in-house with negligible expense.
  4. Clear floor space between the two entrance doors to the men's bathroom is inadequate. *Recommendation*: Remove the second door at no expense.
  5. Chairs stacked in hallway leading to the bathrooms prohibit access. *Recommendation*: Store chairs elsewhere.

This case study showed that ADA guidelines could be readily met in many existing facilities. In this instance, the restaurant had attempted to remodel in an accessible manner prior to the ADA's enactment. Consequently, changes to meet the current requirements were relatively minor.

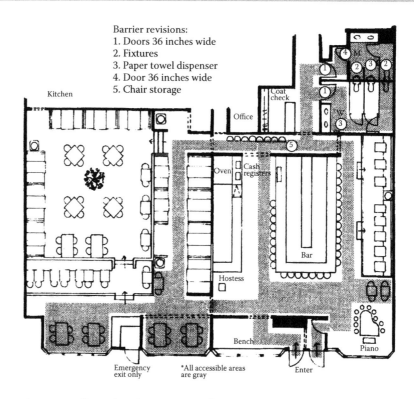

Barrier revisions:
1. Doors 36 inches wide
2. Fixtures
3. Paper towel dispenser
4. Door 36 inches wide
5. Chair storage

FIGURE 11.1   Restaurant floor plan showing modifications for ADA compliance.

ACOUSTICS, THE ADA, AND ANSI STANDARDS

When the Americans with Disabilities Act (ADA) was introduced in 1990, it was considered a major civil rights law with more teeth than prior civil rights laws. In addition to protection against discrimination, this law requires that accommodation be made to remove barriers to full participation by people with physical and mental disabilities. The Access Board is the US Government agency responsible for developing the ADA Accessibility Guidelines to help people meet the law's requirements.

Title III of the ADA, *Public Accommodations and Services Operated by Private Entities*, lists *places of education* as number 10 on its list of 12 categories covered by the Title. Prior to 2002, the United States did not have a Classroom Acoustics Standard. Yet, the original Access Board Guidelines did not address acoustical accessibility in schools for students with hearing loss and other disabilities. This was primarily a reflection of society's priorities about accessibility: a major focus had been to remove barriers for persons with mobility disabilities.

A STANDARD TO REMOVE ACOUSTICAL BARRIERS

We would never teach reading in a classroom without lights. Why then do we teach in "acoustical darkness"? Speaking to a class, especially of younger students, in a room with poor acoustics, is akin to "turning out the lights" (John Erderich, PhD, 1999).

The Acoustical Society of America (ASA), The American Speech–Language–Hearing Association, and others had been urging the Access Board to consider research and rulemaking on the acoustical performance of building and facilities, in particular school classrooms and related student facilities. Other active groups included individual acoustics professionals, parents of children with hearing loss, individuals who are hard of hearing, and a consortium of organizations representing people with disabilities. The effect of poor classroom acoustics had been studied for years and the research data overwhelmingly supported the need for all students, but especially young students and those with hearing loss and other disabilities, to have access to classroom acoustics that supported learning and communication.

In 1997, a parent of a child with hearing loss used a formal legal process of petition. She petitioned the Access Board to amend the ADA Accessibility Guidelines. She asked them to include new provisions for acoustical accessibility in classrooms to remove barriers for children with hearing loss. It was this parental petition that caused the Access Board, in 1998, to publish a Request for Information to gather public input on the issue of classroom acoustics. The Board subsequently actively supported the ongoing efforts to create a standard.

In mid-1997, the ASA commissioned a Working Group on Classroom Acoustics in conjunction with the American National Standards Institute (ANSI) to develop a draft standard for approval by the ANSI committee responsible for noise issues (S-12). This Working Group included audiologists, acoustic engineers, building managers, educators, interior designers, persons with hearing loss, architects, acoustical materials manufacturers, parents, professional organizations, consumer organizations, and governmental organizations.

The ANSI standard was submitted by the Working Group to the ANSI Board of Standards Review for approval at the end of May 2002. The ANSI Board approved this standard on June 26, 2002. The Access Board has proposed this standard to the International Code Council (ICC) for inclusion in a future International Building Code (in process at time of publication). When ICC approves the standard as Code, future school buildings and renovations will have the opportunity to be acoustically supportive to learning and communicating.

Taken by itself, the new ANSI/ASA standard is voluntary unless referenced by a code, ordinance, or regulation. However, school systems may require compliance with the standard as part of their construction documents for new schools, thus making the design team responsible for addressing the issues. Parents may also find the standard useful as a guide to classroom accommodations under IDEA (the Individuals with Disabilities Act). Some parents might include it as part of their student's IEP (Individual Education Plan).

Advocates for barrier-free acoustic environments now have a powerful tool for influencing future school building and renovation plans. The early decades of the 21st century are expected to see major school building activity in the United States. According to the US General Accounting Office, one-third of the nation's schools need major renovation or replacement. Furthermore, census projections indicate that over 400,000 additional students will enter our schools each year for the next 50 years. This growth means we will need about 16,000 new classrooms each year.

Lois Thibault, Coordinator of Research at the US Access Board, said:

> For now, it's my recommendation that advocates focus on getting the message across to the public. It's not an easy task, for we adults have become expert listeners in noise. It's hard for people to understand and be concerned about a problem we ourselves aren't experiencing and can't really model. This is the greatest difficulty faced by advocates: that of connecting the risks of educational delay and failure FOR ALL KIDS to this unseen condition, "noisy" classrooms that don't sound particularly noisy to our listeners.
>
> We know that the invisible hazard of background noise limits access to language skills, reading competency, and effective communication and learning for children everywhere, and at the most critical stage in their education. We know that it's an ADDITIONAL barrier for kids who use English as a second language (a HUGE group), kids with a wide range of learning and attention disabilities (perhaps 10% of school enrollment), kids who have speech impairments (the most common reason kids are referred for special services), kids with ear infections, and kids with hearing loss.
>
> The data are there—young kids need higher Signal-to-Noise for speech intelligibility—but it's not resonating with the public. It should be a concern of every parent and school because it wastes our scarce educational dollars and sentences many kids to reduced potential, even with costly remedial services. My own view is that poor classroom acoustics have been and are continuing to be responsible for a broad host of problems in education for which other remedies are being sought and applied.
>
> If someone can find the message that conveys this, everything could change. The public would demand better schools built to meet the needs of child learners, not adult listeners.

Now is the time for designers to recognize that public awareness of the need for quiet classrooms is growing. Our society is just beginning to develop a priority for quieter listening spaces that support learning and communication. The design team can facilitate that growth by becoming knowledgeable about the standard, understanding the rationale for the standard, and including it in their plans when approaching clients. Designers can work with local acoustic professionals for technical support as well as with community organizations who advocate for good acoustics.

THE STANDARD

Its official name is ANSI S12.60-2002 American National Standards Institute Performance Criteria, Design Requirements and Guidelines for Schools. It is commonly referred to as The Classroom Acoustics Standard.

The two main components of this standard are (1) noise levels should not exceed 35 dBA in an unoccupied classroom and (2) reverberation time should not exceed 0.6 seconds.

This means that in an unoccupied classroom, during regular school hours, with all systems running (including HVAC [heating, ventilation, and air conditioning]), the noise level will not exceed 35 dBA. Also, there will be minimal sounds bouncing off hard surfaces in the room because walls and ceilings will be treated with appropriate tiles and other sound-absorbing materials. When an empty classroom is quiet, the teachers and students can speak at normal levels, without strain, and with good expectations for understanding and being understood.

The ASA sells the standard along with accompanying documents (in 2002, the cost was $35.00). The ASA has a web site, www.asa.aip.org/, or you can email the ASA at asastds@aip.org.

The 6 North Apartments in St. Louis, Missouri, are an example of a universal design ordinance's influence. The ordinance was unique in requiring all the units be accessible instead of just a percentage of them (Figure 11.2) (Stern 2006).

6 NORTH APARTMENTS

6 North Apartments is a three-story, 80-unit residential/mixed-use and mixed-income building located at the corner of Laclede Avenue and Sarah Street in St. Louis's central west end. According to the Center for Universal Design at North Carolina State University, 6 North is the nation's first large-scale example of 100% universal design (UD) in a multifamily residential building. All of the project's one- and two-bedroom apartments—as well as its common spaces, corner coffeehouse, streetfront live/work units, and gated parking lot—are fully accessible by both

FIGURE 11.2   6 North Apartments, St. Louis, corner exterior view.

disabled and nondisabled persons. UD features incorporated at 6 North include stepless entries, open floor plans, front-loading washers and dryers, front-mounted controls, adjustable-height counters and shelves, roll-in showers, offset plumbing controls, lever door handles, rocker light switches, and high-contrast color and texture schemes. The project contains 56% market-rate and 44% affordable units. It is fully leased, and eight units currently are occupied by households with at least one disabled member (Figures 11.3 through 11.5).

6 North is located on an urban infill site one block from the St. Louis University campus (to the east) and about six blocks from the Barnes/Jewish Hospital complex and the Washington University Medical School campus (to the west). It also lies within a few blocks of Forest Park (the second-largest municipal park in the United States), the Center for Emerging Technologies (a business incubator), and CORTEX (the region's first wet labs dedicated to converting biomedical research into usable technologies). Set at the convergence of multiple bus routes and within a mile (1.6 kilometers) of a light-rail station, the site is easily accessible by public transit. Behind the building are a warehouse located across an alley and a post office, while a hardware store and row houses are across Laclede Avenue and a vacant lot and restaurant sit across Sarah Street. A former warehouse that has been converted to condominiums stands on the opposite corner of Laclede and Sarah.

The 6 North project was the brainchild of Paraquad, Inc., a private, nonprofit center founded by disability advocates Max and Colleen Starkloff in 1970 that is dedicated to providing independent living services for those with disabilities. Since 1997, Paraquad had owned and managed the Boulevard Apartments on Forest Park Avenue in St. Louis, which provided subsidized housing for disabled residents. In 2000, the US Department of Housing and Urban Development (HUD) condemned the structure because of safety and other issues. After determining that renovating the structure to meet current HUD requirements would be prohibitively expensive, residents were moved to other buildings, and Paraquad and local politicians began the search for replacement housing that would allow disabled residents to live independently.

Missouri Senator Kit Bond, a longtime housing advocate and chairman of the Senate Appropriations Committee that funds housing programs, was able to secure $1 million for the replacement project in the fiscal year 2002 VA-HUD appropriations bill and an additional $500,000 in HUD funding in 2003. Recognizing MBS's experience in developing urban communities, Bond approached Richard Baron, the company's chairman and CEO, to discuss developing the new project. Baron, in turn, met with the Starkloffs (who had since left Paraquad to form the Starkloff Disability Institute) to discuss how to incorporate accessible design features into the units. Expressing concerns about accessible design—which often results in features that meet rigid design codes but do not always fully address the needs of disabled people and can create an unappealing, institutional image—and explaining the benefits of UD, Colleen Starkloff, the

FIGURE 11.3   6 North Apartments lobby with special acoustic accents.

FIGURE 11.4   6 North Apartments interior living area; notice the wide door opening.

FIGURE 11.5   6 North Apartments kitchen with special adjustable table.

institute's director of education and training, proposed a revolutionary concept: to make the new project 100% UD. Rather than incorporating a specific number of units that would be accessible to disabled residents, the entire structure would be equally accessible to individuals of all ages and abilities. By removing the stigma of "handicapped" or "special needs" housing, UD offers a more mainstream approach to providing housing that meets everyone's long-term needs.

Baron bought into the UD concept from the beginning, encouraging the design and development team to explore a wide range of creative options to ensure that UD features were incorporated into every aspect of the project, often in ways that made them invisible to the able bodied. This open-minded attitude was an important factor in keeping the project on target, on time, and within budget. MBS brought in Trivers Associates, a multidisciplinary architecture, planning, and urban design firm, to design 6 North. The design process began in May 2002, and five months later, MBS applied to the Missouri Housing Development Commission (MHDC)—the state's housing finance agency—for tax credits. Construction, which commenced in October 2003, took 14 months, and the project was completed in December 2004.

PLANNING AND DESIGN

According to Jack Hambene, senior vice president of MBS, the decision to make 6 North the first 100% UD multifamily project created extensive opportunities as well as significant challenges. Throughout the entire development process, UD was the key decision-making driver. The site plan and unit designs had to be accessible to, usable by, and attractive to a wide range of potential users, including those with mobility issues, audio and visual disabilities, and children and adults of all ages. Having worked on Americans with Disabilities Act (ADA) conversion projects that had to conform to strict design codes, Andrew Trivers, president of Trivers Associates, and project architect Greg Zipfel both note that the UD process was a liberating journey, an exciting opportunity to think creatively about how to design spaces that are comfortable, attractive, and easily used by all.

The design and development team—which also included Colleen Starkloff and representatives of Brinkmann Constructors—began meeting early in the design process and worked together throughout the design and development phases. One innovative feature of the apartments—a custom-designed, piston-driven adjustable countertop whose height can be raised or lowered—exemplifies how this collaborative design and development process worked. Trivers's interior designers sent drawings of their original plans for the counter to Starkloff, who reviewed and commented on them. After several series of revisions, Brinkmann representatives discussed the product with prospective suppliers, proposed materials, provided cost estimates, and built a mock-up, which was then further revised before the final version was approved and built in each unit. Although this process was quite time consuming, it ultimately resulted in a product that can be adjusted to meet individual apartment residents' needs. The design team also conducted extensive research on other unit features, including sinks, storage units, appliances, and windows. The UD approach continued throughout the construction process, as various products and ways of using them were tested and reworked as the project was built.

Although several products were designed specifically for 6 North, standard products generally were used throughout the project, which kept costs down. Planning and placement, rather than unique design, are what make UD work. Standard elevator buttons as well as rocker light switches were mounted lower than usual so that they can be easily reached by individuals in wheelchairs; likewise, electrical outlets were placed higher than usual. In kitchens, removable kick plates and sinks with corner drains that allow plumbing and garbage disposals to be offset created legroom under sinks and counters for those in wheelchairs. Standard side-by-side refrigerator/freezers are easily accessible by everyone, as are standard wall ovens (placed lower than usual), dishwashers (placed higher than usual), and smooth cooktops plus front-loading washers and dryers with front-mounted controls. Easy-open hinges, adjustable shelving, and handle pulls (rather than knobs) make cabinet and closet storage spaces more usable and accessible. None of these adaptations strike the casual visitor as "special needs" features; in fact, nondisabled prospective residents typically comment on how user friendly the apartments are.

Large windows bring in copious amounts of natural light, which makes the units more usable for residents with low visual acuity as well as more appealing to everyone, and reduce the need for artificial lighting and energy use. High-contrast colors and textures provide additional cues for those with visual disabilities. Black trim around interior doorways, contrasting countertops and cabinets (light oak/black laminate), boldly contrasting wall colors, low-nap carpets and hardwood floors, and bright yellow trim around exterior doors all help residents orient themselves spatially.

Open floor plans maximize turning and transfer space for those in wheelchairs. Sliding closet doors create more usable space, as do wall-mounted cabinets in bathrooms and kitchens (Figure 11.6).

The bathrooms at 6 North illustrate important differences between standard residential design and UD. They feature wall-mounted cabinets that extend to 12 or 18 inches (30.5 to 45.7 centimeters) above the floor, allowing wall space to be efficiently used for storage while maintaining adequate wheelchair turning space at floor level. Sinks are angled just a bit, making it easier for wheelchair users to reach faucets. Seventeen-inch-high (43.2-centimeter-high) toilets (slightly higher than standard) are not noticeably different but are easier to access from a wheelchair and are placed so that those who find it easier to transfer to a toilet from one side than the other can be accommodated. Prefabricated roll-in shower stalls with adjustable shower heads and center drains do not drain as well as custom-built units would have, but were considerably less expensive, and extra-long shower curtains help keep the slip-resistant bathroom floors dry. Water-mixing valves for bathtubs and showers were installed closer to the outside of the showers/tubs than in standard bathrooms, making them easier to reach and adjust. Perhaps most important, plywood blocking was installed behind the drywall around all toilets and showers, as reinforcement, to allow stable grab bars to be attached (and removed) as required by residents.

The building's common spaces, which include a lobby, a community room, a fitness center, a patio, elevators and elevator landings, and hallways, also incorporate UD features. Clerestory elevator landings bring in natural light and offer wayfinding reference points. Contrasting color/

FIGURE 11.6   Sliding barn door passage and room divider.

texture schemes in the six-foot-wide (1.8-meter-wide) hallways and black-framed doorways help residents orient themselves, and a package shelf by each entry door provides a place to put packages and bags while opening the unit door. A 20-foot-long (six-meter-long) sliding glass door enables the community room to be closed off for private functions. The building shares its outdoor patio, which has become a popular gathering place, with the corner coffee shop. The fitness center, which all residents may use at no additional cost, contains several exercise machines that can accommodate wheelchair users, including a rotating exercise bicycle.

No parking is provided for retail users, although ample street parking generally is available, and two street spaces are reserved for disabled patrons. (Additional spaces for the disabled—more than the number required by code—are found in the secure parking lot.) One parking space per unit is allotted to residents of the income-restricted apartments at no additional charge; other tenants pay $50 per month for a space within the fenced, restricted-access lot.

Although the original building on the site was completely demolished, the design of the new, three-story, red brick structure reflects the industrial vernacular of nearby buildings, and its massing and style complement the surrounding neighborhood. Along the Laclede streetscape, brick and black metal fences create courtyard entrances for the ground-floor apartments, and residents have personalized these courtyards with potted plants and outdoor furniture. Planting strips along the low brick walls help soften the sidewalk edge. Industrial-style, floor-to-ceiling windows; exposed ductwork; 9- to 11-foot (2.75- to 3.35-meter) ceilings; and built-in metal awnings extend the industrial feel throughout the building. Along the Sarah Street facade, a grade difference issue was resolved through the installation of another planting strip, which separates the more public section of the sidewalk from that directly in front of the live/work units.

APPROVALS

The community was quite nervous about what would be built on the 6 North site. Although the existing abandoned building was considered an eyesore, neighbors were concerned about the size and uses of what would replace it. The transitional neighborhood contains both industrial and residential uses (including single-family and multifamily housing), and residents wanted walkable retail uses but worried that housing for low-income and special-needs residents would lower their property values. Recognizing the importance of creating activity on the street, the design team addressed these concerns by placing live/work units (which could be used as retail, office, or a combination of commercial and residential space) at street level, along the Sarah Street facade. These units are occupied by service-oriented commercial uses that currently include a stationary store, a catering service, and an architectural office. Two live/work units at the corner of the building were combined to create the Six North Coffee Company, which has become a popular gathering place for neighborhood residents.

Throughout the design process, members of the development team consulted with the local alderman, the Central West End Association (a nonprofit organization dedicated to maintaining and enhancing the area as a desirable and diverse urban residential, business, and institutional neighborhood), and various other area and neighborhood groups on the building's design and the need for retail services. The tax abatement entitlement process also required public hearings and approval by a subcommittee of St. Louis's board of aldermen. The MHDC/HUD-mandated environmental review was prepared by city staff and approved by HUD, the state historic preservation office (a division of Missouri's department of natural resources), and the federal Advisory Council on Historic Preservation.

FINANCING

McCormack Baron Salazar used a variety of public, private, and nonprofit sources to fund the project. US Bank provided $3.99 million in debt financing through a private first mortgage and $5 million in tax credit equity financing. MHDC provided a $540,000 HUD HOME junior loan, a $693,000 Affordable Housing Assistance Program credit, and $463,000 in federal and state low-income housing tax credits. The use of these credits required 35 of the units to be income restricted for up to 30 years; 5 of the apartments must be rented to residents earning up to 50% of area median income (AMI) and 30 units to those earning up to 60% of AMI.

In addition to the $1.5 million in HUD grants, significant funding ($650,000) came from the St. Louis Affordable Housing Trust Fund (AHTF), which was formed by a city ordinance in 2001 to provide loans and grants "for the rehabilitation, modification, construction, and preservation of affordable and accessible housing." The AHTF is the only housing trust fund in the country that requires all the projects it funds to incorporate UD. Revenues for the fund come from a 2.625% sales tax on out-of-state purchases exceeding $2000. 6 North is the fund's primary demonstration project.

MARKETING AND MANAGEMENT

Although legally the developers could not make units available only to disabled people, McCormack Baron Ragan (MBR) knew that renting the affordable units quickly was key to making the project work financially and recognized that many disabled individuals also have low incomes and are in need of subsidized housing. MBR therefore started promoting 6 North to prospective disabled tenants, holding open houses for key members of this community well before it began its general leasing efforts. The building was quickly leased upon its completion in December 2004. The first resident moved in before Christmas, 6 North celebrated its grand opening in March 2005, and the building was fully occupied by April. The first renters included 12 families with disabled members and eight seniors, all of whom qualified for the affordable units. The market-rate units also leased rapidly, to a wide range of residents that currently includes empty nesters, students at nearby universities, and young professionals and entrepreneurs who work all over the metropolitan area but prefer the neighborhood's urban lifestyle and amenities. While the building has experienced significant turnover (due primarily to its many student residents), vacated units generally have been re-leased quickly.

Removing the blighted building on a key corner in a transitional area of St. Louis and replacing it with residences and street-level businesses have increased redevelopment activity on adjacent land parcels and spurred a substantial increase in the city's income and property tax bases, as new residents and businesses are attracted to the area. And 6 North continues to serve as a demonstration project for UD; designers, developers, and planners throughout the country and around the world still visit the site to learn about the feasibility of and challenges involved in creating UD projects elsewhere.

EXPERIENCE GAINED

- Flexibility was key to obtaining the approval of the community while keeping the project financially and physically viable. Designating the first-floor space as live/work units demonstrated the developer's commitment to including retail space, yet left open the option of these units also being used as residences. After a retail tenant (the coffee shop) was identified, two planned live/work units were combined to create space for this tenant and another was transformed into a fitness center for the building's residents. Although the remaining three live/work units could still be used as both living and work space (since they contain a kitchen/bedroom area and bathroom, separated by a long, sliding door from a loft-type space located at the front of the unit), two are being used as entirely commercial spaces as of August 2006.
- Cost is always a consideration when building affordable housing, but because 6 North was a demonstration of the full range of UD principles, the project averaged about $7000 more per unit than a traditionally designed project. Several of the more expensive options such as the aforementioned piston-driven adjustable countertop could be addressed less expensively, and many of the key design features—such as the open floor plan, wide doorways, lower light switches, contrasting color patterns, and so forth—could either be installed at no additional cost or be purchased as stock items.

- Although preserving and building around the steel farmers market shed was expensive, time consuming, and problematic, the development team believes that retaining it was ultimately beneficial to the project. The structure, now lit at night, serves as a landmark and a reference to the property's past.
- While the design and development team did not specifically plan 6 North as a smart growth or new urbanist project, UD, smart growth, and new urbanist goals clearly complement each other, and 6 North addresses many smart growth/new urbanist objectives through its mix of uses, density, urban infill development, and accessibility.
- Architect Andrew Trivers and other members of the development team say they firmly believe that in the future, all buildings will incorporate UD. Developing 6 North gave them a chance to explore this new type of development.

REFERENCES

Stern, J. D. (2006, July). 6 North apartments. *ULI-development case studies.* Retrieved June 11, 2013, from www.casestudies.uli.org/CaseStudies/C036016.htm.

U.S. Department of Labor–Find It By Topic–Disability Resources–Americans with Disabilities Act. (2013, January 1). *United States Department of Labor.* Retrieved June 11, 2013, from www.dol.gov/dol/topic/disability/ada.htm.

Index

Page numbers followed by f and t indicate figures and tables, respectively.